ADVANCES IN CERAMICS • VOLUME 14

# CERAMIC IN HEAT EXCHANGERS

| | |
|---|---|
| Volume 1 | Grain Boundary Phenomena in Electronic Ceramics |
| Volume 2 | Physics of Fiber Optics |
| Volume 3 | Science and Technology of Zirconia |
| Volume 4 | Nucleation and Crystallization in Glasses |
| Volume 5 | Materials Processing in Space |
| Volume 6 | Character of Grain Boundaries |
| Volume 7 | Additives and Interfaces in Electronic Ceramics |
| Volume 8 | Nuclear Waste Management |
| Volume 9 | Forming of Ceramics |
| Volume 10 | Structure and Properties of MgO and $Al_2O_3$ Ceramics |
| Volume 11 | Processing for Improved Productivity |
| Volume 12 | Science and Technology of Zirconia II |
| Volume 13 | New Developments in Monolithic Refractories |

ADVANCES IN CERAMICS • VOLUME 14

# CERAMICS IN HEAT EXCHANGERS

Edited by

Bryan D. Foster
Norton Company
Worcester, Massachusetts

John B. Patton
U.S. Department of Energy
Idaho Falls, Idaho

The American Ceramic Society, Inc.
Columbus, Ohio

Proceedings of three symposia on heat exchangers, the first at the 86th Annual Meeting of the American Ceramic Society, April 29-May 3,1984, Pittsburgh, Pennsylvania, the second at the Pacific Coast Regional Meeting, October 28-31, San Francisco, California, and the third at the 87th Annual Meeting, May 5-9, 1985, Cincinnati, Ohio.

Library of Congress Cataloging in Publication Data

Main entry under title:

Ceramics in heat exchangers.

(Advances in ceramics, ; v. 14)
"Proceedings of three symposia on heat exchangers, the first at the 86th Annual Meeting of the American Ceramic Society, April 29-May 3,1984, Pittsburgh, Pennsylvania, the second at the Pacific Coast Regional Meeting, October 28-31, San Francisco, California, and the third at the 87th Annual Meeting, May 5-9, 1985, Cincinnati, Ohio."—Copr. p.
Includes Index.
1. Heat exchangers—Materials—Comgresses. 2. Ceramic materials—Congresses.
I. Foster, Bryan D. II. Patton, John B. III. American Ceramic Society. Meeting (86th : 1984 : Pittsburgh, Pa.) IV. Pacific Coast Regional Meeting (1984 : San Francisco, Calif.) V. Series.
TJ263.C463 1985 621.402'5 85-6052
ISBN 0-916094-67-7

Coden: ADCEDE



Printed in the United States of America.

# Preface

This volume attempts to review all known commercially significant applications of ceramics in heat exchangers—ranging from developmental to fully established units.

Many are described in detail by people closely involved, with emphasis on the practical aspects of making currently available technology accomplish economically feasible tasks. Other applications are treated more briefly in appropriate categories and with references to facilitate further study.

Ceramic materials performance criteria and testing—and development efforts for property improvement and cost reduction—are also reviewed.

Representatives of government and other "sponsoring agencies" present their perspectives and goals and motivating forces.

These papers were taken from three ACerS sponsored sessions—the first on May 1, 1984 in Pittsburgh, the second on October 30 and 31, 1984 in San Francisco, and the third on May 9, 1985 in Cincinnati. All papers were invited and all manuscripts were reviewed by the editors and by Dr. Morris Berg at The University of Illinois.

Bryan D. Foster
Norton Company

John B. Patton
U.S. Department of Energy

# Contents

## SECTION I: OPPORTUNITIES/CHALLENGES/ MOTIVATIONS



## SECTION II: APPLICATIONS/APPROACHES/DESIGNS

### RECUPERATORS (TUBE TYPE)



### RECUPERATORS (FINNED PLATES)

# Section I

# Opportunities/Challenges/ Motivations

# A Survey of Ceramic Heat Exchanger Opportunities

SCOTT RICHLEN

Senior Project Manager
Idaho National Engineering Laboratory
Idaho Falls, ID 83415

This paper surveys the areas of opportunity for use of ceramic heat exchangers, presents the basis for their use, and discusses the shortcomings of metallic vs the advantages of ceramic heat exchangers. The categories of heat-exchanger description are also summarized to provide a basis for categorizing developed and emerging ceramic heat exchangers.

In the last few years we have witnessed the full commercial development of a ceramic plate-fin recuperator manufactured by GTE Sylvania, and a ceramic tubular recuperator manufactured by Hague International. Both of them are now in general use for waste heat recovery from a variety of flue gases. More recently, Coors Porcelain Company has begun field-testing another ceramic plate-fin recuperator, and C & H Combustion has commercialized a ceramic tubular recuperator.

Current development efforts include four advanced heat exchangers: a fluidized bed recuperator, by Thermo Electron Corporation; a fluidized bed boiler system, by Aerojet Energy Conversion Company; and two high-temperature tubular recuperator systems, by Babcock & Wilcox and AiResearch Manufacturing Company of California. The fluidized bed concepts yield improved performance, but are limited by the available ceramic fabrication technology from even higher performance. Both high-temperature tubular recuperator projects have overcome ceramic fabrication limitations and will soon be field-tested. Additional development efforts are being conducted by these and other companies to support the development and use of the ceramic heat exchanger.

Allied use areas of ceramics other than recuperators include heat exchangers for process use and for power generation. The use of ceramic tubing in a radiant tube furnace can result in lengthened radiant tube lifetime. Consideration is being given to the use of ceramics in immersion heating for metal melting. Additionally, ceramic's resistance to many gases and liquids which are corrosive to metals should result in its use in chemical process heat exchangers. The use of ceramic will allow the development of power generation cycles incorporating corrosive working fluids, combustion of fuels corrosive to the metal components of available heat exchangers, and higher system temperatures resulting in greater cycle efficiency.

There are relatively few participants attempting to advance the ceramic heat exchanger and that advance is often delayed by the current state of ceramic fabrication technology. However, ceramic heat exchangers have begun to penetrate the market and as progress is made toward dependable performance and as more awareness of the advantages of ceramic occurs, the market penetration is expected to become substantial.

#### Areas of Opportunity for Ceramics

There are three prime candidate industrial areas of the ceramic heat exchanger: process heat exchange, power generation heat exchange, and industrial waste heat recovery.

For chemical process heat exchange, the primary use of the ceramic heat exchanger would be for liquid-to-liquid or liquid-to-vapor heat exchange. This would employ the ceramic heat exchanger in a medium-to-low temperature range where it would have no advantage over metallic heat exchangers with respect to temperature. However, many process streams, because of their corrosive nature, require that the metal heat exchanger have a protective coating. The ceramic heat exchanger could be targeted at these streams. The chemical process heat exchanger has been given little attention, probably because it is a less well-defined market area and may not hold the potential for conserving energy (DOE perspective), or enhancing the position of natural gas (GRI perspective), or enhancing the position of electric power (EPRI perspective). Other than the use of glass coatings and glass-tubed heat exchangers, there has been almost no penetration of the market by the ceramic heat exchanger.

For process heat exchange in the primary metals industry, ceramics could be used in both radiant tube furnaces and in the recently proposed immersion heater. The radiant tube furnace is used by industry in heating processes where the product must not be contaminated by the combustion product of the burner. The combustion flame travels the interior of the radiant tube, which in turn radiates heat to the product. Since ceramic materials have a much higher melting temperature than metals, by replacing the metal radiant tube with a ceramic tube, higher process temperatures or greater throughput could be achieved and the lifetime of the radiant tube extended. The immersion heater is similar to the radiant tube furnace except that the tube is submerged in the bath of a holding or melting furnace. Heat is conducted from the surface of the immersed tube rather than transferred by direct exposure to the flame as in a reverberatory furnace. The immersion heater has the potential of being more efficient than the reverberatory furnace because of the elimination of furnace leakage and of product oxidation from flame impingement. The development of an immersion heater for metal melting will depend on successful development of large ceramic tubes able to withstand the immersion environment.

For power generation heat exchange, the primary use of the ceramic heat exchanger would be the transfer of heat to a working fluid. This use would employ the ceramic heat exchanger in a medium-to-high temperature range, where it would have advantages over the metallic heat exchangers with respect to temperature. Additionally, depending on the nature of the working fluid and on the heat source, the environment could be corrosive to a metallic heat exchanger. The major drawback to use of the ceramic heat exchanger for power generation heat exchange is the lack of high-pressure seals. Because the working fluid can be under very high pressure, development of the ceramic heat exchanger for this market must include development of high-pressure seals and joint configurations. To date there has been almost no penetration of the market by the ceramic heat exchanger.

The industrial waste heat recovery environment is typically medium-to-high temperature and can often be corrosive. In the less corrosive streams, it is a profitable market for the metallic heat exchanger. For industry with corrosive waste streams, little use is made of waste heat recovery, and a large market exists for the heat exchanger that proves viable (e.g., only about 5% of all aluminum remelt furnaces have waste heat recovery devices). Although the industrial waste heat recovery market has been assessed a number of times, there is still some contention about actual market size. Most industrial concerns will not (because proprietary) or cannot identify the temperatures, constituents, and operation cycles of their waste heat streams. Without these values it is very difficult to provide an economic market value. Also, most market studies have addressed the larger industrial concerns but have been unable to account for the smaller firms because of their lower visibility. In spite of these hindrances, significant markets have been identified in the high-temperature range of exhaust gases (see Table I) and penetration by ceramic recuperators has begun.

Examination of Table I reveals that the markets are primarily melting or reheating processes in the metals industry. Even though some of the processes are highly recuperated (e.g., steel soaking pit—85 to 90% recuperated), if recuperation is by a metallic heat exchanger it would probably be protected by a dilution system and provide lower efficiency than a replacement ceramic heat exchanger. If the current system is a tile recuperator, it could be replaced by a more efficient ceramic heat exchanger.

Depending on the temperature of the waste stream and the cost of installation, the economic payback period can be one to three years. Generally, the cost of the recuperator comprises only a small portion (10–30%) of the total retrofit installation cost. Frequently, new burners and control systems are added when the recuperator is installed; they impact payback since they are considered part of the total installation cost. In the case of the commercialized GTE Sylvania recuperator, most economic paybacks have been from one to two years.

The reason for the interest in the higher-temperature range exhausts can be simply stated—the higher the exhaust temperature, the greater the fuel savings on recuperation. This trend is shown in Fig. 1, along with the use areas of metallic and ceramic heat exchangers. Note that the ceramic heat exchanger, because of its ability to operate in the higher exhaust temperatures, will provide fuel savings of more than 30%.

Because recuperation is the lowest cost/highest return method of conserving energy and because the industries listed in Table I are very energy-intensive, the total market for a ceramic recuperator is expected to be very large. The potential amount of energy to be saved in the industries listed in Table I from high-temperature recuperation can be estimated by multiplying the potential fuel savings (30%) by the total consumption. This results in a potential savings of approximately a quarter of a quad (where one quad = $10^{15}$ Btu).

### Categories of Heat Exchangers

Having identified a market for the ceramic heat exchanger, a particular ceramic heat exchanger must be developed for a particular market niche. The ceramic heat exchanger will be similar in many ways to metallic heat exchangers. In discussing heat exchangers, whether they are ceramic or metallic,

Table I. Industrial Waste Heat Recovery Market*

| Furnace | Consumption, kJ ($10^{12}$ Btu) (and year) | Number of Furnaces | Percent of Population Recuperated | Flue Gas Temp., °C (°F) | Average Firing Rate, mm W (mm Btu/h) |
|---|---|---|---|---|---|
| Steel soaking | 160 (152) (1979) | 2275 | 85 to 90 | 815–1371 (1500–2500) | 7.3 (25) |
| Steel reheat furnace | 351 (335) (1979) | | | | |
| —Continuous-type | | 310 | 60 | 1315 (2400) | 72.5 (250) |
| —Batch-type | | 385 | 30 | 815–1371 (1500–2500) | 11.6 (40) |
| Iron and steel forging | 89 (85) (1980) | | 15 | | |
| —Slot forge | | 8000 | | 1315–1426 (1500–2600) | 0.07, 2.1 (0.25, 7.5) |
| —Car bottom and box | | 4500 | | 815–1426 (1500–2600) | 1.5, 14.5 (5, 50) |
| Glassmelting furnaces | 189 (180) (1979) | 800 | 95 | | |
| —Unit melters | | 75 | | 1426–1537 (2600–2800) | 11.6 (40) |
| —Regenerative | | 725 | | 1426–1537 (2600–2800) | 18.8 (65) |
| Aluminum reverberatory furnace | 51 (49) (1978) | 800 | 5 | 760–1260 (1400–2300) | 2.9, 2.6 (10, 90) |
| Copper reverberatory furnace | 18 (17) (1976) | 12 | 8 | 1315–1426 (2400–2600) | 33.3 (115) |

*Ref. 1.

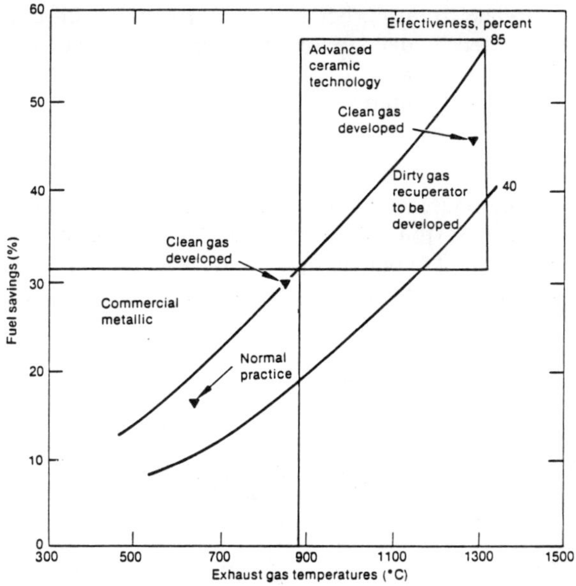


Fig. 1. Fuel savings opportunities with rising exhaust temperatures (Ref. 2, used by permission).

four general categories of description can be used: (1) operational distinction, (2) compactness, (3) configuration, and (4) assembly.

Every heat exchanger can be described in terms of one or all of these categories. The categories do not all describe a heat exchanger to the same level of detail, nor is each category completely independent of any other. However, each category does describe a heat exchanger in a distinctly different way from any other category.

**Operational Distinction**

The operational distinction of a heat exchanger is probably the most basic category of distinction. It addresses the basic question of how a heat exchanger operates with respect to the two streams exchanging heat within it.

In a "recuperator," the two streams at different temperatures flow in spaces separated by a wall and exchange heat by convection and radiation to and conduction through the wall. In the other two types of heat exchanger, the two streams occupy the same space. In the "regenerator," the hot and cold streams pass alternately over the same heat-transfer surface, which accumulates thermal energy from one stream and then transfers it to the second. In the "direct contact" heat exchanger, both streams flow through the same passage simultaneously and transfer energy directly, with no storage medium.

When preheating combustion air to be ducted to a burner, the recuperator is usually termed a "recuperator," but sometimes it is labeled an "air preheater." When used to vaporize a liquid stream it is usually referred to as a "boiler." When condensing a vapor stream, it is referred to as a "condenser." When the regenerator uses a stationary heat-transfer surface, it is frequently referred to as a "checkerwork" or "brickwork." When the heat-transfer surface is composed of a wheel rotating between two streams, such that a portion of the wheel is first heated by one stream and then rotated into the other stream where it is cooled, it is referred to as a "heat wheel" or "rotary regenerator." In utility use, the direct contact heat exchanger is usually referred to as a "cooling tower."

### Compactness

It is rather meaningless to simply state the size of a particular heat exchanger as a way of describing it. A much better method of describing a heat exchanger is to refer to its compactness or its area density. This refers to the heat-transfer surface-area-to-volume ratio. As this ratio increases, the size of

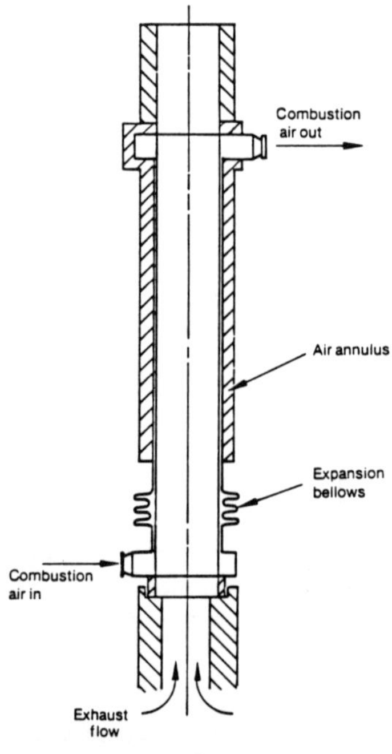


Fig. 2. Shell-and-shell configuration.

the channels through which the heat-exchanging stream passes decreases. This tends to increase the convective heat transfer to the channel wall, increasing the efficiency of the heat exchanger. However, the decreasing channel size also increases the pressure drop; this results in increasing the amount of energy required to move the heat-exchanging stream through the channel. For heat exchangers operating in a convection-dominated environment, the design can be optimized by trading off heat-transfer increases vs pressure-drop increases as compactness is increased.

For the heat exchanger operating at high temperatures, these relationships may not hold. At high temperatures, the heat exchanger can sometimes take advantage of thermal radiation from a waste heat stream. The amount of radiation is proportional to the mean beam length of the volume of radiating gas. The mean beam length in turn is proportional to approximately 3.6 times the radiating gas volume divided by the surface area being radiated. So, other factors aside, radiation will increase as the ratio (compactness) of the heat-transfer surface-area-to-volume is decreased. The pressure drop for the stream will also be decreased. This is in opposition to the compactness effect on convection; therefore, the designer's job becomes more complicated when designing a ceramic heat exchanger for high-temperature environments.

**Configuration**

The physical configurations of recuperators used for industrial waste heat recovery cover a wide range of geometries. The simplest of the recuperator configurations is the shell-and-shell (also termed shell-in-shell, stack type, and radiant). This recuperator consists of two concentric shells making up an inner passage through which the exhaust flows, and an annulus between the two shells through which the combustion air flows (see Fig. 2). It is usually mounted on larger exhaust streams and, because of the size of the center (exhaust) passage, does not require an exhaust fan. The recuperator is frequently run in parallel flow to prevent overheating of the lower portions of the inner shell wall. For this reason, the effectiveness (the percent of energy in the waste stream recovered) of the recuperator is usually less than 50%.

The plate-fin (also termed finned-plate) is a much more compact recuperator than the shell-and-shell. In this type of heat exchanger, the exhaust gases and the combustion air pass through alternate passages of the heat exchanger (see Fig. 3). The passages are separated by flat plates covered with many fins, which significantly increase the amount of heat transferred between

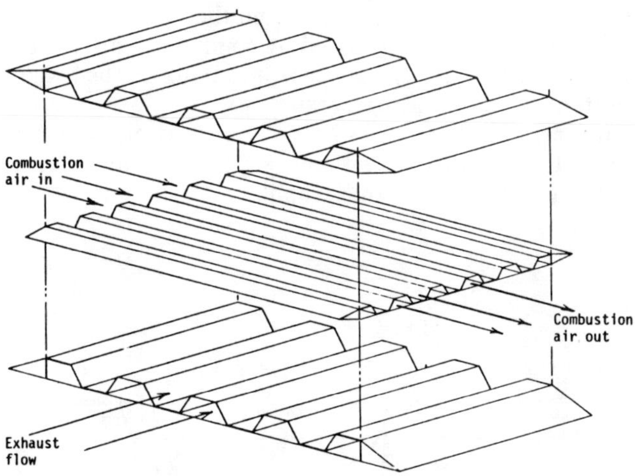


Fig. 3. Plate-fin configuration, exploded view.

the two streams. This type of heat exchanger is usually used on smaller, clean waste streams (recovering 7.3 mm (5 mm Btu/h) or less). The fins, while making the heat exchanger very vulnerable to fouling, also give it a high effectiveness, frequently greater than 50%, and up to as high as 75%.

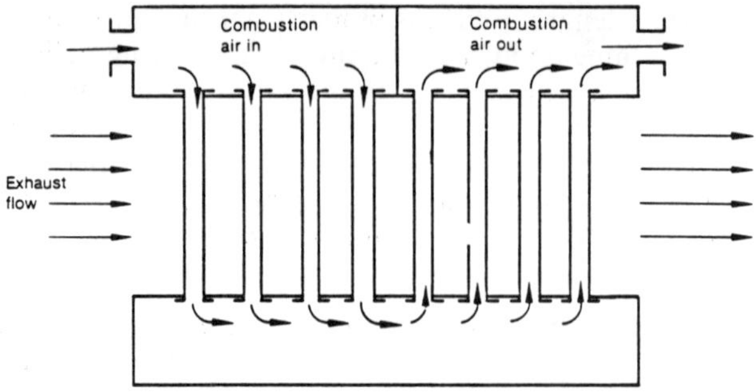


Fig. 4. Tubular configuration.

The tubular recuperator is heavily used by industry. In this type of heat exchanger, the exhaust gas flows over a bank of tubes through which combustion air is flowing (see Fig. 4). The exchanger can be built in a wide range of sizes, competing with the plate-fin in its smallest size, and with the shell-and-shell at its largest. The exchanger is generally resistant to fouling, and in the cleaner waste streams may have external fins on the tubes.

The fluidized bed heat exchanger makes use of the effect of a fluidized bed of particles through which the exhaust passes, to increase heat transfer from the gas (see Fig. 5). An array of tubes carrying combustion air or steam is sometimes immersed in the fluidized bed. The fluidizing effect of the bed increases the convection to these tubes by a factor of up to 10 over tubes in a normal exhaust flow. An alternate method of operation is to transfer heat from the exhaust gas to the fluidized bed, and then transport the bed material to another chamber where it is fluidized by combustion air and its heat transferred to the combustion air. In this case, the fluidized bed would no longer be a recuperator, but rather a regenerator.

The heat-pipe recuperator uses an array of heat pipes to transfer energy from one stream to another. The heat pipe (see Fig. 6) is an evacuated chamber with the inner surface lined with a capillary mesh. The capillary mesh is

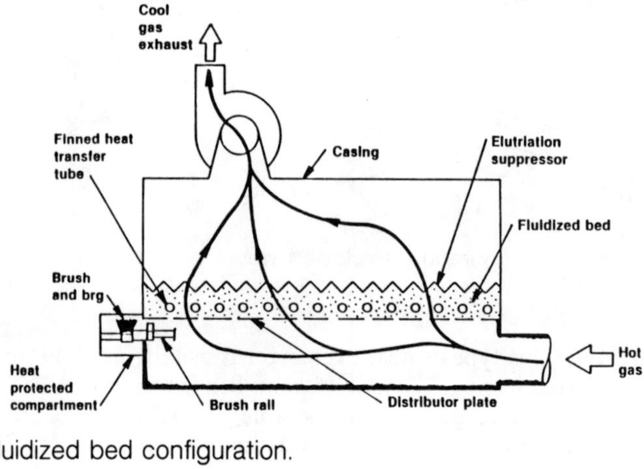


Fig. 5. Fluidized bed configuration.

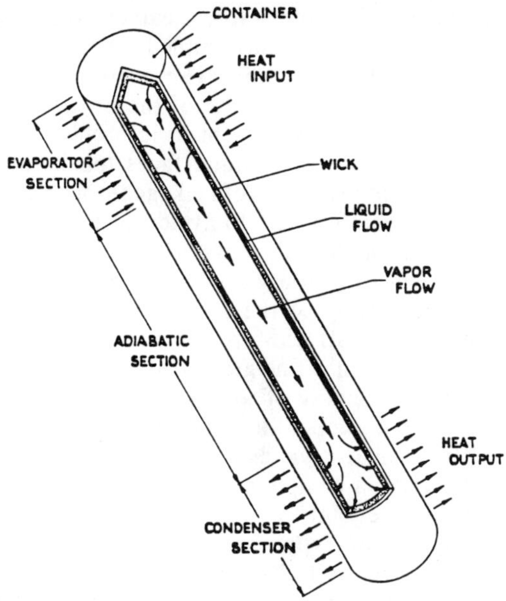


Fig. 6. Heat-pipe element.

saturated with a working fluid, and the remaining internal volume is filled with a vapor of the same fluid. In normal operation, heat is added to one end of the heat pipe and removed at the other end; the working fluid liquid is evaporated from the heated end and the vapor is condensed in the cooled end. The condensate then returns to the heated end via the capillary mesh where it is evaporated again. Because the heat is essentially transferred by latent energy from the working fluid to the container of the heat-pipe device, the individual heat pipe acts as a superconductor of heat, and when configured as a heat exchanger provides levels of effectiveness similar to that of the plate-fin heat exchanger.

**Assembly**

Ceramic heat exchangers currently available, and those being developed, represent either monolithic or nonmonolithic assembly. Both types of assembly have advantages and disadvantages, the selection of assembly method frequently depending on the configuration of the heat exchanger.

Monolithic refers to a process where the parts (either in green state or fired form) are "glued" together with slip, and final-fired to create a singular or one-piece heat-exchanger core. Although ductwork or plenums may not be part of the monolith, if the core of the heat exchanger is monolithic the heat exchanger is termed "monolithic." The major advantage of the monolithic assembly is that there are no joints in the core of the heat exchanger, and the attendant problems of sealing. The major disadvantage of this type of assembly is that stress concentrations under severe thermal operating conditions can significantly shorten the lifetime of the heat exchanger. Additionally, should a portion of the assembly fail, the whole core must be replaced, since individual parts of it cannot.

The term "nonmonolithic" refers to a heat exchanger assembled from individual components that can be disassembled for repair. This is the major advantage of the nonmonolithic assembly. The major disadvantage is the use of joints and their attendant seals. Because of the brittle nature of ceramics, the ceramic-to-ceramic joint has been a historic problem area for the heat-exchanger designer. A joint where ceramic faces touch can be a source of stress for the design. To avoid high rates of leakage through the joint, expensive surface machining would be required. To avoid these two drawbacks, a compliant or yielding material is placed between the ceramic components. This seal can be a cement, a packing, or a glass. Whatever choice is made, the seal must be able to provide sealing so the joint does not leak. It must also be compliant or yielding, so that stresses are relieved at the joint.

The cement seal results in a fine-grained interface that conforms to the ceramic surfaces, so no machining of joint surfaces is required. The cement interface is made up of both crystalline and amorphous phases, and it must be tailored in strength, so that it yields at lower stresses than do the ceramic surfaces around it. Packing, a compliant amorphous fiber material, is usually selected and packed into the joint to obtain high fiber density for the sealing. The major disadvantage of this type of joint is that at very high temperatures it becomes crystalline, bonds to itself, and loses its resilience. The major difficulty of the glass joint design is tailoring the viscosity of the glass so that it does not flow out of the joint at temperature.

## Ceramic vs Metallic

### Disadvantages of the Metallic Heat Exchanger

The metallic heat exchanger is used most by industry, and when used properly it provides relatively efficient service over a long lifetime. However, there are two major limitations of the metallic heat exchanger. The first is temperature. When manufactured of carbon steel, the bulk material temperature in the heat exchanger should not exceed 425°C. When manufactured of stainless steel, the bulk material temperature should usually not exceed 650°C. However, the exhaust gas temperatures of many industrial processes far exceed these temperatures and would heat the heat-exchanger material beyond acceptable limits under optimal operating conditions. The unit must therefore be "protected," which lowers performance. Typically, the heat exchanger is operated in the parallel flow mode, or the exhaust stream is diluted. The preferred mode of operation is the counterflow mode. The counterflow mode provides the highest efficiency since it allows the heat exchanger to heat the cold stream (the stream receiving heat in the heat-exchange process) to a higher temperature than it would be heated in parallel flow. However, the parallel flow mode, because it maintains lower material temperatures in the heat exchanger, and ensures a longer lifetime, is often chosen. In dilution, ambient air is ducted into the exhaust upstream of the heat exchanger. Although this cools the exhaust and ensures lower bulk material temperatures, it also lowers the temperature to which the cold stream can be heated, thus lowering the efficiency of the heat exchange.

The other major limitation of the metallic heat exchanger is its susceptibility to corrosion. Corrosion can occur under various conditions in a waste stream, even under normal conditions, depending on the composition of a waste stream and the material of which the exchanger is made.

Under oxidation conditions, oxygen and other exhaust constituents can attack the material surface. In this case, there is a general migration of oxygen into the material, which causes scaling. The scaling can be protective; however, if the heat exchanger is operated intermittently, causing large variations in surface temperature, the scale can flake off the surface because of differential expansion and expose virgin material. The general attack by the oxygen then resumes. Other exhaust constituents, such as sulfur, diffuse into the grain boundaries, forming eutectics that melt at temperatures much lower than the material melt point (and thus the design point) and can cause disintegration of the material surface. In either case, temperature accelerates the process.

If the process that produces the waste stream is operated such that reducing conditions (caused by fuel-rich operation) exist in the waste stream, carburization of the metal surface can occur. Under high temperature, carbon migrates into the grain boundaries of the metal and forms carbides. The formation of these carbides can cause residual stresses and embrittlement.

Under normal conditions, exhaust components, such as antimony, vanadium, magnesium, or aluminum, can form eutectics with the components of stainless steel, and a process similar to that described above for sulfur under an oxidizing atmosphere may occur. Under low-temperature conditions (when sufficient heat is extracted by the heat exchanger to cause exhaust constituents to condense), acid condensation corrosion can occur.

**Advantages of the Ceramic Heat Exchanger**

The use of ceramic as the material of fabrication for the heat exchanger addresses both of the disadvantages of the metallic heat exchanger. First, with respect to temperature, the ceramics under consideration as heat-exchanger materials do not lose significant strength until bulk material temperatures in the range of 1300°C are reached (see Fig. 7). Second, although ceramics are subject to surface attack by constituents in an exhaust waste stream, coupon testing indicates that generally they are either not attacked or are attacked

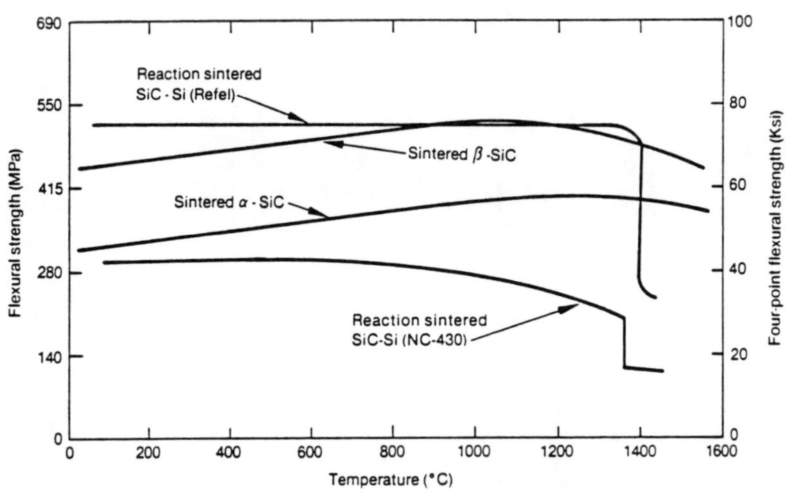


Fig. 7. Strength of SiC ceramics as a function of temperature (Ref. 3, used by permission).

at greatly lessened rates than by the same exhaust constituents that attack the metallic heat exchanger.

Because of the advantages that the ceramic holds over the metallic, the ceramic heat exchanger can be operated in the efficient counterflow mode with no dilution of the exhaust stream. Therefore, it can provide greater energy savings from exhaust streams currently recuperated by the metallic heat exchanger. Additionally, because it is less susceptible to metal corrosives, the ceramic heat exchanger can be operated in many exhaust streams in which the metallic heat exchanger would not survive.

### Conclusions

(1) There are significant markets for the ceramic heat exchanger.

(2) Whereas there is no operational distinction of the ceramic heat exchanger from the metallic, at high temperatures compactness may not show the same trends as it does in the lower-temperature metallics.

(3) Metallic heat exchangers provide a configurational basis on which the ceramic may build.

(4) Both monolithic and nonmonolithic assembly methods have significant advantages and disadvantages.

### References


1. M. Coombs, D. Kotchick, and H. Strumpf, High Temperature Ceramic Recuperator and Combustion Air Burner Programs Semi-Annual Report April–October 1981. GRI Document No. 81–18550, November 1981.
2. W. J. Rebello, PAR Enterprises Incorporated; private communication.
3. J. I. Federer, Presentation to the Waste Heat Recovery Contractors Information Exchange Meeting, Los Angeles, CA, March 3, 1982.

# Ceramics and High-Temperature Waste-Heat Recovery

JOHN B. PATTON

Conservation Technologies Division
U.S. Department of Energy
Idaho Operations Office
Idaho Fall, ID 83401

This paper treats the application of ceramics to the recovery and utilization of waste heat from high-temperature, industrial waste streams and reviews the status of high-temperature waste-heat recovery technology. Factors that are seen as delaying the industrial implementation of ceramic heat exchangers are identified and a brief overview of the Department of Energy's High Temperature Waste Heat Recovery Program and overall approach is provided.

High-temperature waste streams are defined, for this paper, as those industrial waste streams where high temperatures and/or foulants and corrodants preclude the use of metallic heat exchangers. Energy losses from flue gases of many high-temperature manufacturing processes have been estimated to be as high as 70%. Effective recovery and utilization of this waste heat could lead to energy savings as high as 60% of the fuel used. The total amount of energy that could be recovered from high-temperature industrial waste streams in the United States has been estimated to be as high as one quad ($10^{15}$ Btu) annually. In light of the fact that a quad is equivalent to more than 150 000 000 barrels of oil, the need to encourage energy savings in this area is clear.

#### Technology Status

The efficient recovery of waste heat from high-temperature, dirty waste-heat streams has presented industry with a formidable task. Historically, recuperators and waste-heat boilers have been made of metal. However, these conventional metallic heat exchangers usually cannot survive in high-temperature, dirty environments. Where metallic exchangers have survived, severe performance penalties have been incurred (e.g., cooling of the flue gas with dilution air or high fouling factors resulting in reduced exchanger effectiveness (less than 30%) and/or reduced preheat air temperatures (less than 537°C (1000°F))). As an alternative to metals, ceramics offer the potential of significantly higher temperatures and much better resistance to many corrosives. Silicon carbide appears to be the best material currently available to provide reasonable service and cost performance to allow effective heat recovery in these high-temperature industrial waste streams.

The question that must be answered, then, is what has prevented or delayed the application of ceramics in the heat-exchanger area. There appear to be several related factors that have all contributed. These factors include: (1) limited availability of funding for capital equipment; (2) the many and varied types of industrial waste streams, and uses for the recovered energy; (3) the present cost of ceramic components; (4) the lack of documented performance data; and (5) the lack of a design base for ceramic heat exchangers. A brief discussion of each of these factors follows.

#### Lack of Capital Funds

The present scarcity of capital funds in U.S. industry has hurt the implementation of all forms of energy-saving technology. Many companies now require a payback of one year or less before they will even consider a capital investment. Furthermore, many industries such as steel and aluminum are producing at only small fractions of their overall capacity. These industries have traditionally been reluctant to make this kind of capital investment and they are today even more reluctant to commit capital funds. A recent survey indicated that it would be at least five years before there would be a significant improvement in this area.

Given this situation it is extremely difficult if not impossible to sell an unproven waste-heat recovery system. Similarly, because of economic risk and long evaluation periods, there is little motivation on the part of heat-exchanger OEM's to pursue (i.e., pay for) the development of new systems.

#### Varied Types of Industrial Streams

In addition to a shaky market situation, a potential developer or marketer of a high-temperature energy-recovery system is faced with the difficult process of identifying a suitable industrial application, including the use for the recovered energy and the most suitable types of heat exchangers to recover the energy. Major industrial high-temperature waste streams of interest include those from (1) steel soaking pits, (2) steel reheat furnaces, (3) aluminum remelt furnaces, and (4) glass-melting furnaces.

A heat exchanger for each type of furnace would have different design requirements. Many furnaces, even within a specific type of furnace, have different requirements. A heat exchanger developed for a specific application will probably not be readily applicable to another type of furnace and, possibly, not even applicable to other furnaces of the same type (e.g., aluminum remelt furnaces).

There are numerous potential uses for the waste heat recovered from high temperature. Four primary uses are (1) combustion air preheat, (2) steam generation, (3) power generation, and (4) feed-stock preheat.

There is also a great deal of controversy among both potential users and developers as to which use is most appropriate for their specific applications. Not only are the uses for the recovered waste heat varied but the types of heat exchangers that could be used are widely varied. Types of heat exchangers that could be used include shell and tube, plate fin, heat wheel, or fluidized bed.

Failure to select a suitable combination of waste stream(s), heat-exchanger type, and use for the recovered energy could result in the development of a system having a limited market, minimum payback, and/or poor industrial acceptance. Without an adequate design base the technical risk in this area is high.

### The Present Cost of Ceramic Components

A fairly wide variety of different sizes and shapes of heat exchangers has been produced. The costs of these systems have typically been high because of the limited number produced and the initial development and tooling costs for the ceramic components. As more and more components are produced the costs should come down, but until a firm low cost is established, it is difficult to stimulate significant interest in the development of ceramic heat exchangers.

### Documented Performance

The selection of any waste-heat recovery system would be based on three factors: (1) performance, (2) reliability/maintainability, and (3) cost.

These three factors are the primary concerns of the user. The performance of a waste-heat recovery system encompasses more than just energy recovered. Such things as increased throughput, furnace life, and improved product quality are all part of the equation. The system must be reliable. It must provide a minimum level of performance for a minimum lifetime. The cost—which really means payback—must be acceptable to the user. Finally, the user must be confident that a system he selects and installs in his plant or process will not have a detrimental effect on his product, will be reliable and not overwhelm him with maintenance or downtime, will provide the energy savings promised by the systems supplier, and, finally, will have a long enough lifetime to allow him to recover his costs.

Unfortunately, ceramic heat-exchanger technology is in its infancy. We do not yet have adequate answers or responses to the simple concerns of performance, reliability, and cost. The user's risks are presently too high. But we are getting closer. In 1985 the DOE will have three full-scale waste-heat recovery systems on line at a steel soaking pit, an aluminum remelt furnace, and a steel reheat furnace. A fourth full-scale unit is scheduled to go on line in early 1986. These units will provide much of the needed performance information.

### Design Base

There is no apparent technical or economic reason why ceramic materials cannot eventually be used successfully for high-temperature heat exchangers. The key word here is "eventually." Because this technology is extremely young, the designer simply does not have an adequate design base. Characterization of materials is often inadequate. There is not enough assurance that the next batch of ceramic tubes will be the same as the previous batch. Many of the problems faced by the designer are straightforward; unfortunately, they are also being encountered for the first time. Consequently, the development of a ceramic heat exchanger can be a long, tedious, and expensive process.

## High-Temperature Waste-Heat Recovery Program

The Office of Industrial Programs (OIP) has been involved in the area of high-temperature waste heat recovery for almost a decade. In reviewing this OIP program three basic areas will be covered

(1) Available technology—those completed projects that have presently available technology.

(2) Emerging technology—those ongoing projects that are expected to

provide commercially available systems in the next three years.

(3) Advanced technology—those ongoing projects that are addressing critical technology deficiencies.

### Available Technology

The need for advanced heat exchangers was recognized in DOE, specifically the Office of Industrial Programs (OIP), by funding for development and commercialization during the period of 1976 through 1980, of two ceramic recuperators: a plate fin exchanger manufactured by GTE, and a shell and tube heat exchanger manufactured by Hague International. Both ceramic units offer performance advantages over conventional metallic units and are capable of producing 648°–815°C (1200°–1500°F) preheated combustion air (i.e., heat-exchanger effectiveness up to 50%). These recuperators are generally limited to clean or mildly corrosive environments and the GTE unit is quite small (approximately one cubic foot in volume) and applicable only to small furnaces. Other companies are now pursuing or marketing similar exchangers.

### Emerging Technology

High-temperature environments (above 815°C) can both decrease the structural strength and increase the corrosion rates of the heat exchanger. Fouling degrades the performance and corrosion decreases the lifetime as well as the structural strength of the exchanger. Recognizing the need for improved fouling and corrosion resistance, other concepts such as the fluidized bed heat exchanger and alternate tube-type recuperators have been introduced. They appear to have the potential of performing satisfactorily in the waste streams of interest (e.g., steel soaking, aluminum remelt, and glass-melting furnaces). Fluidized bed heat exchangers are of interest because of their compactness, enhanced heat transfer, and self-cleaning capability. Ceramic tube recuperators are of interest because of their ability to survive in high-temperature, corrosive streams without excessive fouling. In 1981 the Office of Industrial Programs funded the development of one fluidized bed recuperator with Thermo Electron, one fluidized bed boiler system with Aerojet, and two high-temperature burner duct recuperators (HTBDR) with Babcock and Wilcox (B&W) and AiResearch.

Thermo Electron is pursuing a "moving" fluidized bed and a "raining" fluidized bed concept for preheating combustion air. The Aerojet fluidized bed concept will generate steam and features a special system for cleaning the distributor plate. Both fluidized bed waste-heat recovery systems (FBWHRS), although providing improved performance, are limited by available (metal) distributor plate materials. AiResearch is developing a ceramic shell and tube exchanger; B&W is developing a ceramic tube-in-tube exchanger. All four systems have been or are being installed and evaluated in actual industrial streams.

### Advanced Technology

OIP recognizes the necessity of assisting the private sector in establishing the technology base necessary to develop, produce, and market high-temperature waste-heat recovery equipment. OIP is supporting a broad-based technology development effort in the area of high-temperature waste-heat recovery. The objective of this effort is to develop a technology base in *critical areas* to enable the private sector to successfully exploit high-temperature, industrial waste-heat streams.

DOE efforts here are limited to those areas where the private sector alone cannot or will not participate. This part of the program is funded at approximately $2 000 000 per year and is entering its second year. The program is field-managed by the Idaho Operations Office.

Critical technology areas that have been identified include advanced heat-exchanger concepts, fouling and corrosion, fabrication methods, NDE and lifetime prediction, sealing and joining of ceramic components, and less expensive ceramic materials.

This part of the program presently consists of the following R&D projects;

(1) Advanced heat-exchanger concepts
  (a) Vortex heat exchanger (United Technology Research Center)
  (b) Ceramic composite heat exchanger (Babcock & Wilcox, Thermo Electron Corp.)
  (c) Spouting fluidized bed (Southwest Research Institute)
  (d) Ceramic tube distributor plate for fluid beds (Aerojet Energy Conversion Co.)

(2) Waste-heat environmental effects—fouling, corrosion, waste stream characterization
  (a) Assessment of waste-stream characterization methods (Idaho National Engineering Lab)
  (b) Fouling probe development (Jet Propulsion Lab)
  (c) Effects of specific waste stream corrodants on ceramic materials (Oak Ridge National Lab)

(3) Fabrication methods for lower cost and/or improved ceramic components
  (a) Laser drilling of ceramic components (United Technologies Research Center)
  (b) Medium-diameter SiC tube fabrication (Oak Ridge National Lab/Carborundum)

(4) NDE and lifetime prediction of ceramic components
  (a) NDE/test method development of ceramic tubes (Idaho National Engineering Lab/X-ray Products Co.)
  (b) Assessment of strength-limiting flaws (Babcock & Wilcox)

(5) Methods to effectively and economically seal and join ceramic components
  (a) Ceramic tube-to-header data base (AiResearch Manufacturing Co. of California)
  (b) Composite materials for low cost, reliable joining, and sealing (United Technologies Research Center)

(6) Low-cost ceramic materials resistant to hostile waste streams
  (a) Improved SiC powder development (Oak Ridge National Lab/ART, Inc.)

Many of the OIP projects mentioned here are the subject of other papers presented in the Ceramics in Heat Exchanger sessions.

## Summary

Much work remains to be done in the move to gain industrial acceptance of advanced heat-recovery systems. DOE has provided and will continue to provide funding in key areas where private funding is not available.

Documented performance is essential for industrial acceptance. The design base will continue to grow as more and more ceramic heat exchangers are developed, installed, and operated; their operation for extended periods of time will build confidence and help establish cost performance factors.

DOE's Emerging Technology projects and Advanced Technology projects are designed to provide a strong foundation for expanding the technology base.

Cheaper materials—including new precursor techniques—are being developed. Increasing quantities of ceramic components are being produced annually, leading to unit cost decreases. Again it will take time to reach truly economical costs, but there is little doubt that ceramic heat exchangers will eventually gain widespread industrial acceptance.

# Advanced Heat-Transfer Systems: GRI Strategy and Accomplishments

CHARLES J. DOBOS

Gas Research Institute
Chicago, IL 60631

Existing high-temperature industrial systems are currently limited in efficiency and productivity by inadequate material temperature capability, corrosion resistance, and durability. GRI's objective is to improve the performance and capability of gas-fired heating and melting systems through the development of high-temperature ceramic heat-transfer and combustion systems. A matrix of development projects involving ceramic materials and component development, heat-exchanger developments, and industrial system field testing has been established to accomplish this objective. Development success to date has included both metallic and ceramic tubular recuperators for large-volume clean or corrosive flue gases and compact ceramic recuperators for smaller, clean industrial applications. Ceramics development projects have resulted in low-cost manufacturing technology for recuperation system components, and will be expanding to develop the diverse components required for high-temperature radiant tube heating systems, immersion tube melting systems, and gas-fired prime movers. Development strategies and technical accomplishments are described.

## Industrial Ceramic Needs

High-temperature, gas-fired industrial processes are limited in thermal efficiency, productivity, and temperature capability. The thermal efficiency of these processes, typically 10 to 25%, results in part from lack of combustion and heat-transfer control, heat loss from insufficient insulation, and poor operating practices. The dominant factor, however, is energy loss from flue gases discharged at 871°–1426°C (1600° to 2600°F). Substantial gains in the efficiency of gas use can be realized if the energy in these gases can be used to preheat combustion air, preheat process feedstock, or generate steam. Development of an advanced recuperator system, schematically depicted in Fig. 1, can provide 40 to 60% energy savings by preheating combustion air to 871°–1093°C (1600° to 2000°F).

The outstanding high-temperature heat-transfer capabilities of natural gas for indirect heating, melting, or preheating processes are currently constrained by material-temperature and corrosion-resistance limitations. Despite the recognition of these limitations, development of advanced heat-transfer systems has been slow due to concerns about their complexity, high initial cost, and reliability at high operating temperatures. Performance and cost effectiveness of those systems presently in use are degraded by poor corrosion resistance, thermal/mechanical failures, and high maintenance requirements. R&D is needed to (*a*) increase the thermal efficiency of existing and new gas-fired processes through cost-effective waste heat utilization; (*b*) improve productivity and product quality through the use of higher furnace operating temperatures

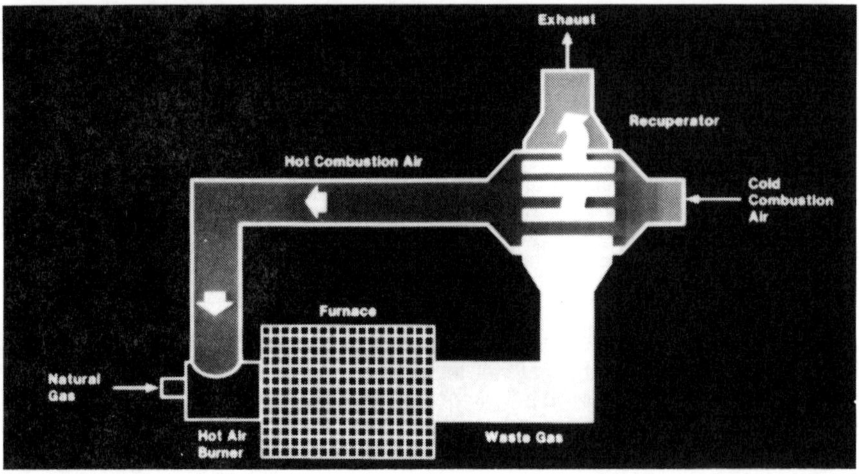


Fig. 1 Schematic of industrial recuperation system.

and more precise heat-transfer control; and (*c*) expand the cost-effective use of gas combustion to new, higher-temperature heating and melting systems. Development of acceptably priced and reliable ceramic components is the major factor in successfully addressing these needs.

### Heat-Transfer System Development Strategy

GRI's tactical objective is to improve the efficiency, productivity, and temperature capability of gas-fired heating and melting systems through the development of high-temperature ceramic heat-transfer and combustion-system components.

For waste heat utilization opportunities, specific technical improvements are needed to (*a*) enhance the durability of ceramic components in extended high-temperature service; (*b*) improve the corrosion resistance of ceramics for use in harsh, corrosive environments; and (*c*) reduce component costs and installed-system payback periods. Extending the temperature capability of heat exchanger and materials technology to 1315°C (2400°F) is the key to efficiency improvements (see Fig. 2) and is our technical goal.

To successfully accomplish this goal, a systematic integration of ceramic material and component development, heat-exchanger development, and industrial field testing must be accomplished. GRI will continue to capitalize on the interest and support of qualified heat-exchanger manufacturers to accomplish these activities; but a matrix of organizations is required to enable industrial recognition and acceptance of these new technologies. System field-test activities must be conducted with qualified system and furnace manufacturers as well as industrial end-users, as shown in Fig. 3. Material development and long-term ceramic material testing must also be integrated with research by ceramic manufacturers, universities, or government laboratories where appropriate.

Due to the diverse industrial furnace population (Fig. 4) and the require-

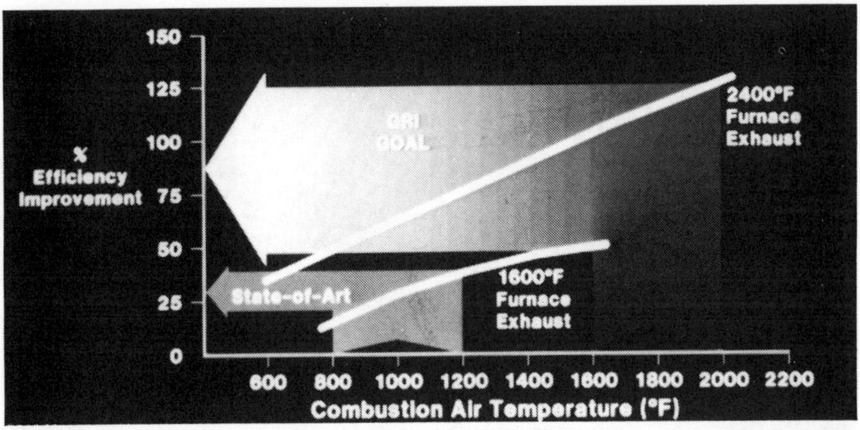


Fig. 2. GRI's efficiency improvement goal.

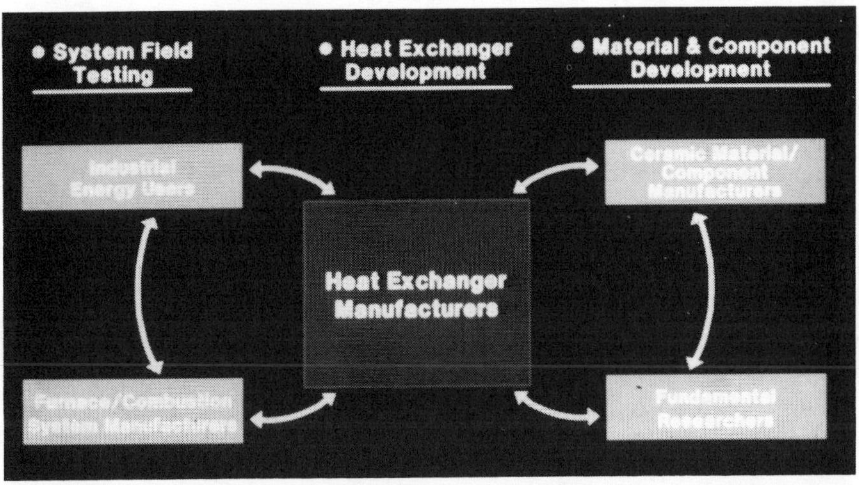


Fig. 3. Required R&D cooperation.

ments for multiple organization involvement and cooperation, a matrix of development projects was initiated in 1981 through 1983. Heat-exchanger development to date has centered on (*a*) both metal and ceramic tubular recuperators for large volume 2.9 m W (10 mm Btu/h and up) clean or corrosive flue gases; (*b*) compact ceramic recuperators for small, clean flue gas applications 0.3 to 8.8 mm W (1 to 30 mm Btu/h); and (*c*) high-temperature radiant tube recuperators for indirect-fired furnaces.

Material and manufacturing technology development activities have focused on methods for low-cost manufacturing of reliable ceramic components of the diverse sizes and shapes required for use in advanced heat-transfer systems. The most noted accomplishments are listed below.

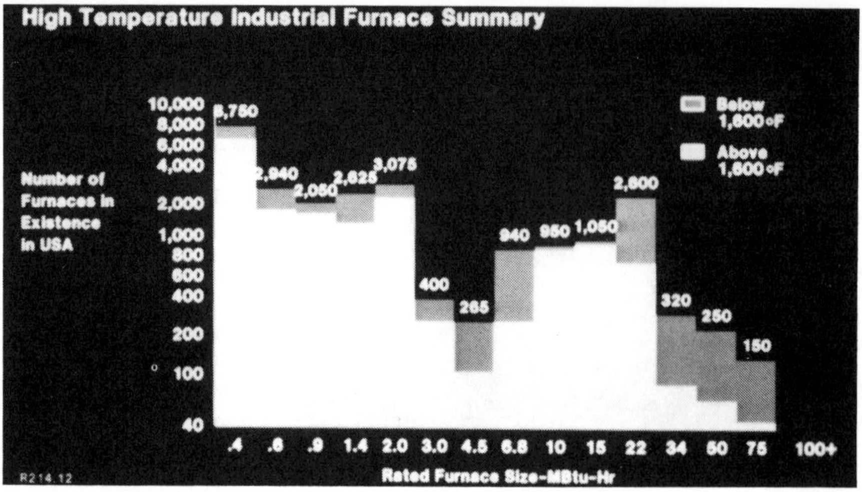


Fig. 4. Waste heat recuperator applications.

**Development Accomplishments**

AiResearch Manufacturing Co. completed proof-of-concept testing of a tubular ceramic recuperator prototype which produces 1093°C (2000°F) combustion air from 1315° (2400°F) exhaust inlet temperatures. A recuperation system based on this technology will be field-tested (with DOE funding) during 1985 to quantify the benefits of 1093°C combustion air in operation of a steel rotary reheating furnace.

Norton Co. developed a low-cost silicon carbide ceramic tube (2 inches in diameter and 4 feet long) and has applied it to a low-maintenance recuperator for corrosive flue gases designed and developed by C&H Combustion. Two new, efficient gas-fired processes (a heat-treatment/casting-wax incinerator system for turbine blade manufacturing and an aluminum melter/holder for aluminum ingot recycling) have been designed, constructed, and installed by C&H with integral ceramic recuperator, advanced controls, and high-temperature refractory and fiber insulation. Field testing is currently under way to quantify improvements over the existing heating/melting technology used by the industrial partners.

Coors Porcelain Co. initiated field-testing of compact ceramic matrix recuperators 0.3 mm W (1 mm Btu/h) based on the development of a low-expansion, mixed-oxide ceramic material. Crossflow matrices have been retrofitted to three heat-treating furnace systems with exhaust temperatures up to 1093°C. Data collected to date indicate excellent matrix durability, decreased fuel consumption (34, 42, 48%, respectively), and enhanced heat-treatment quality.

Norton Co. successfully developed SiC precision casting and sintering techniques for 2 foot by 2 foot ceramic plates for AiResearch's finned-plate recuperator concept. The unit incorporates advanced techniques for mechanical plate sealing and for complex silicon carbide ceramic plate fabrication. System proof-of-concept testing 1343°C (2450°F) will be initiated in 1985 to permit the design and installation of a field test system in a steel industry application by 1986.

Coors Porcelain Co. developed a process for producing siliconized silicon carbide ceramic which successfully meets the manufacturing cost target of less than \$20/foot for 1-inch-diameter recuperator tubes. Thermal and mechanical evaluations are currently under way by heat-exchanger and thermal-system manufacturers to determine long-term performance, and by industrial end-users to determine flue gas compatibility and corrosion resistance.

Carborundum Co. has developed an advanced plastic-extrusion technology for fabricating complex ceramic shapes from their alpha-sintered silicon carbide material. Various finned tubular components and large-diameter (4-inch OD) radiant tube components have been manufactured using the low-cost extrusion method. These components will be tested in advanced thermal systems up to 1537°C (2800°F) to determine the feasibility of new indirect heating and melting concepts.

**Future R&D Strategy**

These heat-exchanger and ceramic material development projects have provided the opportunity for a number of companies to develop and test product designs and concepts with substantially reduced financial risk. The majority of these development efforts have been successful through proof-of-concept testing in the laboratory and initial system field testing with cooperative industrial partners.

While financial risks involving product development have been tempered through GRI support, the anticipated business risks faced by the heat-exchanger and ceramic-component manufacturers remain high. Decisions in terms of capital investments, business partnerships, and commercialization strategies must be made. These decisions can be made wisely only if a market need is developed and can subsequently be assessed and addressed. To develop this market, technology recognition and acceptance are required by industry. GRI strategy to accomplish this market development (shown in Fig. 5) will involve two critical activities in 1984 through 1986:

(1) System field test to determine the cost/performance benefits of advanced heat-transfer technology.

(2) Process compatibility testing of newly developed ceramic materials and long-term testing of ceramic components' performance and reliability.

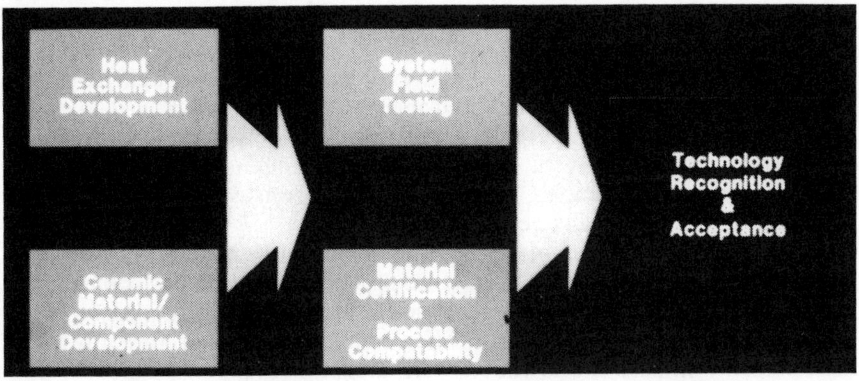


Fig. 5. Advanced heat-transfer systems project area R&D strategy.

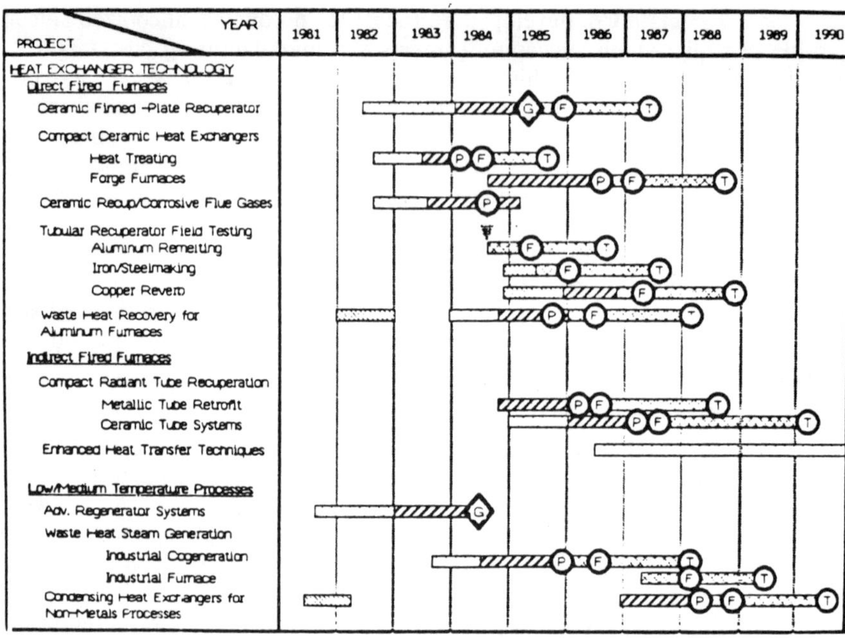


Fig. 6. Advanced heat-transfer systems network diagram (part 1). Lab-scale testing (///, prototype testing; xxx, full-scale testing). G=go/no go decision. P=proof of concept. E=field test program completion. T=technology transfer.

Planned project specific activities are summarized in part 1 of the network diagram (Fig. 6).

While field testing and quantification of system performance and efficiency improvement will continue for these systems in 1984 and 1985, a shift in development focus is required to address the material needs and opportunities in three new specific areas:

(1) *High-Temperature Indirect Heating*: Application of ceramic materials and components to radiant tube heating systems can extend gas-fired system temperature and productivity capabilities (Fig. 7)

(2) *High-Temperature Indirect Melting*: Immersion melting concepts reliant on advanced materials and components can extend current efficiency and productivity limitations of direct-fired melting systems (Fig. 8).

(3) *Gas-Fired Prime Movers*: Development of new materials is required to improve the versatility of gas-fired movers through (*a*) higher operating temperatures and efficiencies, (*b*) consolidation of waste heat into one usable form, and (*c*) improved engine operating life.

As a result, the focus of GRI's high-temperature material development activities will be reordered (see Fig. 9) to provide impetus to development and acceptance of these new, higher-temperature technologies. The plan for a balanced program of ceramic heat-transfer component projects to address this need is shown in part two of the network diagram (Fig.10).

These longer-range activities will consider advanced materials technology including ceramic matrix composites, carbon/carbon composites, ceramic coat-

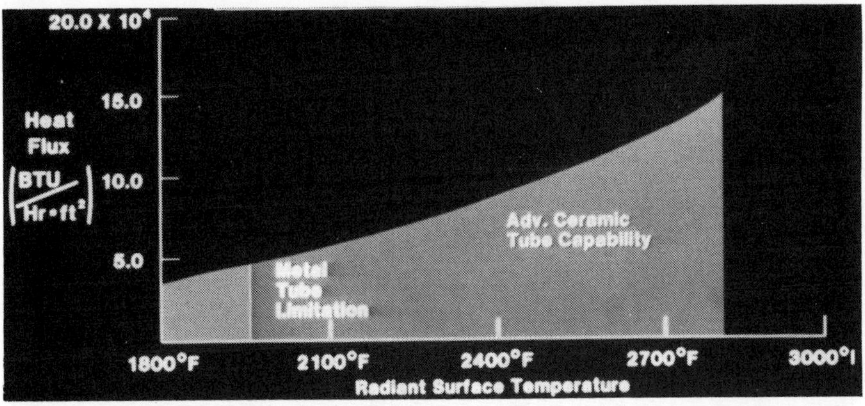


Fig. 7. Indirect radiant tube heating system.

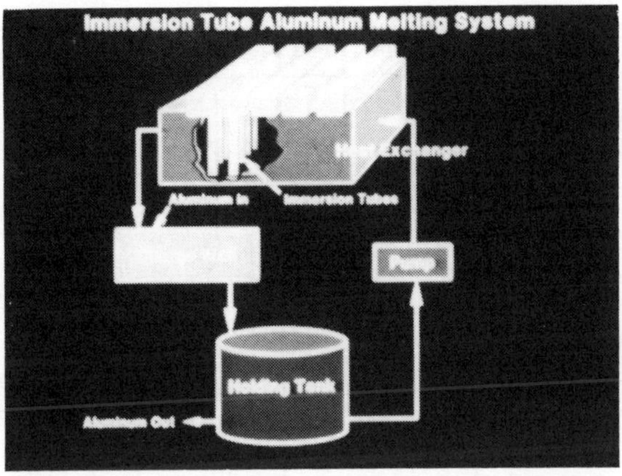


Fig. 8. Indirect metal melting process.

ing, transformation-toughened ceramic, monolithic structural ceramic, etc., to extend the applicability of natural gas to new industrial systems. Subsequent to identification of a feasible application, GRI will evaluate the potential benefits to capitalize on these emerging technologies. In addition, GRI will develop and continually update a high-temperature materials technology data base to classify and prioritize GRI's materials R&D requirements, opportunities, and options.

GRI must continue to capitalize on the interest and support of qualified ceramic-component manufacturers to accomplish these goals. System field-test activities with qualified system or furnace manufacturers as well as industrial end-users remain critical to the acceptance of all advanced material technologies. Ceramic component and long-term ceramic material testing will feed back into research by ceramic manufacturers, government laboratories, and university professionals. Response to the need for special materials and com-

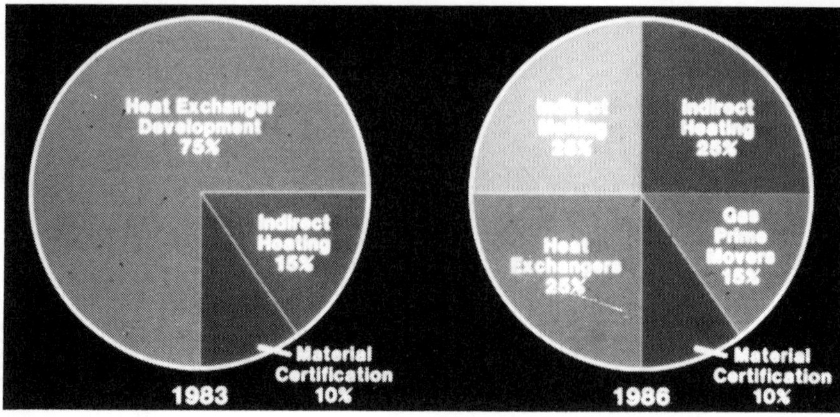


Fig. 9. Reordered R&D focus, 1983–1986.

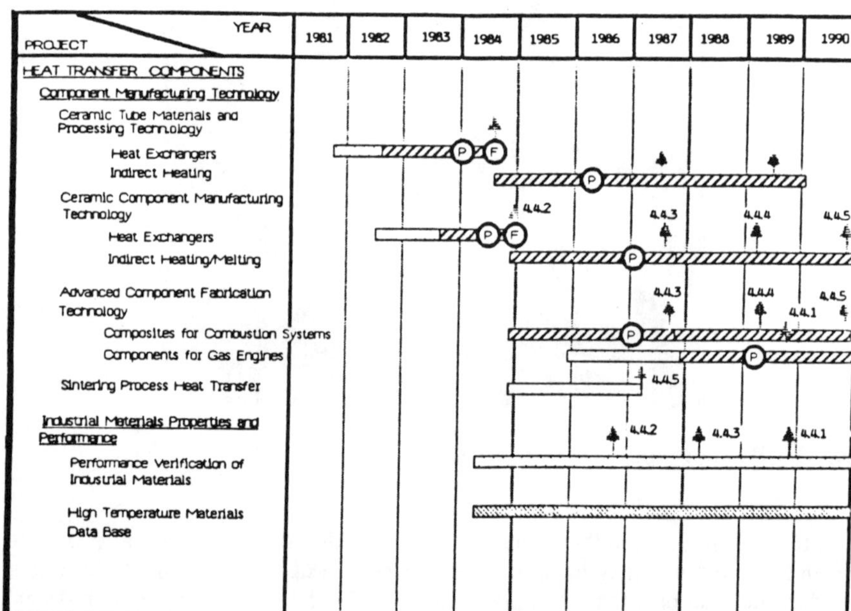


Fig. 10. Ceramic-component development network diagram (part 2). Technical market feasibility paper study.

ponents for advanced systems being developed in GRI's other industrial utilization project areas will continue to broaden the scope and application of R&D pursued in this project area.

Section **II**

# Applications/Approaches/Designs

## RECUPERATORS (TUBE TYPE)



## RECUPERATORS (FINNED PLATES)

# Ceramic Tube Heat Recuperator— A User's Experience

PATRICK W. WARD

Aluminum Company of America
Davenport, IA 52808

The Aluminum Company of America has actively pursued the development of recuperation on aluminum melting furnaces. Since 1974, these installations have included radiation, high-efficiency metallic, and ceramic heat wheel recuperators. These early recuperators were disappointing economically due mainly to poor performance and excessive downtime. This paper describes an improved recuperator design which is able to tolerate the severe environment of an aluminum melting furnace and give a consistent 30+% fuel savings.

In proposing a recuperator for Davenport Works, Alcoa's largest heat-treated plate and sheet mill, three problem areas were addressed:

(1) *Space:* In Davenport's Ingot Plant, there is little additional space around the melting complexes. The recuperator must be an integral part of the furnace.

(2) *Reliability:* The recuperator must operate a minimum of five years with only minor annual repairs. The unit should be designed to operate in undiluted products of combustion without benefit of cooling combustion air.

(3) *Operation:* The furnace must operate the same as a nonrecuperator-equipped furnace. No special consideration to scrap cleanliness or scrap type shall be required.

In 1980, Davenport and Corporate Engineering felt the Hague Cerhx* Ceramic Tube Recuperator could best address the before mentioned concerns. However, the proposed unit would be three times larger than any previous Hague recuperator and would be the first on an aluminum melting furnace. This paper describes Alcoa's experience with this recuperator since its installation in November of 1981.

The first section of the report describes the design of the heat-recovery system. The second section discusses the recuperator's construction and how its ceramics tolerate a production-furnace environment. In the third section, the recuperator's performance is summarized and recommendations are given.

Information pertaining to the *measured* fuel savings resulting from this installation will not be discussed in this report. Although the heat-recovery system's contribution is very significant to these savings, additional savings are due to furnace design that Alcoa considers proprietary.

---

*Cerhx, Hague International, South Portland, ME.

### Heat Recovery System Design

Although the recuperator is the heart of a heat-recovery system, its performance and reliability can be positively or negatively enhanced by the design of the ductwork, blowers, burners, and controls.

Several important features have been designed into the heat-recovery system associated with the ceramic tube heat recuperator. These design features include energy input sizing, combustion system, furnace pressure control, data acquisition, and maintainability.

A conservative approach was applied in each of these areas due to the unknowns associated with a "first of its kind" installation. Less costly approaches to control are available; however, these control systems might not offer the control flexibility and reliability expected of mill-duty, production equipment.

#### Energy Input Sizing

Hague's recuperator was designed to develop a 815°C (1500°F) preheated air temperature with a flue gas temperature of 1371°C (2500°F), an effectiveness† of 66%. The refractory-lined, hot air ductwork and burners were sized based on that predicted preheat at a 7 mm W (24 mm Btu/h) *new* fuel input. The fuel train and blower were sized for a 9.9 mm W (34 mm Btu/h) new fuel input, in the event a catastrophic recuperator failure returned the furnace to a cold firing mode. No provisions were made to operate the furnace with the installed recuperator "off line."

#### Combustion System

The dual-fuel combustion system features a fuel-primary electronic air-fuel ratio controller that measures the preheated combustion air after the recuperator. In addition, the system can be switched from natural gas to No. 2 fuel oil in minutes, if required. Provisions to fire oil at 10% and natural gas at 90% to produce a luminous gas flame have been installed, but not evaluated to date.

#### Furnace Pressure Control

Since the pressure drop on the flue side of the recuperator was predicted at one inch of water column, an eductor was installed to provide the necessary mechanical draft to the unlined furnace stack. The mechanical draft is controlled by an electronic furnace pressure control. A flue thermocouple in the unlined stack controls a dilution air damper above the recuperator to maintain a stack temperature of 371°C (700°F) (Fig. 1).

#### Data Acquisition

The recuperator and furnace are monitored by a computer data-acquisition system. The software, developed and written by plant engineers, collects pertinent information on the furnace complex on a 24-h basis. This tool has provided valuable operating and diagnostic information to production, maintenance, and engineering. Sample real-time displays can be found throughout the report.

#### Maintainability

Since the recuperator weighs 12 tons and must be set on top of the

†Effectiveness is a measure of a recupertor's performance at the designed firing rate:

$$\text{Effectiveness (in percent)} = \frac{\text{Preheated air temperature} - {}^\circ\text{R}}{\text{Inlet flue gas temperature} - {}^\circ\text{R}} \times 100$$

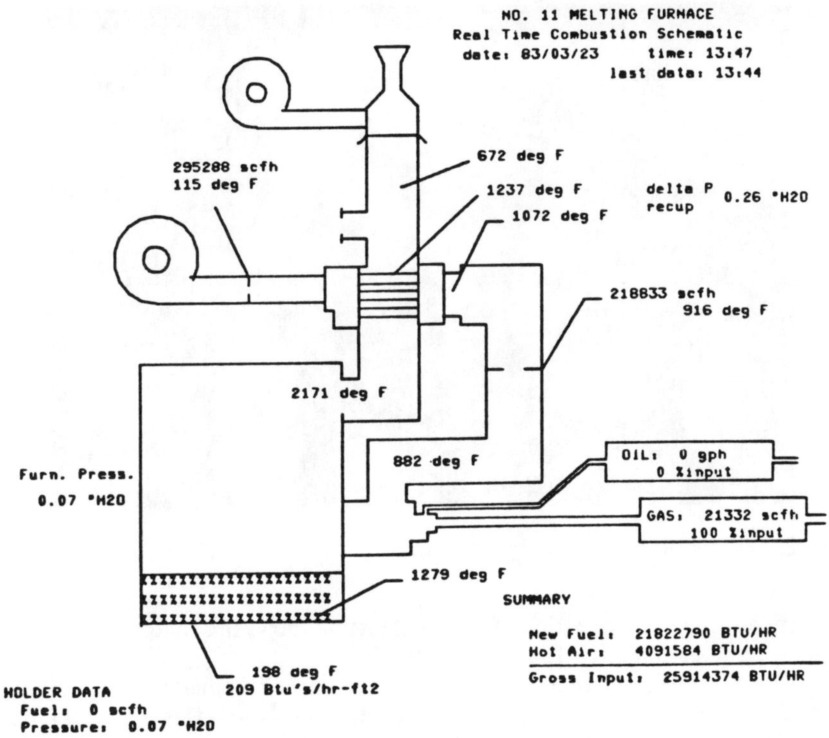


Fig. 1. Real-time furnace schematic.

refractory flue, directly underneath the building crane girder, a recuperator handling system was developed. A portable stand attaches to the back side of the flue, forming a railway to roll the recuperator out into the lift of the 30-ton overhead crane. With this system, the recuperator can be removed in one shift and reinstalled in two shifts (Fig. 2).

## Recuperator Performance and Maintenance History

In over two years of operation, no significant furnace down time has been associated with the ceramic tube recuperator. The heat-recovery system reduces overall fuel consumption an estimated 35% when operating near designed effectiveness. However, due to combustion air leakage, the recuperator has operated near designed effectiveness only 50% of the time. This section will review the construction of the recuperator, and the unit's chronological maintenance history and measured operating performance over the last 29 months.

### Recuperator Construction

The recuperator consists of 8 rows of 24 ceramic tubes. Each tube is held by two silicon carbide hex adapters. A set of ceramic turbulators is inserted in each tube to increase the convective heat transfer of the tube. The entire assembly is held under spring tension by a spring pack assembly (Fig. 3).

Three types of recuperator tubes have been used to date:

(1) HT–35 radiation tube: A nitrite-oxynitride bonded silicon carbide cast

Fig. 2. Recuperator handling system.

in a smooth tube configuration. Analytical testing shows this material's bonding system remains intact to 1537°C (2800°F).

(2) HT–25 finned tube: A high-purity calcium aluminate-bonded silicon carbide in a finned configuration; analytical testing showed this material failed exponentially over 1093°C (2000°F).

(3) HT–35 finned tube: A prototype nitride-oxynitride bonded silicon carbide cast in a finned configuration.

From the flue side of the recuperator, the initial installation consisted of three rows of HT–35 radiation tubes, one row of HT–35 finned tubes, and four rows of HT–25 finned tubes. This combination was selected at a joint meeting of Hague and Alcoa personnel and was based on Hague's predicted mean tube temperature of a recuperator with a mean flue gas inlet temperature of 1371°C (2500°F). The HT–35 material in the finned configuration was unproven on a production recuperator; however, the group felt the predicted increase in recuperator effectiveness warranted the use of the prototype tube.

It should be noted that Hague designed-in a 15% combustion air leakage to cool the spring pack assemblies. Therefore, all subsequent leakage figures must take this initial condition into consideration.

*New to Six Months:* Initially, the performance of the recuperator was close to Hague's predictions, with a measured effectiveness of 62% (767°C (1250°F) preheat from 1260°C (2300°F) flue gas temperature). After two months of stable operation, combustion air leakage began to increase from 18 to 23% in a 9-week period (an increase of 5%). This leakage reduced the combustion air preheat to 454°C (850°F) with a 1232°C (2250°F) flue gas temperature and reduced the effectiveness to 48%. Data taken by Mr. Roger Woodward, Hague's Marketing Director, confirmed the leakage in April 1982 and plans were made to repair the unit during the 1982 plantwide maintenance shutdown in July. Prior to repairing the recuperator, the leakage was measured at 45% and the effectiveness had been reduced to 41%.

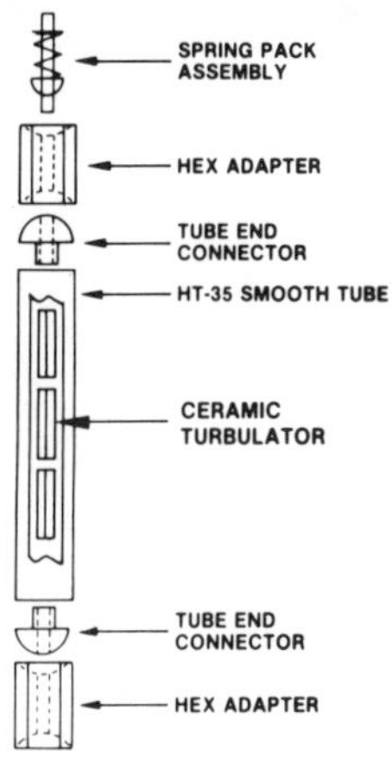


**TYPICAL SMOOTH TUBE ASSEMBLY**

**HAGUE CERHX RECUPERATOR**

Fig. 3. Recuperator spring pack assembly.

After disassembling the recuperator, the damage was found to be mainly confined to the fourth row of HT–35 finned tubes and fifth row of HT–25 finned tubes. Almost all tubing in these rows had longitudinal cracks, some of which were one-half-in. wide and ran three-quarters of the length of the tube (Fig. 4). Hague's Mr. Woodward determined that the finned HT–35 tubes failed due to thermal expansion. This failure probably provided air to sustain secondary combustion during periods of charging oily scrap, which in turn overheated the HT–25 material, resulting in the failure of row five.

Based on Hague's negative experience with HT–35 finned tubes, in several other applications, the recuperator was returned to production with four rows of HT–35 smooth tubes and four rows of HT–25 finned tubes.

*Six to Fourteen Months:* Returning to production on July 12, 1982, the recuperator operated with 18% leakage, well within Hague's spring cooling leak-

Fig. 4. Finned tube failure.

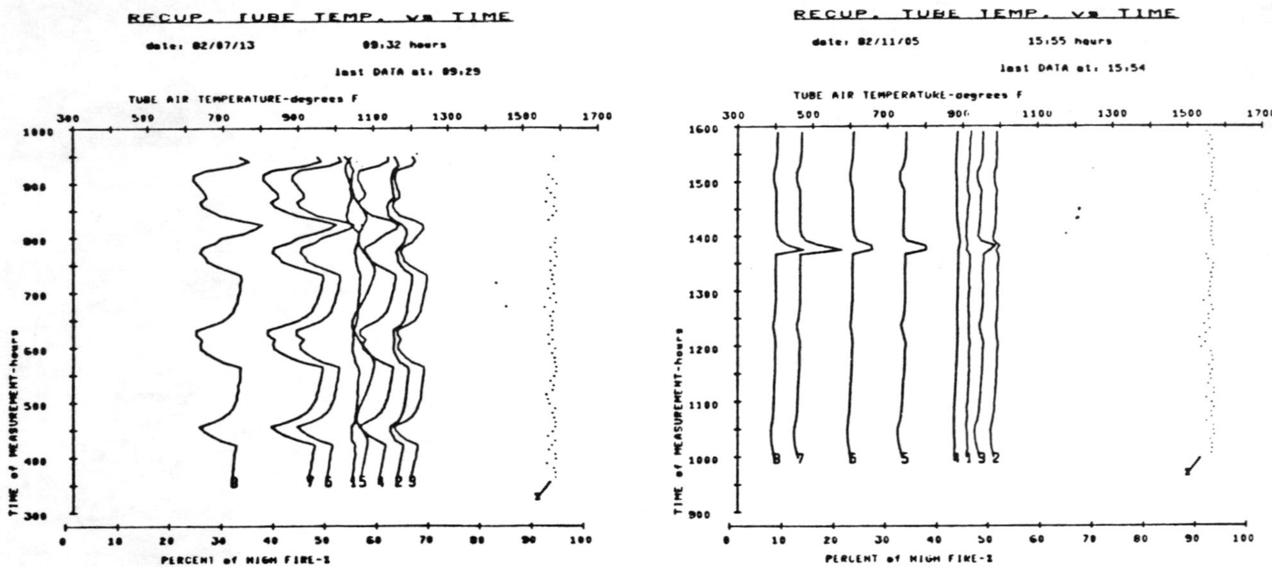


Fig. 5. Combustion air temperature by tube row vs time.

age prediction. Under these conditions, the effectiveness was restored to 62% (676°C (1250°F) preheat from 1260°C (2300°F) flue gas temperature).

After only six weeks of operation, the combustion air leakage rapidly increased to 40%, with the leakage appearing to be located at row 5, the first row of finned HT–25 tubes (Fig. 5). For the next five months, the leakage remained near 45%. The recuperator's effectiveness was again reduced to 42%.

Apparently, the problem had been incorrectly diagnosed during the previous repair. It is possible that the operation of the HT–25 material above 1093°C (2000°F), in conjunction with the finned tube design, was causing the tube failures.

It appears that the computer prediction of recuperator performance overestimated the amount of energy removed from the flue gas by the smooth tubes via radiation. This phenomenon would result in a higher-than-predicted mean operating temperature for the HT–25 tubes. The higher temperature would push the first rows of HT–25 tubes over their critical 1093°C limit. Possibly the leakage stabilized after it became great enough to cool the remaining HT–25 tubes below 2000°F.

Probably more important, the finned tube design provides the perfect "flame holder"; therefore, any secondary combustion resulting from charging oily scrap would occur in the first two rows of finned tubes. Also, over time, a dry aluminum oxide would collect on the finned tubes, restricting the flow of flue gases, providing the potential for hot spots in the recuperator. This plugging was apparent on all rows of finned tubes, but was most obvious on the first two rows.

On the other hand, the HT–35 smooth tubes have performed better than expected. Visual and analytical inspections reveal no apparent breakdown of HT–35 smooth tubes after one year of operation in the hottest portions of the recuperator.

Based on this hypothesis, Davenport management approved the purchase of materials to revise the recuperator to eight rows of HT–35 smooth tubes. This repair was set for the first week of February 1983.

*Fourteen to Twenty-Four Months:* The disassembly of the recuperator revealed that the fifth and six rows, which are the first and second rows of finned HT–25 tubes, were 70% destroyed. However, most of the smooth HT–35 tubes appeared to be in excellent condition.

During installation of new tubes, the assembled dimensions of the new parts were inadequate to engage the spring pack assembly. This problem was resolved by the addition of two standard 1½-in. washers between the spring and the cold air strong back.

An air flow check of the recuperator prior to bringing the furnace up to temperature revealed a 31% air leak. This leakage was found to be spring cooling leakage. It appears that the addition of the washers doubled this leakage over Hague's estimate.

During this segment of the recuperator's campaign, an average flue gas temperature of 1287°C (2350°F) provided a 648°C (1200°F) combustion air preheat (59% effectiveness). Based on 10 months of continuous operation near the original designed preheat specification of the recuperator, the writer speculates that the convective heat-transfer advantage of the finned tube was exaggerated.

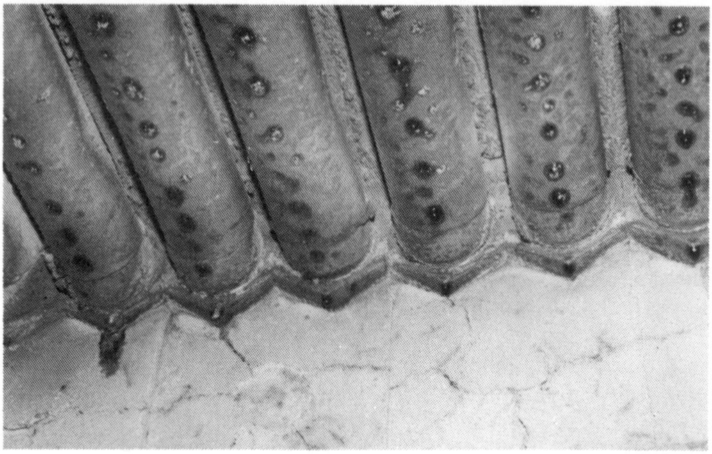

Fig. 6. Flue-side tubes, first row, and cracked support block.

The combustion air leakage stabilized at approximately 21% after two months of operation. The installation of all turbulator-equipped tubes has increased the combustion air pressure drop to 20 in. water column vs 12 in. water column predicted for the original designed input. The retubed recuperator's flue side pressure drop was reduced to 0.30 in. of water column at high fire at installation vs a 1.00 in. water column drop with the previous design. This pressure drop steadily increased to 0.90 in. of water column during the eight months of continuous service. A compressed air blowoff, worked from the top side of the tube bundle, was sufficient to remove the aluminum oxides that collect on the tubes. However, this cleaning technique appears to be effective only on the top two rows of tubes. After back-blowing the recuperator with compressed air, the flue-side pressure drop was reduced to 0.70 in. of water column.

In November 1983, the two-year-old furnace was shut down for two weeks for miscellaneous refractory repairs. During this outage, the bottom row of recuperator tubes was inspected by Davenport Engineering personnel. Although the recuperator tubes appeared to be in good condition, with no obvious pitting or discoloration, a significant amount of glasslike residue was found on the flue floor beneath the tube bundle. In addition, severe deterioration to the cast refractory support blocks on the hot air side was noted during this inspection. The support block failure could be categorized as a classic dissimilar material expansion failure between a castable and its anchor (Fig. 6).

A phosphate-bonded plastic was used to repair the damaged support blocks since enough original material appeared to remain to support the tube bundle. No recuperator tubes were replaced during this outage.

*Twenty-Four Months to Present:* During the November furnace repair, a revision was made to the furnace, increasing its internal surface area. Since the furnace input was limited to 6 736 240 mm W (23 000 000 Btu/h) due to a pressure-constrained blower, an average flue gas temperature of 1204°C (2200°F) was all that could be maintained. Consequently, the average preheat from the

COMBUSTION AIR

Air--IN: 373833 scfh
140 degrees F

Air--OUT: 300236 scfh
1019 degrees F

Leakage: 20 percent

PRODUCTS of COMBUSTION

POC--IN: 2349 degrees F

POC--OUT: 1395 degrees F

Pressure Drop: 1.41 inches of water

RECUPERATOR EFFICIENCY

Energy in Flue Gas: 14654014 BTU/HR

Energy in Combustion Air: 5155571 BTU/HR

Efficiency: 35 percent

STACK SIDE

Row No. 8: 727 degrees F

Row No. 7: 816 degrees F

Row No. 6: 817 degrees F

Row No. 5: 897 degrees F

Row No. 4: 978 degrees F

Row No. 3: 1058 degrees F

Row No. 2: 1065 degrees F

Row No. 1: 1020 degrees F

FURNACE SIDE

Fig. 7. Real-time recuperator performance data taken with input at 8 m W (28 m Btu/h).

recuperator dropped to 482°C (900°F), 51% effectiveness. The furnace's productivity improved slightly after this furnace modification.

In April 1984, the furnace was taken down for one week to install a new combustion air blower and repair the furnace flue. During this outage, an inspection of the recuperator produced one broken tube in the top row of the unit. This tube was identified as a first-generation smooth tube and was easily replaced.

After returning to production, the new blower provided the means to take advantage of the before-mentioned furnace revisions with a higher new fuel input. As a result, the recuperator's effectiveness was restored to 55%, 604°C (1120°F) preheat from 1315°C (2400°F) flue gas as compared to the 59% effectiveness measured after the February 1983 repair. The 4% loss in effectiveness can be attributed to an 18% increase in combustion air flow and recuperator fouling, which is reflected in the higher flue-side pressure drop,‡ 0.45-in. w.c. new to the current 1.41-in. w.c. (Fig. 7).

### Conclusions and Recommendations

This recuperator installation has provided much insight into the potential for ceramic tube recuperation on large aluminum melting furnaces. This installation had addressed the space, operation, and, to a large extent, the economic requirements of this project's scope. The success of this installation

‡Calculated flue-side pressure drop at 8.2 m W (28 m Btu/h) using base of 0.30-in. w.c. at 6.7 m W (23 m Btu/h).

after its February 1983 repair enables Alcoa to install future recuperators based on their economic justification.

### Performance Summary

Several methods are available to provide the user with an indication of performance of a heat-recovery system. The best method would be to measure the energy used by the furnace prior to the recuperator installation vs the energy used after the retrofit. Although this provides the true savings of the new installation, the recuperator often is not the sole provider of the savings, as is the case here.

However, an analytical method is available to predict fuel savings potential based on burner air preheat temperature and furnace outlet temperature. The North American Manufacturing Co. provides a handbook supplement entitled "Thinking Man's Guide to Preheated Air Systems," which contains a table that can be used to predict the percent natural gas saved (Fig. 8).

Based on the table in Fig. 8, the estimated average fuel savings of the recuperator vs time in service was plotted (Fig. 9).

### Recommendations

In proposing future ceramic tube recuperators, the following guidelines are offered:

(1) Process temperatures must exceed 982°C (1800°F) to warrant the cost of ceramics.

(2) In aluminum melting furnaces, use only nitride-oxynitride bonded silicon carbide tubes.

(3) The benefits of finned tubes are questionable and fin tubes should not be used in an aluminum melting furnace installation.

| % Natural Gas Saved | $T_2$, Air Preheat Temperature, F | | | | | | | | | | |
|---|---|---|---|---|---|---|---|---|---|---|---|
| $T_3$, Furnace Outlet Temperature, F | 600 | 700 | 800 | 900 | 1000 | 1100 | 1200 | 1300 | 1400 | 1500 | 1600 |
| 1000 | 13.4 | 15.5 | 17.6 | 19.6 | – | – | – | – | – | – | – |
| 1100 | 13.8 | 16.0 | 18.2 | 20.2 | 22.2 | – | – | – | – | – | – |
| 1200 | 14.3 | 16.6 | 18.7 | 20.9 | 22.9 | 24.8 | – | – | – | – | – |
| 1300 | 14.8 | 17.1 | 19.4 | 21.5 | 23.6 | 25.6 | 27.5 | – | – | – | – |
| 1400 | 15.3 | 17.8 | 20.1 | 22.3 | 24.4 | 26.4 | 28.4 | 30.2 | – | – | – |
| 1500 | 16.0 | 18.5 | 20.8 | 23.1 | 25.3 | 27.3 | 29.3 | 31.2 | 33.0 | – | – |
| 1600 | 16.6 | 19.2 | 21.6 | 24.0 | 26.2 | 28.3 | 30.3 | 32.2 | 34.1 | 35.8 | – |
| 1700 | 17.4 | 20.0 | 22.5 | 24.9 | 27.2 | 29.4 | 31.4 | 33.4 | 35.3 | 37.0 | 38.7 |
| 1800 | 18.2 | 20.9 | 23.5 | 26.0 | 28.3 | 30.6 | 32.7 | 34.6 | 36.5 | 38.3 | 40.1 |
| 1900 | 19.1 | 21.9 | 24.6 | 27.1 | 29.6 | 31.8 | 34.0 | 36.0 | 37.9 | 39.7 | 41.5 |
| 2000 | 20.1 | 23.0 | 25.8 | 28.4 | 30.9 | 33.2 | 35.4 | 37.5 | 39.4 | 41.3 | 43.0 |
| 2100 | 21.2 | 24.3 | 27.2 | 29.9 | 32.4 | 34.8 | 37.0 | 39.1 | 41.1 | 43.0 | 44.7 |
| 2200 | 22.5 | 25.7 | 28.7 | 31.5 | 34.1 | 36.5 | 38.8 | 40.9 | 42.9 | 44.8 | 46.6 |
| 2300 | 24.0 | 27.3 | 30.4 | 33.3 | 36.0 | 38.5 | 40.8 | 42.9 | 45.0 | 46.9 | 48.7 |
| 2400 | 25.7 | 29.2 | 32.4 | 35.3 | 38.1 | 40.6 | 43.0 | 45.2 | 47.2 | 49.2 | 51.0 |
| 2500 | 27.7 | 31.3 | 34.7 | 37.7 | 40.5 | 43.1 | 45.5 | 47.7 | 49.8 | 51.7 | 53.5 |
| 2600 | 30.1 | 33.9 | 37.3 | 40.5 | 43.4 | 46.0 | 48.4 | 50.6 | 52.7 | 54.6 | 56.4 |
| 2700 | 33.0 | 37.0 | 40.6 | 43.8 | 46.7 | 49.4 | 51.8 | 54.0 | 56.1 | 58.0 | 59.7 |
| 2800 | 36.7 | 40.8 | 44.5 | 47.8 | 50.8 | 53.4 | 55.8 | 58.0 | 60.0 | 61.9 | 63.5 |
| 2900 | 41.4 | 45.7 | 49.5 | 62.8 | 55.7 | 58.4 | 60.7 | 62.8 | 64.7 | 66.4 | 68.0 |
| 3000 | 47.9 | 52.3 | 56.0 | 59.3 | 62.1 | 64.6 | 66.7 | 68.7 | 70.4 | 72.0 | 73.5 |
| 3100 | 57.3 | 61.5 | 65.0 | 68.0 | 70.5 | 72.7 | 74.6 | 76.2 | 77.7 | 79.0 | 80.2 |
| 3200 | 72.2 | 75.6 | 78.3 | 80.4 | 82.2 | 83.7 | 85.0 | 86.1 | 87.1 | 87.9 | 88.7 |

Fig. 8. Table for predicting percent natural gas saved (reproduced by permission of the North American Manufacturing Co., Cleveland, OH).

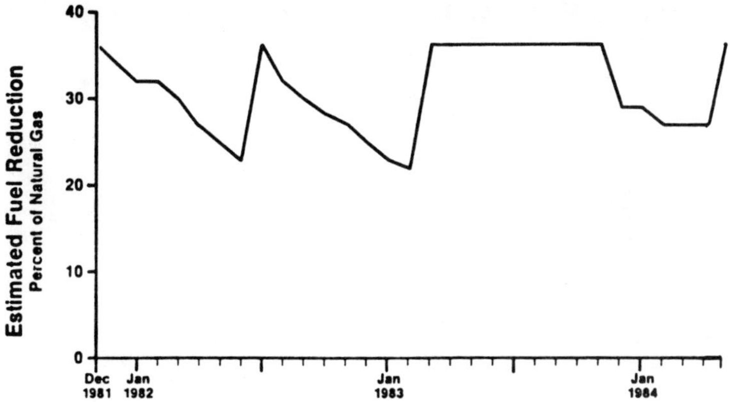


Fig. 9. Estimated fuel reduction vs time in service.

(4) Cold-air metering shall be used for combustion control. This would minimize hot-air ducting length, thereby reducing ductwork energy loss. A recuperator design that minimizes cooling air leakage would make this type of combustion control more desirable.

(5) A ceramic-fiber, lined stack would provide sufficient draft for a smooth tube recuperator, thereby eliminating the need for mechanical stack draft in most installations. In any event, a 2.0 safety factor should be utilized when sizing stack draft in anticipation of recuperator fouling.

(6) The use of a two-pass recuperator design to boost recuperator effectiveness should be considered.

The writer would like to acknowledge the support and technical guidance of Mr. J. W. McIntee, Alcoa Corporate Engineering. In addition, the refractory analysis by Ms. E. M. DeLiso, Alcoa Technical Center, and repair summaries by Mr. J. B. Gorss, Corporate Engineering, and Mr. D. J. Ritter, Davenport Engineering, have been very helpful in generating this report.

# Development of an Air-to-Air Heat Exchanger with All-Ceramic Internals

ROBERT G. GRAHAM

C&H Combustion
Troy, MI 48083

Design criteria, comparison of ceramics vs metals, prototype development, five years of test data, recent commercial operating experience, and materials performance and cost requirements for various potential markets are reviewed.

## History

C&H Combustion is an industrial furnace company that custom designs high-temperature process furnaces and incinerators. Over the years, systems have been supplied with heat recovery, usually metal air-to-air heat exchangers or boilers. These devices, in many cases, could not be used when there was acid, slag, or abrasive particulate in the flue gas.

In 1975 a test stand was built and a program started to develop an all-ceramic heat exchanger. This stand would accommodate three zones with tubes up to two feet long (Fig. 1). From 1975 through 1982, different designs were constructed and evaluated. Extensive heat-transfer data were obtained, using both smooth and internally finned tubes. After testing, some designs were scrapped, including those that required springs, tube collars, external tube seals, or elaborate components.

In early 1982 a new test stand was built. The stand would handle four-foot-long tubes, and the tube diameter could be varied up to four inches. The air-side temperature could be set from ambient up to 871°C (1600°F), and the furnace-side temperature varied from 315°C (600°F) to 1426°C (2600°F). With these controls, it was possible to duplicate any zone in a multipass exchanger with furnace temperatures as high as 1426°C and preheat the air to as high as 1093°C (2000°F). By March of 1984, 14 tests had been run, including one 2400-h test. The main purpose of this last battery was to test different ceramic tile, tubes, and seals.

Improved ceramic materials and knowledge gained from testing helped C&H set the following design criteria:

(1) *Heat exchanger*: The heat exchanger was to be of the shell and tube type with the furnace flue gas passing outside the tubes and the air to be heated inside the tubes.

(2) *Tubes*: The design was to incorporate tubes with smooth exteriors and, in some applications, internal finning.

(3) *Tube sheets*: The tube sheets were to be built of C&H's proprietary structural ceramic shapes with removable inserts and internal tube seals.

(4) *Pressure drop*: The furnace-side pressure drop was to be held at less than ¼ in. w.c. to ensure that an induced draft fan would not be needed in most applications. The air-side pressure drop was to be held at less than 1½ in. w.c. per bank. In a normal three-pass heat exchanger, this meant only a

Fig. 1. Overview of tube bundles in a three-pass test stand heat exchanger.

few inches had to be added to the combustion air fan to obtain optimum heat-exchanger performance.

(5) *Soot blowing*: The casing was to be designed with doors or ports to allow either manual or automatic soot blowing.

(6) *Maintenance*: It was important to build into the unit a method for removing individual tubes and replacing any tube or seal.

(7) *Leakage*: Leakage was to be from the air side to the flue gas side, and not exceed 2% per zone.

**Results**

The results were excellent and many ceramic materials were tested successfully. They varied from standard high-alumina up to siliconized silicon

Fig. 2. Overview of test bundle in a single-pass test stand heat exchanger.

carbide tubes. The program culminated in the building of three field units, two for the investment castings industry and one for an aluminum scrap remelt furnace. These exchangers are either in service or presently under construction. The 2400-h test was completed in March of 1984, and best demonstrates to date the successful application of ceramics to a high-temperature heat exchanger.

**2400-H Test**

*General:* In November 1983, a test bundle was constructed using 16 four-foot-long Norton extruded siliconized silicon carbide tubes. The tubes were assembled into two C&H plug-style tube sheets with internal gaskets.* The unit was brought to 426°C (800°F) on November 28, and soaked at temperature until December 3, 1983, at which time the furnace temperature was elevated in regular increments to 982°C (1800°F) by December 6. For the next three months, furnace temperatures were varied up to 1037°C (1900°F), with the occasional excursion to 1315°C (2400°F). The air-side-in temperature was varied from ambient to 426°F. During February, one of the lead tubes on the furnace side was subjected daily to compressed air soot blowing. On March 4, 1984, the stand was shut down, and on March 6 opened up and inspected (Fig. 2).

*Proof of Concept*: The system was assembled as described above, and there were no difficulties in installing tubes, seals, plugs, and tile. Two months into the test, the unit was cooled down, a plug removed, work done on the gasket thermocouple, reassembled, and brought back up to operating temperature, all within eight hours. There was no difficulty in removing the plug, gasket, and tube, or in reassembly. After the test was completed, inspection revealed that no tile had cracked, no joints had opened, and no tubes were broken.

*Soot Blowing*: The Norton tube did not break, even though the soot was blown when the furnace temperature was 1315°C with the preheated air temperature

*Fiberfrax, Carborundum Co., Niagara Falls, NY.

Fig. 3. Four-pass, 1315°C (2400°F), air-to-air heat exchanger with all-ceramic internals. Photo shows tubes installed in zone one only.

at 648°C (1200°F). The tube has been returned to Norton for laboratory analysis. However, there were no cracks, checking, or other visible signs of deterioration.

*Leakage*: The attached curve shows that there is no appreciable change in leakage from start to finish of the test. The leakage includes the tube seals, mortar joints, and tube sheet periphery. The curve has been corrected to exchanger design flow. The Fiberfrax seals stayed compliant.

*Ceramic Life*: After the test was completed and the tube bundle dismantled, it was noted that all of the ceramic components were reusable. Some of the tile and tubes have been cleaned and are now being used in a new series of tests.

*Summary*: The test was successful and there were no surprises. The tubes and tube sheet tile are full-sized, when compared to proposed commercial units. This met one of the primary goals, which was to test four-foot tubes and thermally duplicate field conditions.

**Field Test Units**

Three field test units have been contracted for, with one installed and operating and two under construction. The first unit is an exchanger attached to an investment castings furnace. The furnace burns off wax and develops flue temperatures as high as 1315°C. With the style of burners and minimum mass flow requirements, it was not feasible to supply preheated air above 900°F. The furnace was installed in the spring of 1984, has been fired, and the heat exchanger works within the design parameters. Leakage has been less than 5%, and, with furnace temperatures varying from 760°C (1400°F) to 1204°C (2200°F), remained constant.

The second unit is a modernized version of the first exchanger, also for an investment castings furnace (Fig. 3). The third unit is part of a dry hearth aluminum melter and was scheduled to be installed in mid-summer of 1984.

Prior to selling the three field units described above, C&H installed a tube in a highly corrosive environment. The biggest problem in predicting the life of a ceramic exchanger has not been thermal, since this can be duplicated in the lab, but has been acid and particulate attack in field units. C&H placed a Norton four-foot-long NC-430 slipcast tube in the flue of a large aluminum scrap melter in the winter of 1982. The furnace ran around the clock for a year with daily fluxing. The tube was subjected to considerable acid and particulate attack, along with temperatures up to 1093°C (2000°F). At the end of a year, the tube was returned to Norton for analysis. There were no cracks, structural damage, or other problems with the tube, and the concensus was that the tube would have lasted in that environment for several more years.

Since this paper was presented to the American Ceramic Society, C&H Combustion has contracted for 10 all-ceramic heat exchangers. They include the three described above plus units for two aluminum scrap melting furnaces, a slot drop forge furnace, two copper scrap remelt furnaces, and one titanium oxide kiln. Considerable field test data have been obtained and will be published at a later date.

**Conclusions and Future Plans**

C&H Combustion is now building several high-temperature ceramic exchangers. A typical exchanger, designed to handle acidic, particulate-laden flue gas at 1315°C (2400°F) that preheats the combustion air to 760°C, would be designed as follows:

(1) Three pass, counterflow.

(2) Zone one would use Norton's Crystalon CX–283 tube sheets with Cryston CN–178A plugs. The tubes would be Norton's extruded siliconized silicon carbide or slipcast CS–101K.

(3) Zone two would use Norton's Crystalon CN–163 tile with Cryston plugs. The tubes would be Norton's nitride-bonded silicon carbide.

(4) Zone three would have Harbison-Walker or Coors mullite wall tile and Norton's nitride-bonded silicon carbide tubes.

(5) The setting would be constructed of Harbison-Walker Alamo, backed with B&W 1260°C (2300°F) insulating firebrick.

The test facility is being set up to fire different combinations, using Carborundum silicon carbide and Coors mullite internal-fin tubes. The exchanger allows for varying both tube and tile throughout, so the best material at the most economical price can be selected for each zone.

Exchangers are being quoted with flue gas temperatures as high as 1537°C (2800°F) and preheated air temperatures up to 1093°C (2000°F). As more field experience is obtained, it is anticipated that the more-expensive, high-performance ceramics will continue to be used in the first zone, with less-expensive materials in the latter zones.

It has been demonstrated that development of a high-temperature ceramic heat exchanger has been mainly a problem of selecting, testing, and proving the ceramics. Certain features had to be worked out to use ceramics and have low pressure drop, minimum leakage, and easy maintenance. The materials listed above have now been proven and, after more field service, sufficient data will be obtained to project service life accurately.

# A High-Temperature Ceramic Recuperator for Industrial Applications

M. Coombs, D. Kotchick, and H. Strumpf

AiResearch Manufacturing Co.
Torrance, CA 90509

The design and development of a ceramic recuperator and high-temperature burner are described. This system is designed to operate in industrial flue gas temperatures of up to 1371°C (2500°F) and provide preheated combustion air of up to 1093°C (2000°F). Such a system offers the potential of reducing furnace fuel consumption by 60%. The design, fabrication, and operation of a 1.5 mm W (5 mm Btu/h) prototype unit is described and test data presented. The design of a full-size system, slated for a one-year industrial host site demonstration test, is also described. This program is jointly sponsored by the Gas Research Institute and the U.S. Department of Energy.

Furnace recuperation represents an effective means of conserving fuel and reducing furnace operating costs. The energy savings theoretically obtainable with recuperation are shown in Fig. 1.

In this figure, the percent of fuel savings is plotted against the flue gas temperature and the air preheat temperature for systems operating at 20%

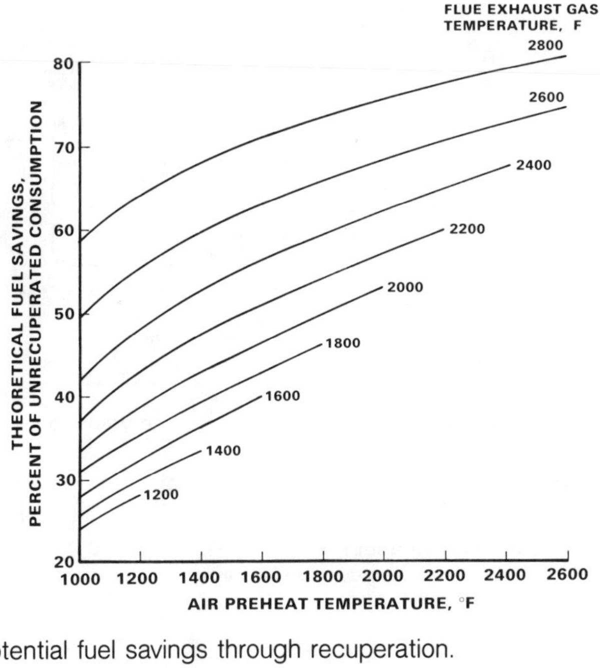


Fig. 1. Potential fuel savings through recuperation.

excess air. Thus, for a flue gas temperature of 1371°C (2500°F) and an air preheat temperature of 1093°C (2000°F), the theoretical savings is approximately 65% of the fuel consumption of an unrecuperated furnace operating at the same temperature levels. While certain considerations may limit the degree of recuperation acceptable with a given furnace (e.g., process control, $NO_x$ formation, etc.), high levels of recuperation can offer a dramatic improvement in process efficiency for high-temperature industrial furnaces.

As a result, the Gas Research Institute (GRI) and the Department of Energy (DOE) are jointly sponsoring a program to develop a ceramic recuperator capable of operating in flue gas temperatures of up to 1371°C. The objectives of this program are:

(1) Develop a ceramic recuperator system capable of producing 1093°C preheated combustion air.

(2) Develop an industrial burner to operate with up to 1093°C combustion air.

(3) Provide a proof-of-concept prototype system demonstration.

(4) Provide a one-year industrial demonstration of a full-size system.

**Concept Description**

The recuperator concept selected for development by AiResearch is illustrated schematically in Fig. 2. It consists of a ceramic recuperator and a metallic recuperator operating in series, and is consequently referred to as a hybrid system.

An air dilution box located between the ceramic and the metallic recuperators is used as a control element to ensure that the exhaust gas entering the metallic recuperator remains below 815°C (1500°F), and that the combustion air delivered to the burner never exceeds 1093°C. The air dilution system eliminates the need to limit the gas temperature exiting the ceramic

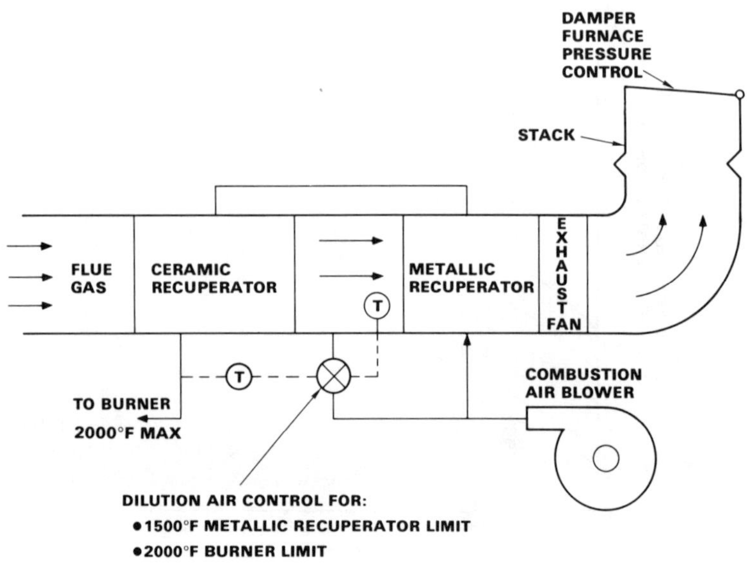


Fig. 2. Recuperator concept schematic.

recuperator. As long as the final air preheat temperature is 1093°C (2000°F), the dilution scheme does not affect the fuel savings available through recuperation, but simply shifts heat load from the ceramic recuperator to the metallic recuperator. This approach minimizes the size of the ceramic unit while maintaining the maximum fuel savings under a wide range of operating conditions.

An extensive research effort was carried out to develop the technology base required for the fabrication of the recuperator. Details of this work are given in Ref. 1 and 2. Following this extensive component development effort, a 1.5 mm W (5 mm Btu/h) prototype recuperator/burner system was designed, fabricated, and tested.

### Prototype Recuperator Design and Fabrication

The construction details of the ceramic recuperator core are illustrated in Fig. 3. As shown there, the ceramic heat exchanger is a two-pass, tubular, crossflow unit with silicon carbide tubes arranged in two bundles. Ceramic

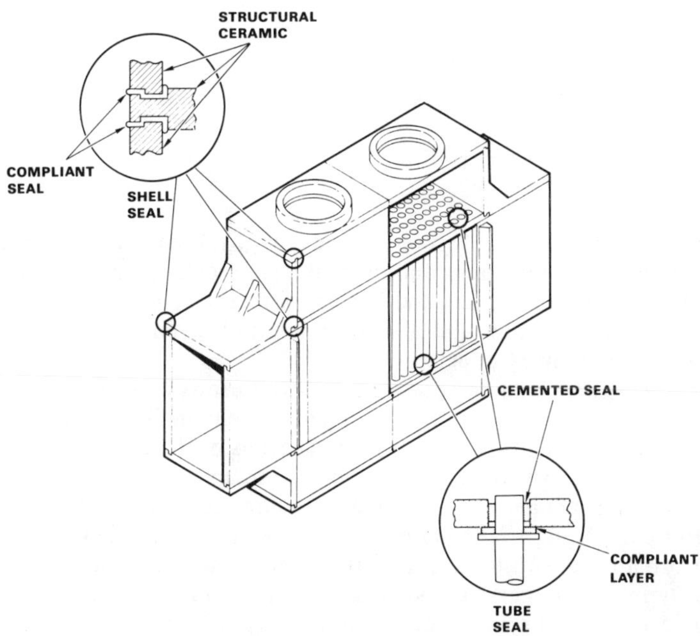


Fig. 3. Core/plate assembly.

header plates and walls (silicon nitride-bonded silicon carbide) form a boxlike structure around the tube bundles defining the flow path for the exhaust gases as well as for the combustion air circuit. These plates also form the inlet, exit, and turning plenums for the combustion air circuit. They are located on the top and bottom of the ceramic core. The tubes are cemented into the ceramic header plates with a viscous cement that allows for differential thermal growth between the tubes and the plates. Ceramic fiber seals (Carborundum 880J film insulation paper) are employed in the boundaries where the structural ceramic plates mate in a tongue and groove fashion.

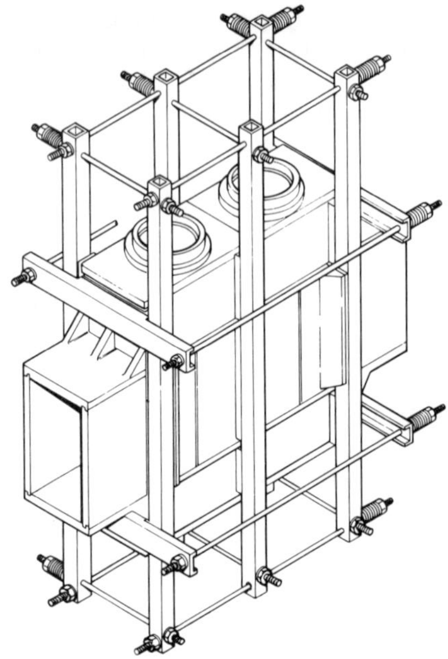

Fig. 4. Core support assembly.

The core support assembly is illustrated in Fig. 4. It consists of four vertical high-temperature steel columns and four horizontal cross members, which are tied together at the top, sides, and bottom with spring-loaded tie rods. The vertical columns are hollow and a small amount of cooling air flows through them due to the chimney effect.

Surrounding the boxlike ceramic core structure is a 1-ft layer of board-type insulation, and around this, a steel containment shell. A combination of springs and pressure loads is used to keep the walls of the core in compression and load the seals to reduce both air and gas stream leakage.

The metallic recuperator (AiResearch) is an industrial plate-fin flue gas recuperator comprising triangular end sections and a counterflow core section. High-performance offset fins are used throughout the recuperator.

The complete recuperator system is illustrated in Fig. 5. This figure shows the ceramic and metallic units, the air dilution system, and the interconnecting ducting. Figure 6 illustrates the progression of the actual buildup of the ceramic recuperator which was assembled in the AiResearch laboratories in Torrance, California.

The high-temperature preheat burner used in the AiResearch prototype system was developed by the Bloom Engineering Company and is illustrated in Fig. 7. The burner is a modification of a standard Bloom industrial baffle burner with the extensive use of internal insulation to protect the casing from the high-temperature combustion air.

**Testing**

A special test facility was designed and fabricated to support the development of the recuperator and burner. This test facility is shown schematically

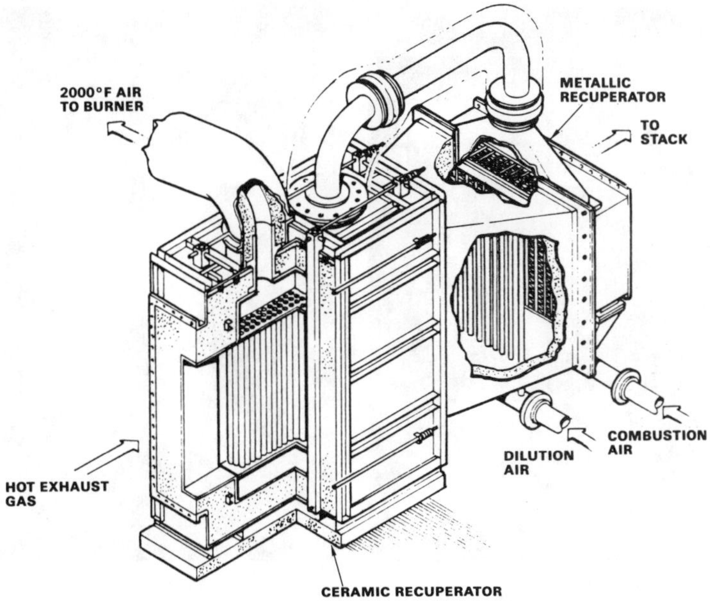


Fig. 5. 1.5 mm W (5 mm Btu/h) prototype recuperator.

in Fig. 8. It consists of two small furnaces: a workhorse furnace and a burner test furnace. The workhorse furnace is pressurized and can provide a gas stream over a wide range of flows at temperatures of up to 1371°C (2500°F). This hot gas stream is used to simulate the hot flue gases entering the ceramic recuperator. The second furnace, the burner test furnace, is the test bed for

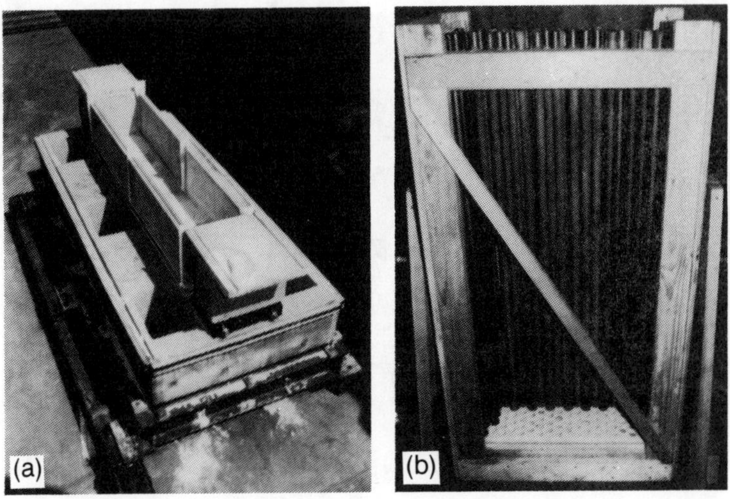


See page 54 for Fig. 6 caption

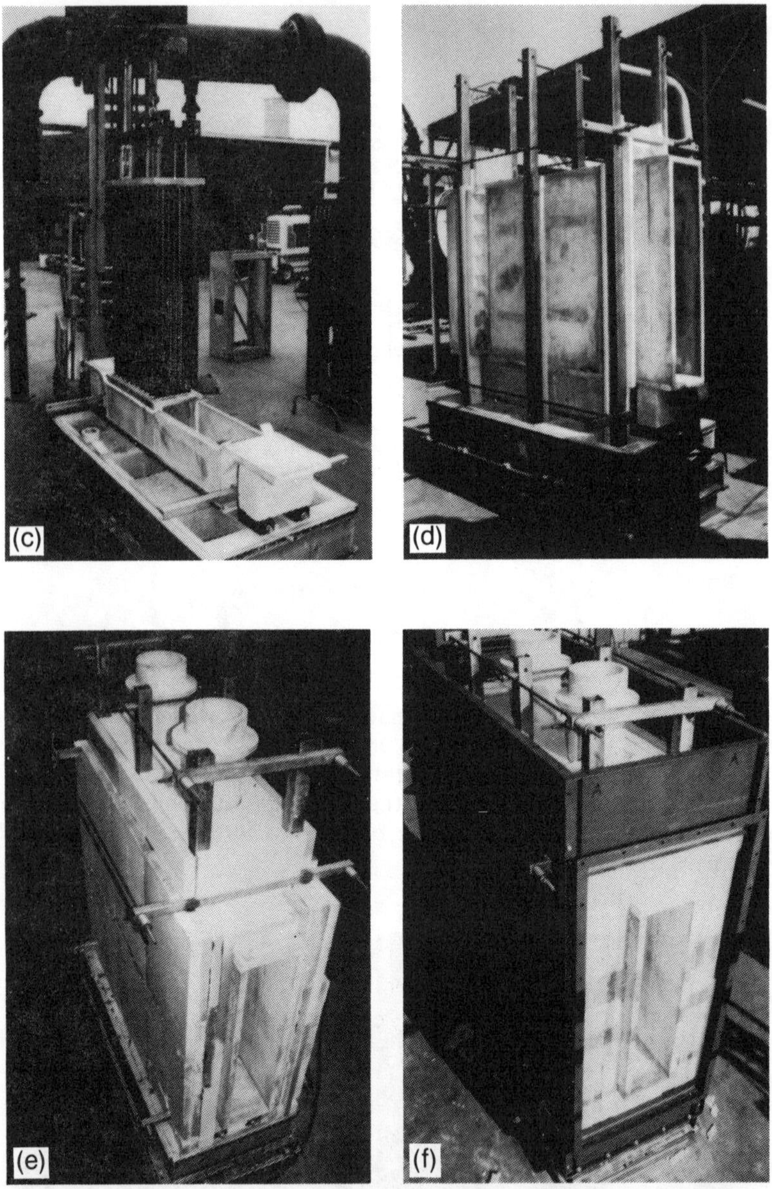


Fig. 6. Buildup of prototype unit. (*a*) Bottom air plenums; (*b*) core buildup; (*c*) first core module loaded; (*d*) metallic air-cooled structure; (*e*) insulation buildup; (*f*) complete ceramic recuperator.

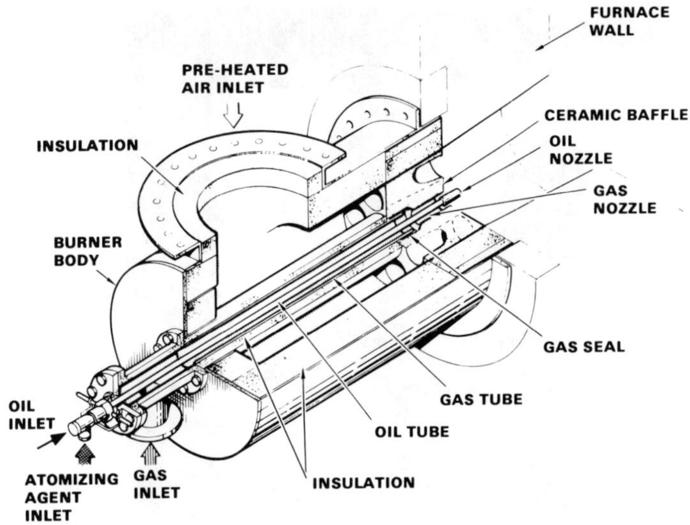


Fig. 7. High-temperature preheat air burner.

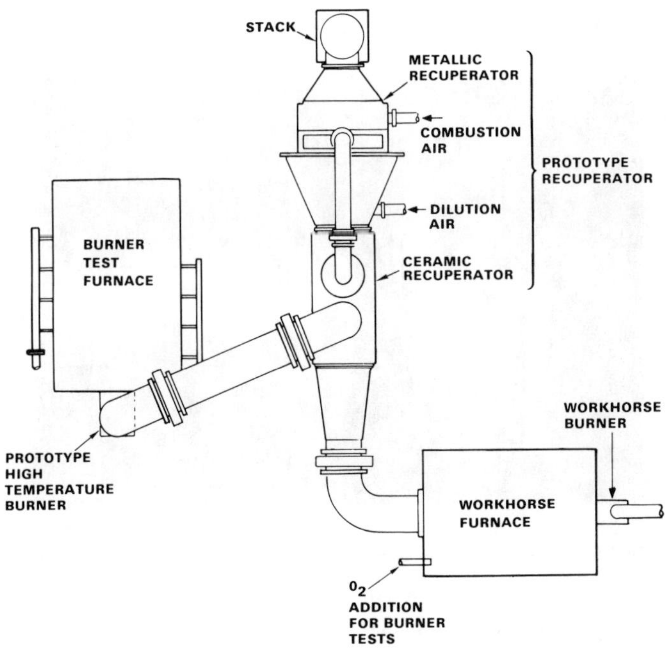


Fig. 8. Test facility schematic.

Fig. 9. Bloom high-temperature burner installed on test furnace.

Fig. 10. Recuperator/burner on test.

the high-temperature burner. This furnace has an internal water-cooled coil that simulates a heat load for the test burner. Figure 9 shows the high-temperature Bloom burner installed on the test furnace.

Testing of the recuperator/burner took place in January 1984. The unit was operated over a wide range of flows and inlet temperatures. (Figure 10 shows a photo of the system on test.) Figure 11 is a summary of some of the data taken during the test period. With a 1371°C (2500°F) inlet temperature

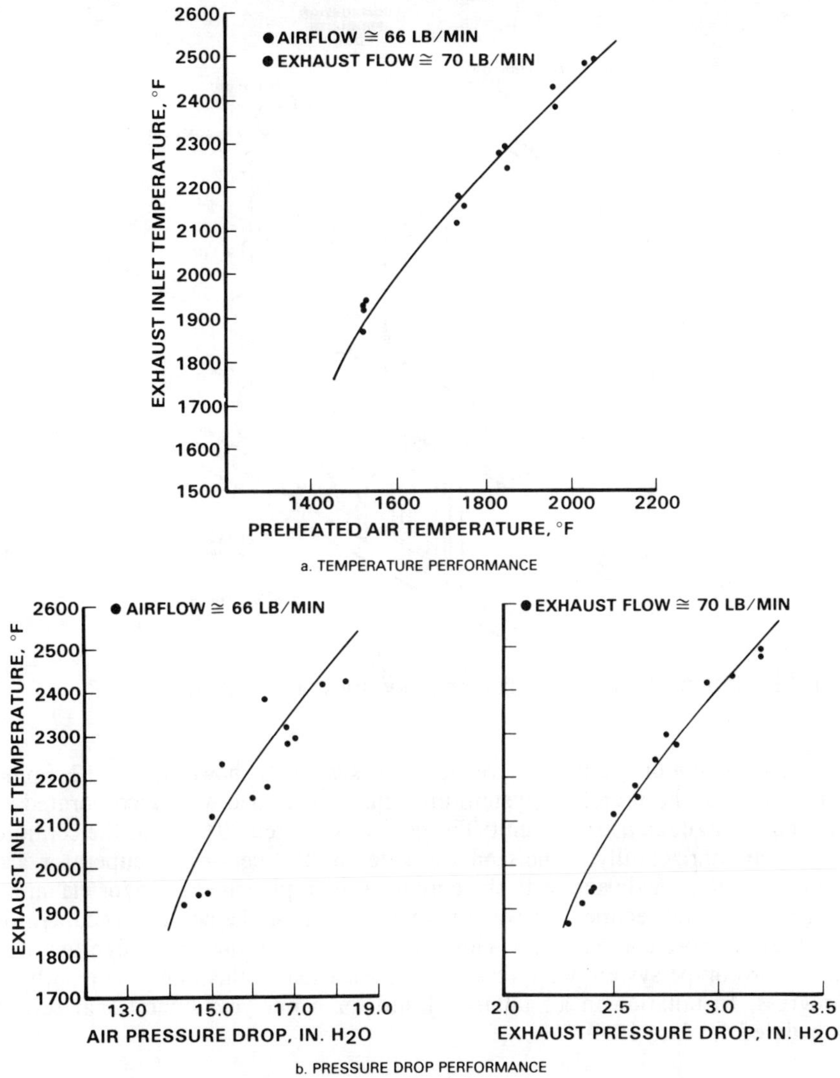


Fig. 11. Test points for integrated recuperator burner tests.

to the recuperator, the combustion air was preheated to 1121°C (2050°F). Performance was close to predicted values and stable operation was observed over the range of test conditions.

#### Industrial Demonstration System

The last phase of the program involves the design, fabrication, installation, and operation of a full-size system at a typical industrial site. This will provide performance, economic, and operational data in a more realistic environment than can be provided in laboratory tests.

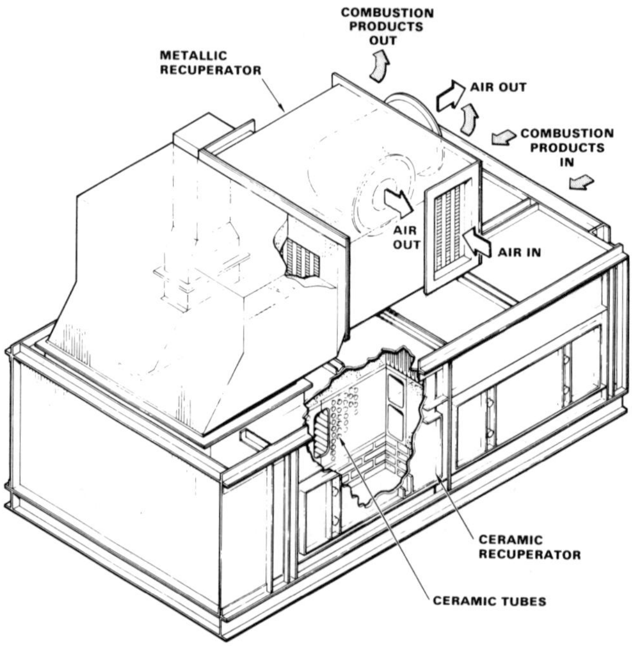


Fig. 12. Industrial demonstration recuperator configuration.

The design of the full-size recuperator system is shown in Fig. 12. Some changes from the prototype system previously described were incorporated in the industrial demonstration unit: The core was rotated 90° so that the ceramic tubes run horizontally; panels on the sides of the ceramic recuperator are removable to provide access to the core for tube replacement and/or cleaning; and the metallic recuperator was mounted on top of the ceramic recuperator to obtain a more compact unit. However, most of the internal design features of the prototype system were retained. Fabrication of this unit is currently in progress. Installation in an industrial site for a one-year operational test is scheduled for late 1984.

**References**

[1]M. Coombs, H. Strumpf, and D. Kotchick, "A High-Temperature Flue Gas Heat Recovery System"; presented at the 1983 International Gas Research Conference, London, UK.

[2]M. Coombs, D. Kotchick, and H. Strumpf, High-Temperature Burner-Duct-Recuperator (HTBDR) Program, AiResearch Report 82–19445, October 1982.

# Design and Materials Selection for a High-Temperature Burner-Duct Recuperator

JOSEPH E. SNYDER AND DANIEL R. PETRAK

Babcock & Wilcox
Lynchburg Research Center
Lynchburg, VA 24506

Babcock & Wilcox is participating with the U.S. Department of Energy in the development of a high-temperature burner-duct-recuperator system which is to be capable of delivering 982°C (1800°F) preheated air to the combustion system of a steel mill soaking pit (Cooperative Agreement DE–FC07–81ID12296). The recuperator designed for the project is composed of ceramic heat-exchanger elements supported by ceramic fiber-insulated, air-cooled, metallic tube sheets. A project overview is presented, with emphasis on the design of the ceramic components.

Babcock & Wilcox is participating with the U.S. Department of Energy in a project entitled "High Temperature Burner-Duct-Recuperator System Evaluation." The project involves the design, construction, installation, and evaluation of a system which incorporates a high-temperature recuperator, burner, and attendant ducting. The recuperator must be capable of operating in the flue of an industrial furnace which contains gases of 1343°C (2430°F) maximum temperature and contaminants from a steel soaking pit process. The recuperator system must also be capable of delivering 982°C minimum preheated air to the combustion system of the industrial furnace.

Since the recuperator requirements are beyond the capabilities of metals for materials of construction, the concept selected by B & W uses ceramic components to address the high-temperature application and metallic components in the lower-temperature areas. This article presents an overview of the project, a description of the ceramic materials activities, and a summary of the system specifications.

### Project Overview

There were a total of eight major project activities divided into two phases. Phase I incorporated the system design and verification using a prototype recuperator and the exposure of candidate ceramic recuperator materials to the industrial flue environment. The approach used to accomplish the design was to use an interactive, multidisciplinary team to develop the recuperator concept shown in Fig. 1. North American Manufacturing Co. designed the system's high-temperature burner with input from the operating and staff engineers at the B & W Tubular Products Group steelmaking site in Koppel, Pennsylvania, which is also the host site for the system.

The activities over the term of the project are listed below by task.

*Phase I*

Task 1. Energy audit.

RECUPERATOR CONCEPT

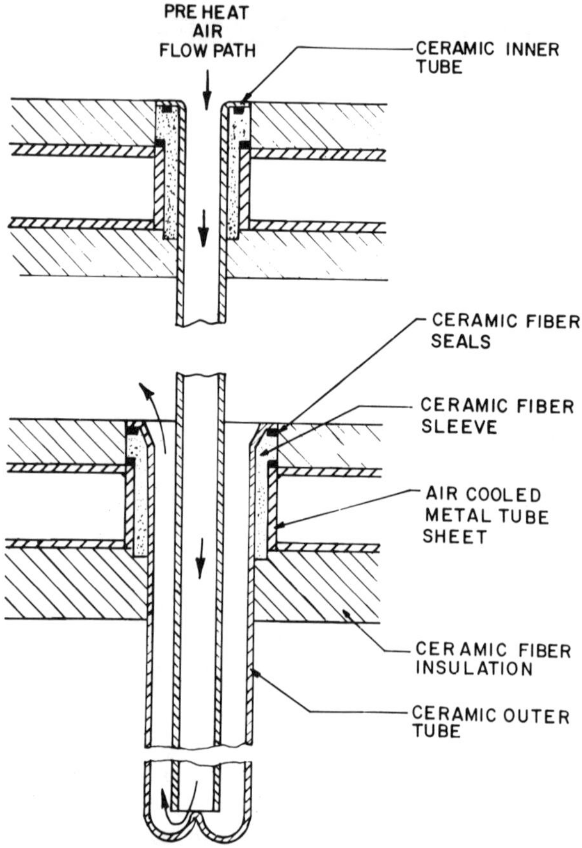


Fig. 1. Recuperator concept.

Task 2. HTBDR system design and prototype evaluation.

*Phase II*

Task 3. HTBDR specification, fabrication, and installation.
Task 4. Environmental evaluation.
Task 5. System shakedown and check-out.
Task 6. System evaluation.
Task 7. Technology transfer.
Task 8. Project management and reporting.

The project team consisted of individuals representing ceramics, system design, process control, thermal and fluid flow, applied mechanics, dynamics, chemical engineering, and tubular products group staff.

**Design Concept**

The concept developed by the design team to recover the heat in the waste-gas stream consists of two ceramic tubes: one tube open on both ends, inserted into another tube which is closed on one end, i.e., bayonet style (see

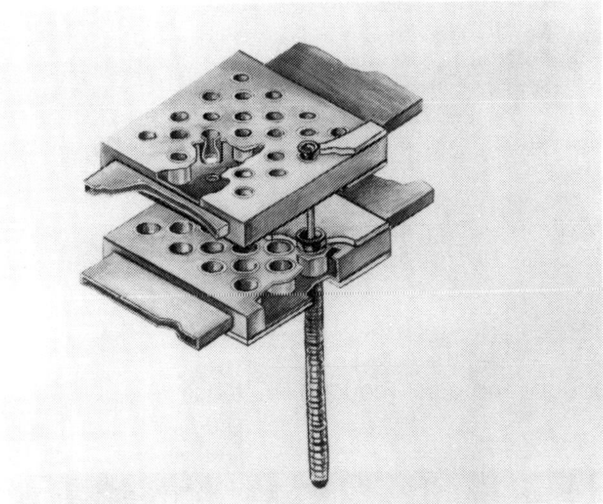

Fig. 2. Air-cooled metallic tube sheet arrangement.

Fig. 1). The tubes are suspended vertically from the plenums into the flue of the soaking pit using air-cooled, metallic tube sheets. As seen in Fig. 1, combustion air is blown into the upper plenum where it begins the heat-exchange process with the hot gases by flowing down the inner tube, turning around at the bottom of the outer tube, and flowing up the outer tube to the lower plenum and then out of the recuperator. Air flow direction through the recuperator could, of course, be reversed. The system is described as a cross-counterflow heat exchanger.

Ceramic fiber insulation is used to line the walls of the plenums and the surfaces of the tube sheets as well as the ducts and burner. Ceramic fiber products are also used to provide individual support sleeves and seals for the tubes. The tubes in this concept rest on the sleeves and seals and are not constrained from moving due to thermal expansion. This design reduces or eliminates mechanical stresses on the ceramic tubes.

Figure 2 illustrates how the tubes will be supported by the metallic tube sheets. Figures 3 and 4 show the ceramic tubes and the ceramic fiber-based sleeves and seals used in the prototype test evaluation, respectively.

The benefits of the B & W concept described above are: no dilution air needed, high operating temperature, low flue-side pressure drop, unconstrained structural ceramics, and ease of maintenance.

Since the heat-exchanger elements are all ceramic, the dilution air required for metallic heat exchangers is not required as long as the temperature of the flue gases does not exceed the use temperature of the ceramic tubes. Also, the use of high-quality SiC components permits a higher system operating temperature than metallic and other ceramic heat exchangers.

The design team had as goals to incorporate the last three benefits. The concept chosen employs large-diameter tubes spaced on 6-in. centers, thereby creating large passageways on the flue side. This translates into a low flue-side

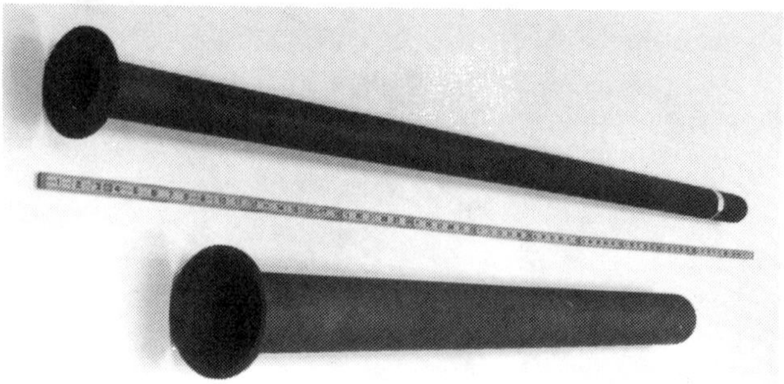

Fig. 3. Inner (top) and outer (bottom) SiC tubes.

Fig. 4. Ceramic-fiber sleeves and seal for the outer (left) and inner (right) tubes.

pressure drop which means there is no need for an energy-wasting, expensive, high-temperature induced draft fan.

The sleeve-seal tube supports and the vertical suspension of tubes were developed to eliminate mechanical constraints of the tubes, thereby minimizing mechanical stresses on the tubes. The tubes are "floating," i.e., they are free to move due to thermal expansion.

A major concern from a user perspective is that any piece of operating equipment has to be maintainable. The recuperator was designed to allow both plenums to be removed so that any tube can be replaced if necessary.

**Host-Site Description (Before HTBDR)**

The evaluation site for the HTBDR system is a 120-ton capacity steel soaking pit manufactured by the Loftus Engineering Corp. The soaking pit

(Pit 17) is located at the Koppel, Pennsylvania steel-melting facility of the B & W Tubular Products Group whose headquarters are in Beaver Falls, Pennsylvania. Pit 17 is one of three pits in a battery. Each pit is top-fired by three Bloom No. 12 burners (each rated at $2.5\times10^6$ W ($8.5\times10^6$ Btu/h)) using preheated combustion air. There are horizontal flue sections for each pit which join together before going to a common natural draft stack which is 150 feet high. Each flue section contains an American-Schack canal-type recuperator which provides 537°C (1000°F) maximum preheated combustion air to the pit combustion system.

The inside dimensions of the pit are: 9 ft wide by 27 ft long by 14 ft high. The ingots are placed in the pit and leaned against each long wall. The burners fire down the length of the pit from the top of the burner wall. The hot gases then circulate around the pit, heat the ingots, and exit through two openings in the bottom of the burner wall. When exiting the soaking pit, the flue gases are immediately diluted with ambient air to lower the flue-gas temperature to about 898°C (1650°F) to protect the metallic recuperator.

The combustion system for the soaking pit operates under low-fire/high-fire and short-flame/long-flame conditions. High-fire is used when the temperature in the pit is being raised from a lower to a higher temperature. Under this condition all three burners are used. Low-fire is used when the pit has reached its soaking temperature (maximum temperature) and the pit temperature is just being maintained. Under this condition only one burner is used. The long-flame/short-flame operational mode is when the combustion system oscillates the length of the flame to establish a temperature uniformity between the ingots on opposite ends of the soaking pit.

**Materials Study**

The materials selection process for the tubular heat exchanger (HX) elements consisted of five steps. A similar process was used for the ceramic fiber-based sleeves and seals. The steps were: determine materials requirements, select candidate materials, choose and conduct appropriate testing, conduct flue exposure tests, and select material and supplier.

Material requirements for the outer tubes were: thermal shock and corrosion resistance, impermeability, adequate strength for calculated stresses, and ease of fabrication.

Because of the thermal excursions anticipated in the industrial process, the materials were limited to the silicon carbide (SiC) class of materials due to this material's excellent thermal shock resistance.

Candidate silicon carbide materials were: siliconized, recrystallized, silicon nitride-bonded, and sintered.

Tests performed on ceramic tubes before and after flue exposure testing included C-ring strength, microstructure, thermal expansion (ORNL), X-ray, and chemistry.

Two flue exposure tests were conducted in the project. The first lasted 36 days and was used to screen materials. The second was conducted by Oak Ridge National Laboratory technical staff and lasted 144 days. The purpose of the second test was to develop long-term corrosion data.

Both flue-exposure tests were considered to be "worst case" evaluations since the tubular specimens were uncooled and located in the mouth of the flue opening under the burner wall. The actual recuperator will use tubes cooled on the inside and will be located in the furnace flue, which is two 90° turns and about 20 ft from the flue opening.

Fig. 5. Appearance of siliconized SiC before and after flue exposure testing.

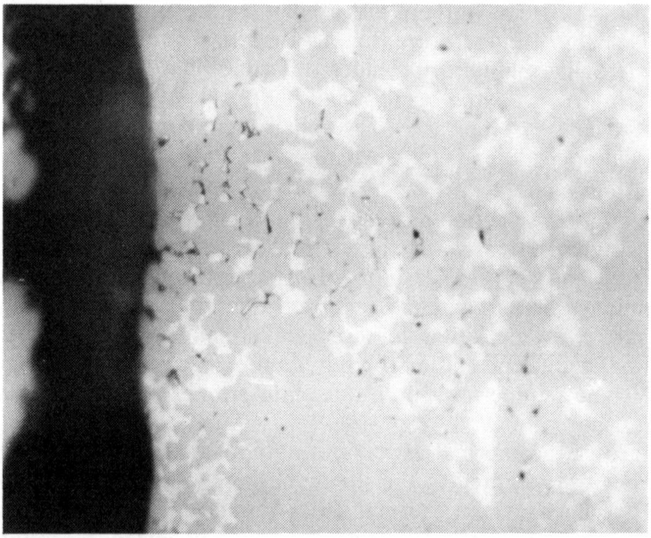

Fig. 6. Microstructure of siliconized SiC from flue exposure testing.

The results of both flue tests have been presented elsewhere.[1,2] Figure 5 shows the effect of the exposure on the SiC tube material selected for the recuperator. Note the glassy appearance and the nodules on the upstream side of the exposed tube. Both conditions were attributed to surface oxidation ($SiO_2$ and CO formation) and reaction with flue-stream contaminants.

Figure 6 shows the microstructure of the NC-430* tube removed from the first flue exposure test. Little surface modification was noted and the results

*Norton Co., Worcester, MA.

Table I. Tube Tensile Stresses and Strengths (psi)

| | Outer Tube, MPa (psi) | Inner Tube, MPa (psi) |
|---|---|---|
| Anticipated maximum tensile stress | | |
| Seal region | 22 (3 190) | 26 (3 800) |
| Along tube | 66 (4 530) | 24 (3 440) |
| Tube strength (C-ring tensile) | | |
| Ambient temp. | | |
| Exposed | 215 (31 200) | 140 (20 400) |
| Unexposed | 226 (32 700) | 118 (17 100) |
| 2370°F | | |
| Exposed | 201 (29 200) | 139 (20 100) |
| Unexposed | 208 (30 200) | 111 (16 100) |

Table II. Short-Term Compressive Creep Behavior and Modulus of Elasticity Data for Candidate Seal Materials

| Material | Temp. °C (°F) | Maximum Stress, MPa (psi) | Modulus of Elasticity, MPa ($10^3$ psi) | Creep Rate $10^{-5}$ (min) | Time (min) |
|---|---|---|---|---|---|
| B & W mill board | 649 (1200) | 0.2 (30) | 1.6 (0.23) | 3.6 | 60 |
| (loading ⊥ fiber) | 649 | 0.2 (30) | 9.8 (1.42) | 7.1 | 30 |
| | 649 | 0.35 (50) | 13.8 (2.00) | 6.2 | 60 |
| | 649 | 0.35 (50) | 17.8 (2.58) | 3.7 | 30 |
| Pressed board | 649 | 0.2 (30) | 53.1 (7.70) | N.D. | 60 |
| (loading ⊥ fiber) | 649 | 0.2 (30) | 96.6 (14.0) | 0.37 | 30 |
| | 649 | 0.35 (50) | 153.2 (22.2) | 0.69 | 60 |
| | 649 | 0.35 (50) | 164.9 (23.9) | 0.62 | 30 |
| Pressed board | 649 | 0.2 (30) | 71.1 (10.3) | 0.43 | 60 |
| (loading ‖ fiber) | 649 | 0.2 (30) | 113.2 (16.4) | N.D. | 30 |
| | 649 | 0.35 (50) | 184.9 (26.8) | 0.82 | 60 |
| | 649 | 0.35 (50) | 222.9 (32.3) | 0.79 | 30 |
| Pressed board | 1093 (2000) | 0.25 (37) | 59.3 (8.59) | 12.5 | 60 |
| (loading ⊥ fiber) | 1093 | 0.3 (44) | 200.1 (29.0) | 11.4 | 30 |
| | 1093 | 0.4 (63) | 217.4 (31.5) | 13.2 | 60 |
| | 1093 | 0.44 (64) | 296.0 (42.9) | 12.2 | 30 |
| Pressed board | 1093 | 0.23 (33) | 60.9 (8.84) | 6.1 | 60 |
| (loading ‖ fiber) | 1093 | 0.29 (43) | 167.6 (24.3) | 6.2 | 30 |
| | 1093 | 0.36 (53) | 289.8 (42.0) | 4.7 | 60 |
| | 1093 | 0.38 (55) | 312.6 (45.3) | 3.5 | 30 |

*All pressed board data in this table are for 800 kg/m³ (50 lb/ft³) density.
N.D.=not determined.

of the second flue exposure test indicated that wall thinning should be expected but at a rate that cannot be determined from the exposure tests because of the test conditions. It appears that the siliconized silicon carbide tubes will survive the flue environment. However, the life of the tubes cannot be ascertained with existing data.

Table I summarizes the strength testing of the siliconized SiC material, before and after flue testing, compared to the anticipated tube stresses cal-

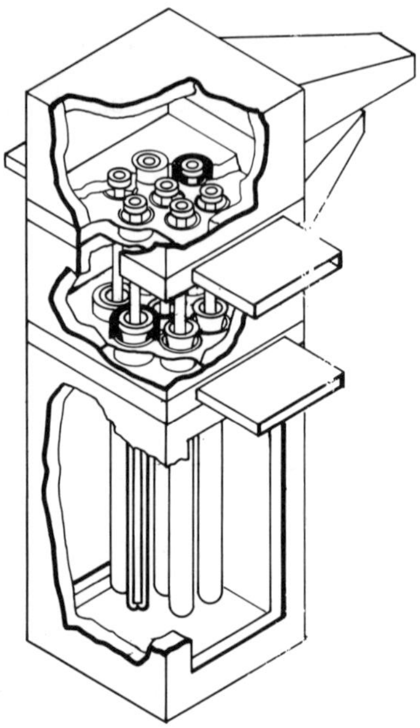

Fig. 7. Schematic of prototype recuperator.

culated by applied mechanics techniques. It will be noted that the strength of the ceramic tube exceeds the anticipated stresses.

The testing of the ceramic fiber-based sleeves and seals was limited to the evaluation of compressive strength and creep resistance since these are the chief concerns. Table II summarizes the creep testing of the sleeve and seal materials. Both materials had sufficient strength and creep resistance to withstand the use conditions. Details of the test results are described in Ref. 1.

Based on the results of laboratory testing and an analysis of the materials after flue exposure testing, it was decided to use the siliconized silicon carbide material for the outer tubes, a recrystallized silicon carbide for the inner tubes in the hotter recuperator sections, and AISI 446 stainless steel for the inner tubes in the cooler section of the recuperator.

**Prototype Testing (Fig. 7)**

As part of the Phase I effort, a seven-tube assembly prototype recuperator was constructed and evaluated to verify the design concept, i.e., the materials selection, the heat transfer and pressure drop computer codes, and to measure system leakage. The prototype was in operation for 500 h at a range of temperatures up to 1260°C (2300°F). The exhaust gases from a gas-fired kiln were used to simulate the flue gases and the prototype was constructed using full-size components except that the tubes were 30 inches shorter than the full-size tubes.

As seen in Figs. 8 and 9, the pressure drop and heat transfer codes (computer model) were verified. The materials showed no evidence of degradation. However, the metallic tube sheets were slightly discolored. The leakage was

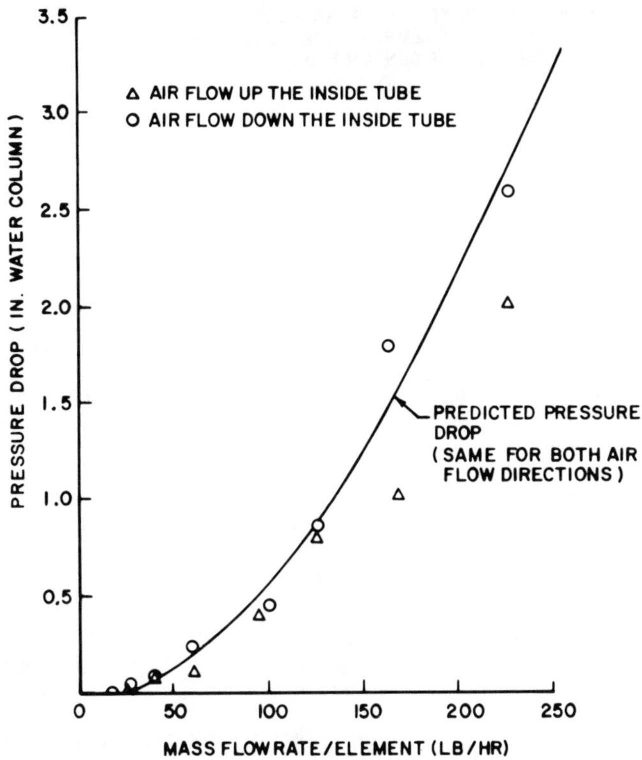


Fig. 8. Comparison of predicted and measured pressure drops from the prototype test with all tube assemblies open.

determined by pressurizing the recuperator and measuring the volume of air flow required to maintain pressure.

The air side-to-flue side leakage was measured at 9.0%, which is too high. Also, the system heat losses through the walls and to the tube sheets were felt to be excessive. To correct the leakage and heat loss, several modifications were made to the design. The outer tube sleeves were changed to incorporate the same seal design as the inner tube sleeves whose leakage was measured at 1.0%, which is an acceptable value. Four modifications were incorporated to reduce the heat loss to the tube sheets and through the plenum walls: (1) the tube sheet thickness was reduced from 3 to 2 inches, (2) insulation was added to the inside of the tube sleeves, (3) the bottom tube sheet insulation was closed around the tubes, and (4) the inside plenum insulation was increased from 3 to 8 inches.

**Recuperator Specifications**

All the technical disciplines described earlier participated in the specification of the recuperator. During the course of the project the system was changed from an in-series three-stage unit employing air-cooled metallic tube sheets and ceramic tubes to a one-stage unit in series with the two-stage metallic recuperator that exists in the flue.

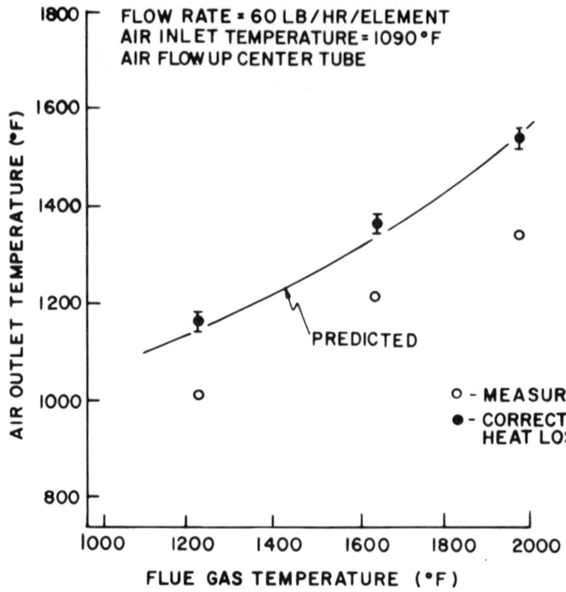


Fig. 9. Thermal performance of the prototype recuperator at 587°C (1090°F) air-inlet temperature. Comparison of predicted and measured temperatures.

The decision to retain the metallic recuperator and eliminate one of the ceramic stages was based on system economics. The cost to fabricate a recuperator from ceramics compared to metals dictates that cne should use a ceramic unit only when the temperature and corrosion prohibit the use of metals. The decision to eliminate one of the ceramic stages was also influenced by the inability to obtain enough ceramic tubes in a timely fashion.

The following tables describe some of the recuperator specifications. Table III is the general description of one stage. In Table IV are the dimensions for

Table III. General Specifications for One Ceramic Stage

| |
|---|
| 50 tube assemblies/pass |
| Tube arrangement: 5-4-5 hexagonal array |
| Tube spacing: 6 in. on center |
| Tube-tube gap: 3/8 in. |
| Overall dimensions (in.): 36W by 76L by 40H |
| Plenum inside dimensions (in.) 28W by 68L by 10H |

Table IV. HTBDR Tube Dimensions

| | Outer Tube | Inner Tube |
|---|---|---|
| Length (in.) | 76 7/8 | 96 |
| Outside diameter (in.) | 3 1/2 | 2 1/4 |
| Wall thickness (in.) | 1/4 | 1/4 |

the ceramic tubes. The anticipated air-side and flue-side pressure drops are shown in Tables V and VI. The mass flow rates at low and high fire are shown in Table VII. Table VIII illustrates the predicted flue temperature changes at the maximum preheat air condition for one ceramic stage condition, and Table IX shows how the fuel savings can be expected to change when the number of stages is reduced from two to one.

Table V. HTBDR Air Side Pressure Drop*

| | High Fire | Low Fire |
|---|---|---|
| Existing recuperator | 7.2 | 0.6 |
| Ceramic stage | 12.3 | 0.7 |
| Plenums | 1.2 | 0.1 |
| Burner | 8.0 | 0.4 |
| Total | 28.7 | 1.8 |

*Inches of water column.

Table VI. HTBDR Flue Side Pressure Drop*

| | High Fire | Low Fire |
|---|---|---|
| Existing recuperator | 0.2 | 0.02 |
| Ceramic stage | 0.3 | 0.02 |
| Total | 0.5 | 0.04 |

*Inches of water column.

Table VII. HTBDR Mass Flow Rates (lbs/h)

| | High Fire | Low Fire |
|---|---|---|
| Natural gas | 690 | 170 |
| Combustion air | 12 000 | 3 000 |

Table VIII. Air-Side and Flue-Side Fluid Temperature Change With One Ceramic Stage °C (°F)

| | Air Side | Flue Side |
|---|---|---|
| Metallic recuperator | | |
| In | 81 (179) | 871 (1600)* |
| Out | 638 (1182) | 427 (802) |
| Ceramic recuperator | | |
| In | 638 (1182) | 1232 (2250) |
| Out | 940 (1725) | 955 (1752)* |

*After dilution.

Table IX. Fuel Savings Change From Two to One Ceramic Stage*

| Flue Gas Temp. °C (°F) | Preheat Air Temp. °C (°F) 1 Stage | | 2 Stages | | |
|---|---|---|---|---|---|
| | 871 (1600) | 940 (1725) | 982 (1800) | 1010 (1850) | 1093 (2000) |
| 1204 (2200) | 46.6 | | 49.9 | | 52.8 |
| 1232 (2250)† | | 49.8 | | 51.7 | |
| 1260 (2300) | 48.7 | | 52.0 | | 54.9 |

*North American Handbook.
†B & W condition.

**Project Status**

The design phase of the project, which included the construction and evaluation of a prototype recuperator, the successful development of a high-temperature burner by North American Manufacturing Co., and the resolution of critical problems, has been successfully completed. Enough tubes have been received from the tube supplier to complete one stage. The fabrication of the stage is 90% complete. It is anticipated that the installation phase will begin in mid-1985, followed by a one-month shakedown period and a six-month evaluation period.

**References**

[1]J. E. Snyder, et al., "High Temperature Burner-Duct-Recuperator System Evaluation," Annual Report, October 1982, D.O.E. Project DE–FC07-81ID12296.

[2]J. I. Federer, "Corrosion Resistance of SiC Ceramics in Industrial Combustion Environments"; for abstract see *Am. Ceram. Soc. Bull.*, **63** [3] 520 (1984).

# Ceramic Recuperator Design for an Aluminum Reclamation Furnace

M. E. WARD and A. D. RUSSELL

Solar Turbines Incorporated
San Diego, CA 92138

W. W. LIANG

Gas Research Institute
Chicago, IL 60631

The aluminum reclamation furnace stack environment represents a major challenge for energy recovery devices because of its use of a continuous salt flux. Solar Turbines Incorporated (Solar) under contract to the Gas Research Institute (GRI) is developing a new ceramic recuperator that has been designed for corrosive and fouling environments. This new modular design recuperator uses both metal and ceramic components in this environment and uses a Solar-developed ceramic-to-metal joining process which facilitates in-process cleaning or removal of modules for maintenance.

Waste heat recuperation has a significant potential for saving energy in fossil-fuel-fired furnaces in the aluminum industry. Although published data on the use of recuperation in the aluminum industry are scarce, a recent survey indicated that only about 5% of the approximately 1000* holding and remelt furnaces in the United States were recuperated and that these units provided preheated combustion air in the range 304°C (580°F) to 537°C (1000°F). It was estimated that, if 648°C (1200°F) preheated air were provided to the entire furnace population through recuperation, an annual energy savings of $2.8\times10^{13}$ Btu's could be realized. This represents an approximate fuel savings of 34% compared to the nonrecuperated furnace fuel utilization. Based on a price of $\$5/10^6$ Btu's, this fuel savings represents an economic savings of $140 000 000 per year.

Looking at a specific installation where a 5.8 mm W (20 mm Btu/h) reclamation furnace is typically operated on a 24-h, 7-day a week basis, the savings for 648°C combustion air system would be $229 500 for a 7500-h year (see Table I for details). This would appear to be an attractive incentive to recycle the waste energy in these installations. However, there is a hurdle to overcome, and that is the corrosiveness of the waste gases.

### Application Environment

The above furnace information was for the most severe environment to be encountered in the aluminum industry. In these furnaces, molten halide

*Only approximately—or less than—this number are reclamation furnaces, to which this paper specifically applies.

Table I. Example Application Economics

| | | |
|---|---|---|
| Furnace Size | 5.8 mm W (20 mm Btu/h) | |
| Fuel usage at 90% of high fire | 18 000 scf/h or 5.2 mm Btu/h | |
| Hours/year | 7500 | |
| Hourly fuel cost at $15/mm W (mm Btu) | $90/h | |
| Yearly fuel cost | $675 000 | |
| Year fuel savings | | |
| Preheat, °C (°F) | % Savings* | Dollars Saved |
| 537 (1000) | 30 | 202,500 |
| 648 (1200) | 34 | 229,500 |
| 760 (1400) | 38 | 243,000 |

*Furnace at 1260°C (2300°F)

fluxes are used to remove surface oxides, and/or chlorine gas is used for alloy sparging. A typical salt flux will contain potassium and sodium chloride to provide a low-melting base material (melting temperatures 593°–648°C (1100°–1200°F)) in which to dissolve fluoride salts. Fluorides are the active solvent material in effecting oxide removal from aluminum melts. A typical low-cost flux can have the following composition: 50% NaCl, 45% KCl, 5% $Na_3AlF_6$; other materials such as LiCl and $MgCl_2$ can be added to lower the melting temperature, and LiF and $AlF_3$ can be added to increase the oxide fluxing capability.

Vapors of the salt flux and high concentrations of $AlCl_3$ and other volatile chlorides formed in sparging are present above the bath and in the stack gases. These vapors will condense on heat-transfer surfaces (tubes) and accelerate the oxidation process. Fluoride salts carry-over, due to their solvent action on the normally protective tube-alloy oxides, are extremely detrimental.

The $Na_2SO_4$ can also be expected to condense on tubes and have the potential to cause sulfidation attack at temperatures of 648°–926°C (1200° to 1700°F) with salt concentrations in the stack gases of only 10 to 1000 ppb.

Use of metallic recuperators in the stack of fluxed aluminum melts may therefore be impractical at high temperatures, depending on metal oxidation and recession rates. Initial tests indicate that recuperators constructed with high-temperature SiC ceramic materials have the potential to provide satisfactory service.[2]

**Ceramic Recuperator**

A ceramic recuperator has been designed for the corrosive-fouling environment of the aluminum reclamation furnace. The design uses a self-contained module, (Fig. 1) as the basic building block. The module is made up of 12 1¾ OD by 3/16 wall Hexoloy[†] silicon carbide tubes that are five feet long.

[†]Hexoloy, SA Tube, Carborundum Co., Niagara Falls, NY.

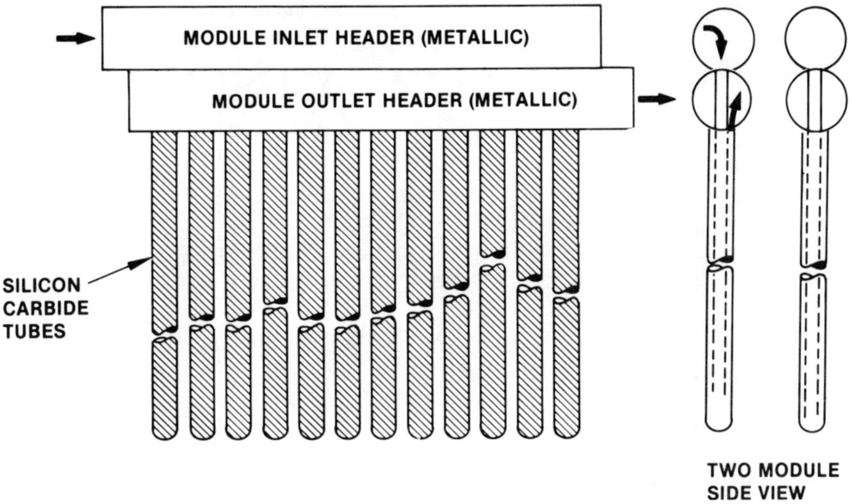


Fig. 1. Ceramic tube recuperator module.

Each tube is sealed at one end before final sintering.

The tubes are permanently bonded to stainless steel sleeves by a Solar Turbines Incorporated-developed proprietary joining process. This bond is non-permeable to air at pressures as high as 690 MPa (100 psi) and has temperature limits as required by the metal sleeve material. After a tube has been bonded to a metal sleeve, standard metal joining techniques can be used to join the tube to a metal header.

Referring to Fig. 1, air enters the module through the upper section of the header and travels down a 304 stainless tube inside the ceramic tube. The air exits the stainless tube into an annulus formed by the stainless and ceramic tube, travels up the annulus while being heated, and exits into the lower section of the metal header.

Modules would be arranged one behind the other to form a single-pass recuperator. Figure 2 shows two of these multimodule components configured into a two-pass recuperator.

Attributes of this design are:

(1) A standard-size factory-assembled module which can be used to form recuperators of any capacity.

(2) An individual module can be easily removed for cleaning, inspection, or replacement.

(3) An individual tube on a module can be easily field-replaced.

(4) Each tube can freely expand independent of each other and the structural metal components.

(5) Each tube is permanently bonded to a metal sleeve which gives a nonchanging leak-free condition. The bond allows the module to be an independent structure, free of sealing material between ceramic and metal components that age and deteriorate with time. This allows the use of a simple double-diaphram regulator control system to maintain proper fuel-air ratios.

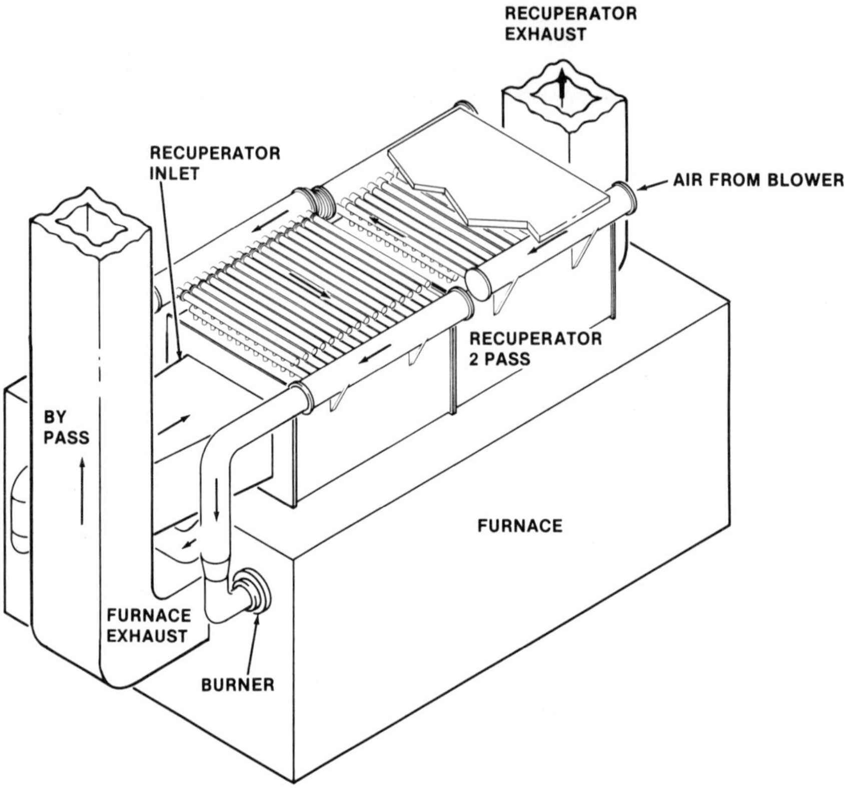


Fig. 2. Possible recuperator-furnace configuration.

**Full-Size Module Tests**

Recently completed performance tests of a full-size module have permitted the design of a full-size recuperator for field testing on a 5.8 mm W (20 mm Btu/h) reclamation furnace in 1985. The tests performed on the module, shown in Fig. 3, included measurements of the ceramic/metal joint and preheat temperatures. These are summarized in Fig. 4. Also, air-side pumping requirements were measured and correlated into a flow parameter (Fig. 5).

One hundred hours of endurance tests have been completed. Included in this test were 20 start and stop cycles and 400 thermal cycles. These thermal excursions were set up to quickly cycle the module and not as a performance check. A typical cycle is illustrated in Fig. 6. These tests verified the soundness of the Solar bonding technology which had undergone extensive laboratory tests previous to this full-size demonstration.

**Future Applications**

In addition to the planned full-size field demonstration of this ceramic recuperator technology in 1985 on an aluminum reclamation furnace, other applications are being examined, including the glass industry and applications requiring high-pressure preheated air.

Fig. 3. Full-size ceramic recuperator module.

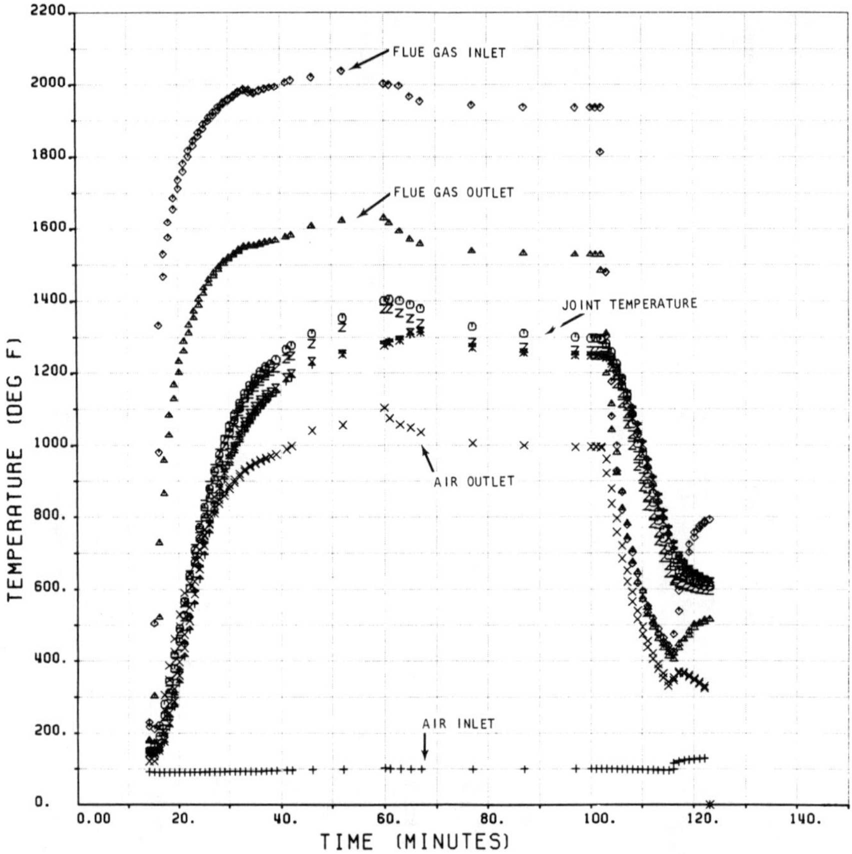


Fig. 4. Results from full-size ceramic recuperator module performance test.

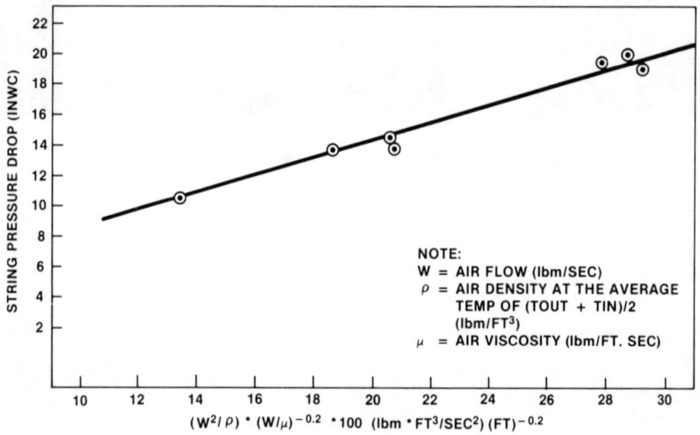


Fig. 5. Pressure-drop relationship for ceramic heat-exchanger module.

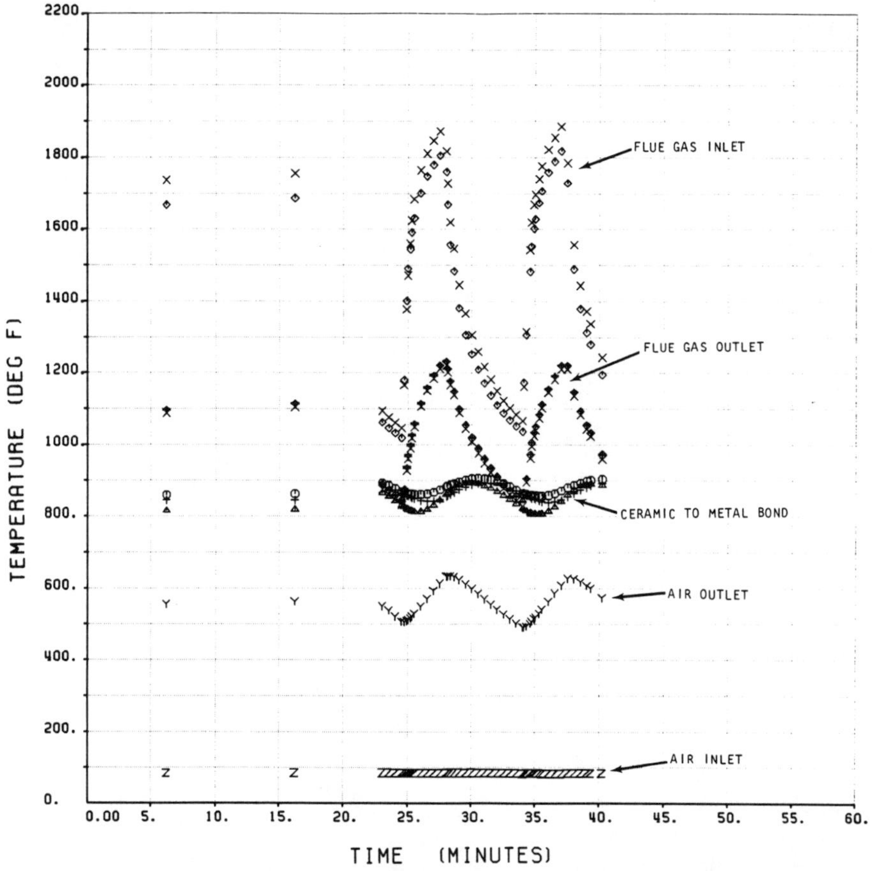


Fig. 6. Typical thermal cycle for the ceramic recupcrator module endurance test.

## References


[1]V. J. Tennery, "Economic Application, Design Analysis, and Material Availability for Ceramic Heat Exchanger," Oak Ridge National Laboratory. ORNL/TM-7580; January 1981.

[2]A. D. Russell, M. E. Ward, and C. E. Smeltzer, "Waste Heat Recuperation for Aluminum Furnaces," GRI 5081-342-0493, GRI Final Report, April 1983.

# Experimental Assessment of an Axially Finned Ceramic Tube Heat Exchanger

KENNETH O. SMITH

Solar Turbines Inc.
San Diego, CA 92138

TIMOTHY E. EASLER AND ROGER B. POEPPEL

Argonne National Laboratory
Argonne, IL 60439

A high-pressure, ceramic tube heat exchanger of an axial, counterflow configuration was tested. The test rig contained heat-exchanger tubes composed of 3-m (10 ft) long axially finned ceramic sections and 1.5-m (5 ft) long metallic sections. The ceramic components were fabricated using sintered $\alpha$-SiC. Testing was conducted using coal and natural gas to fire the heat exchanger to investigate tube fouling characteristics and thermal performance, respectively.

Over the past eight years, Solar Turbines Incorporated has been advancing ceramic heat-exchanger technology through a series of programs supported by DOE and EPRI.[1-3] These efforts have included fundamental technology development applicable to a wide range of ceramic heat-exchanger devices and the performance testing of a reduced-scale ceramic heat-exchanger unit. The specific heat-exchanger configuration adopted for development through these programs is a high-pressure, ceramic tube heat exchanger of an axial, counterflow design. The unit can accept ceramic heat-exchanger tubes up to 4.6 m (15 ft) in length and utilizes a tubular ceramic header at the top to collect the high-temperature working gas. This axial counterflow configuration has the advantages of higher heat-transfer efficiency and lower fouling potential on the shell side than other configurations using tube-in-tube and cross-flow geometries.

The work described below was one of three tasks comprising Solar Turbines' most recent ceramic heat-exchanger development work supported by DOE.[4-7] This effort was an extension of earlier rig testing of the Solar unit using ceramic tubes with axial fins to enhance heat-transfer rates. Earlier testing examined the performance of bare tubes. Specific goals of the test plan included:

(1) Rig operation at shell-side temperatures on the order of 1590 K (2400°F) and tube-side pressures up to $8\times10^5$ Pa (120 psia).

(2) Assessment of a hybrid ceramic/metal heat-exchanger tube configuration as a means of reducing the ceramic content of the heat exchanger. The tube assemblies were composed of 1.5 m (5 ft) metal sections and 3 m (10 ft) finned ceramic tube sections.

(3) Experimental verification of the heat-transfer enhancement achieved through finning of the ceramic tubes.

(4) Documentation of the fouling potential of the axially finned tubes when the test rig was fired at 1267 K (1820°F) using coal. These tests were to be directly comparable to earlier fouling tests conducted using nonfinned heat-exchanger tubes.[3]

(5) Assessment of a modified ceramic header design that allowed joining of the header to downstream piping with a V-band clamp.

(6) Evaluation of the suitability of sintered $\alpha$-SiC as a material for use in the heat exchanger.

**Facility Description**

The ceramic heat-exchanger test facility is of an axial-counterflow design (Fig. 1). The rig can accommodate 4.6 m (15 ft) long heat-exchanger tubes and has a rectangular flow cross section of 23.5 by 11.7 cm (9.25 by 4.625 in.).

With the adopted rig configuration, the high-temperature shell-side gas stream enters the heat exchanger at the top. This shell-side gas stream flows downward over the heat-exchanger tube bundle and exhausts at the bottom of the unit. Low-temperature, high- or low-pressure air enters at the bottom

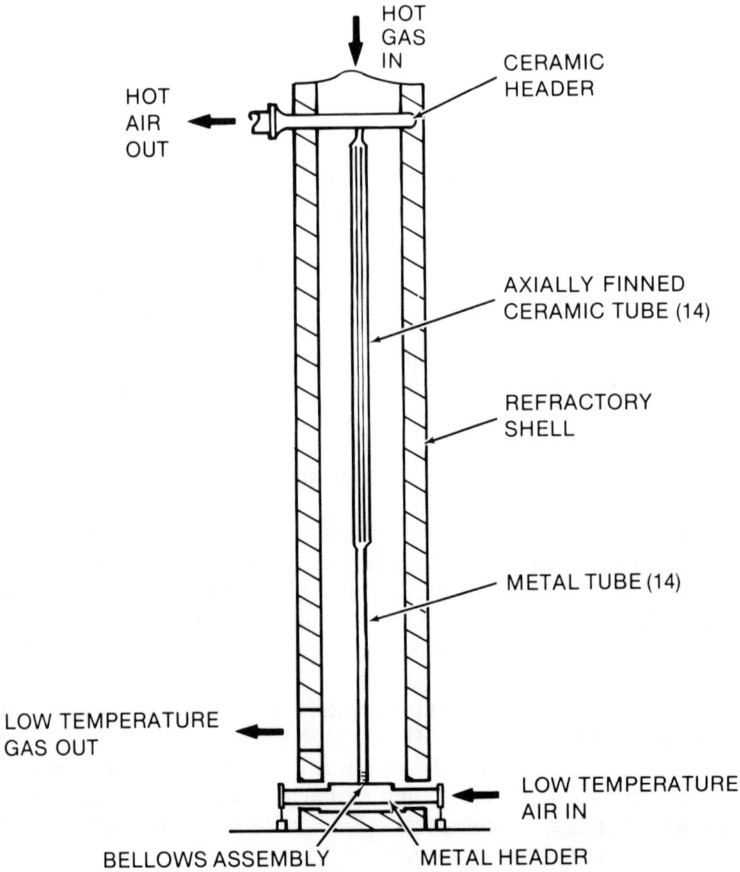


Fig. 1. Ceramic tube heat exchanger.

of the heat exchanger and is distributed to the heat-exchanger tubes through a metallic header. The air flows upward inside the tubes and discharges into a tubular ceramic header located at the top of the heat exchanger.

The ceramic header is supported in a fixed position within the heat exchanger. The lower, metallic header is supported from below using a series of pneumatic pistons which allow the header to "float" vertically. In this way, the gross thermal expansion of the tube bundle can be accommodated. The pistons are operated so as to maintain a constant compressive force on the header and tube bundle to ensure good sealing of the spherical compression joints used to mate the ceramic components of the heat exchanger.

The shell side of the heat exchanger is fired with an atmospheric pressure combustor using either natural gas or pulverized coal as the primary fuel. Tube-side air flows can be preheated using a natural gas-fired, vitiating air heater upstream of the metallic header.

### Heat-Exchanger Tube Assembly

A hydrid ceramic/metal heat-exchanger tube assembly design was selected for evaluation. The concept involves the use of metal tubes in low-temperature regions, thus minimizing the extent to which ceramics are employed. This approach has the potential economic advantage of substituting lower-cost metals for ceramics. Testing in earlier programs was limited to all-ceramic heat-exchanger tubes.

A series of heat-exchanger performance calculations indicated that, to maintain metal temperatures below approximately 1144 K (1600°F), the tube assemblies should be comprised of 1.5 m (5 ft) metal sections and 3 m (10 ft) ceramic sections.

### Finned Ceramic Tubes

The as-designed and as-fabricated ceramic tube cross sections are illustrated in Fig. 2. The original design reflects the incorporation of 12 0.64 cm

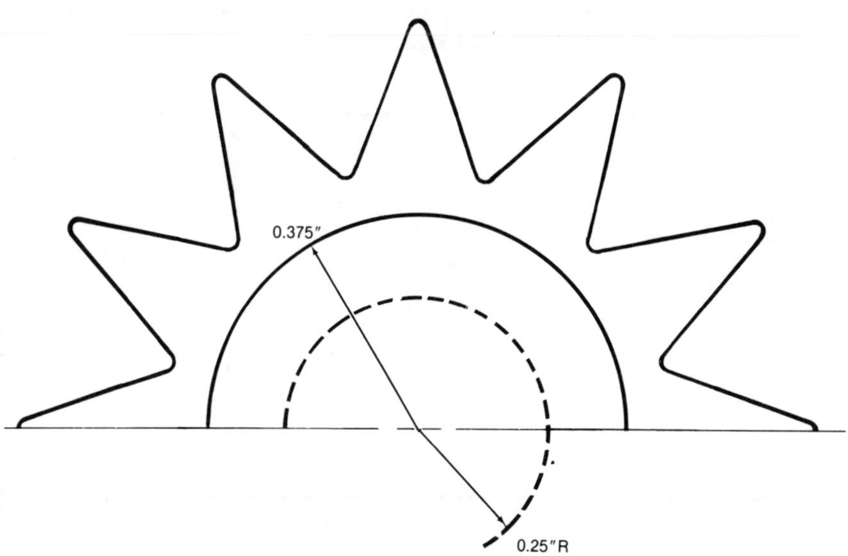


Fig. 2. As-designed and as-fabricated ceramic tube cross sections.

(0.25 in.) triangular-shaped fins on a 2.54 cm (1 in.) outer diameter (OD) and 1.27 cm (0.5 in) inner diameter (ID) ceramic tube. Testing in earlier programs employed bare ceramic tubes of the same OD and ID. As discussed below, manufacturing constraints prevented the tubes from being fabricated to this thick-walled design.

The triangular fin shape was selected as a result of thermal shock tests performed in earlier studies.[2] Inductively heated, finned ceramic tube specimens were subjected to rapid cooling to investigate the thermal shock resistance of both triangular and rectangular axially finned tubes. The triangular configuration was superior and was chosen for the present study.

Figure 3 illustrates the tube bundle geometry for the selected finned tube design. The available test rig cross-sectional area limited the overall finned tube OD to 3.8 cm (1.5 in.) if a reasonable number of tubes (14) was to be accommodated in the rig. For comparison, only eight tubes could have been installed if the fin height was increased to 1.27 cm (0.5 in.). A tube bundle of 28 bare tubes was tested in previous programs.

The design details of the ceramic tube ends are shown in Fig. 4. A 10 cm (4 in.) length of bare tube was left at each end. This served two purposes. First, it facilitated the machining of the tube ends as required for the spherical compression joint. In addition, it was expected that the lack of finning on the tube ends would reduce the level of fouling just downstream of the ceramic header. Earlier tests with coal firing showed this region to be the most rapidly fouled.[3] The bare tube ends were intended to provide more free space in the wake region below the header.

### Metallic Tube Sections

The 1.5-m (5 ft) long metal heat-exchanger tubes were fabricated using 1.9 cm (0.75 in.), Schedule 80, Incoloy 800H pipe. The top end of each tube was machined to provide the female half of the spherical compression joint used for the metal-to-ceramic joint. The lower ends of the metallic tubes were fitted with a bellows assembly to accommodate any tube-to-tube variation in thermal expansion of the heat-exchanger tubes.

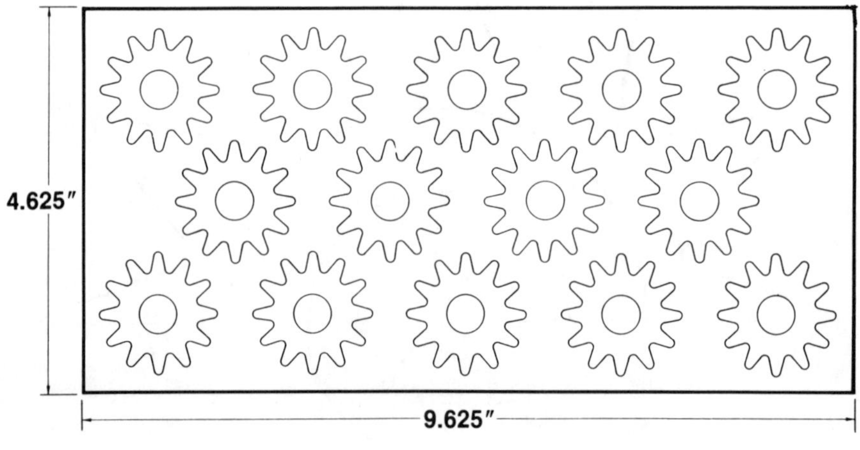


Fig. 3. Ceramic heat-exchanger tube bundle geometry.

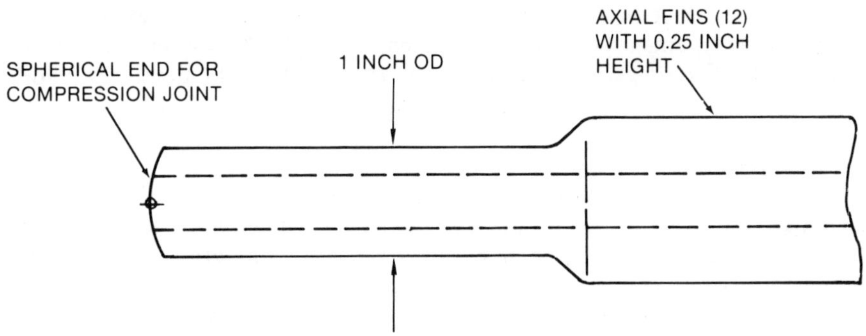


Fig. 4. Details of ceramic tube ends.

Fig. 5. Flanged ceramic header.

**Ceramic Header**

The ceramic header used in this program (Fig. 5) was basically similar to the designs adopted for earlier studies. The ceramic tubes mate with the header by means of ball-and-socket-type compression joints. Modifications to the earlier header design for this program included a different pattern for the header stubs to accommodate the new tube bundle geometry. In addition, a flanged conical section was added to the header exhaust end. Thus, a V-band clamping arrangement could be used to maintain a backpressure orifice plate in place. Earlier programs had employed an externally mounted pneumatic piston to hold the backpressure plate up against the ceramic header.

### Material Selection

Of the ceramic materials currently available, only sintered $\alpha$-SiC and siliconized SiC were deemed viable candidates for the ceramic heat-exchanger components. The major factor in selecting the $\alpha$-SiC material was the fabricator's ability to meet design specifications. Only the $\alpha$-SiC supplier indicated the capability to produce 3-m (10 ft) long axially finned ceramic tubes. An additional factor favored the selection of the sintered $\alpha$-SiC material. Earlier programs had evaluated siliconized SiC. Thus the sintered $\alpha$-SiC represented a new material for testing.

### Ceramic Tube Fabrication

Fabrication of the ceramic tubes proved to be the pacing activity in terms of program performance. An indication of the effort involved in tube fabrication can be seen in the delivery schedule of the tubes. An initial tube delivery of four pieces was received from the manufacturer approximately five months after the initial scheduled delivery date. A total of 14 tubes, the number needed to fill the tube bundle, had been delivered 3 months later. However, 6 of these 14 tubes contained cracks (to be discussed below) and were not suitable for high-pressure operation. Fourteen acceptable tubes were eventually received approximately 8 months after the first tube delivery. Additional tubes continued to arrive after completion of the program test plan.

The finned tubes were fabricated by extrusion. Initial attempts to extrude the tubes using a water-based slurry were unsuccessful. The material failed to fill the extrusion die, with resulting poor fin definition. Subsequently, a resin-based binder was employed with better success. However, use of the resin-based binder required that a binder removal step be added to the fabrication process prior to sintering.

Because of the need for efficient binder removal, it was determined by the manufacturer that 0.64 cm (0.25 in.) tube wall thicknesses could not be provided. Thus, a design variance was permitted to reduce the tube wall to 0.32 cm (0.125 in.). It was clear that the wall thickness reduction would result in higher than design-level stresses in the tubes. In addition, the integrity of the spherical compression joints was at risk because of the reduction in joint contact area.

As finned ceramic tubes were received, a number of fabrication flaws and characteristics were found. They included: (1) a variable degree of tube straightness, (2) spiraling of the tube fins, (3) surface pits in the tube material, (4) tube wall thickness variations, and (5) axial cracks in the tube ends (Fig. 6).

The impact of these flaws and tube characteristics on heat-exchanger assembly and performance is detailed below:

(1) Tube straightness deviations proved only a minor problem in terms of rig assembly. Ceramic tubes had to be selectively located within the tube bundle to prevent tube-to-tube interference.

(2) Fin spiraling resulted in no performance or assembly problems.

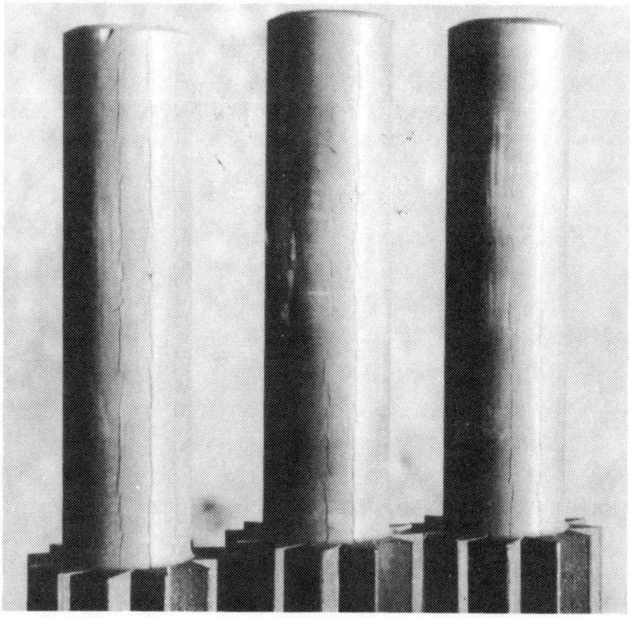

Fig. 6. Cracked ceramic tubes.

(3) Surface pits were found in only a single tube and had no apparent impact on tube performance over the achieved test duration.

(4) Tube-wall thickness variations were of a magnitude that resulted in walls as thin as 1.6 mm (0.0625 in.). These large deviations from the nominal 3.2 mm (0.125 in.) wall further increased material stresses from the design levels originally established for the 6.4 mm (0.25 in.) tube walls and further threatened joint integrity.

(5) The axial cracks made the affected tubes unsuitable for high-pressure operation as the cracks acted to reduce tube strength and contributed to tube-side leakage.

**Ceramic Header Fabrication**

The ceramic headers were fabricated by slip casting. Throughout the fabrication activity, significant problems were encountered with cracking of the header castings at the bases of the stubs used for the spherical compression joints. The cracking was attributed to the conical exhaust section of the header. This conical geometry prevented the SiC from shrinking freely in the mold as the slurry dried. Consequently, the header was placed in tension with subsequent cracking. Eventually two headers of excellent quality were delivered for use in the heat exchanger.

**Fouling Tests**

The first test series examined the fouling tendency of the heat exchanger. The nominal test conditions were similar to those of earlier bare tube fouling tests:

Shell-side flow rate: 0.23 kg/s (0.5 lb/s)

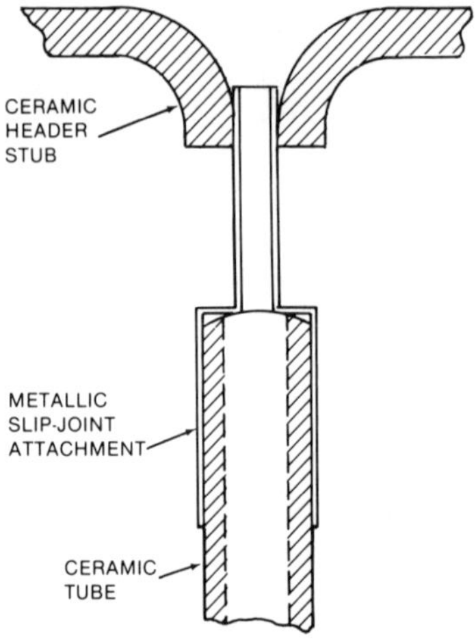


Fig. 7. Slip joint used for low-pressure tests.

Shell-side inlet temperature: 1261 K (1810°F)
Coal flow rate: 32 kg/h (70 lb/h)
Initial coal ash fusion temperature: 1722 K (2640°F)
Tube-side inlet temperature: 322 K (120°F)
Tube-side exhaust temperature: 1106 K (1530°F)
Tube-side pressure: $1.2 \times 10^5$ Pa (3 psig)

A high-volatile, bituminous coal identical to the coal employed in earlier testing was used.[3] The low-pressure fouling tests were conducted using a slip-joint configuration to mate the ceramic tubes to the ceramic header (Fig. 7). Rig shakedown tests had demonstrated that, without high internal pressures, the metal bellows in the tube assemblies were not sufficiently stiff to support the heat-exchanger tubes. As these bellows tended to collapse, they were unable to accommodate any tube-to-tube differential in thermal expansion. Consequently, the slip joints were employed at low pressure to serve this purpose.

The first fouling test was conducted using a full tube bundle of 14 finned tubes for 14 h. Progressive fouling of the heat exchanger was monitored through measurement of the shell-side pressure drops across the ceramic header and across the entire tube bundle (metal and ceramic tubes) (Fig. 8). A rising pressure drop implied increased fouling. These data show rapid fouling of the header and bundle over the first 10 hours of operation. At the 10-h mark it appears that ash agglomerates became dislodged from the header (causing a decrease in pressure drop) and dropped into the tube bundle. This additional blockage of the bundle resulted in an extremely rapid fouling rate during the remaining test period. During this same period the header pressure drop remained relatively constant. The test was terminated after 14 h when it appeared that no steady state fouled condition was being approached.

A post-test inspection of the heat exchanger showed a high degree of fouling. In general, the level of fouling did not appear as severe as had been observed in the earlier bare tube test. However, bare tube testing had been continued for twice as long as this first finned tube test.

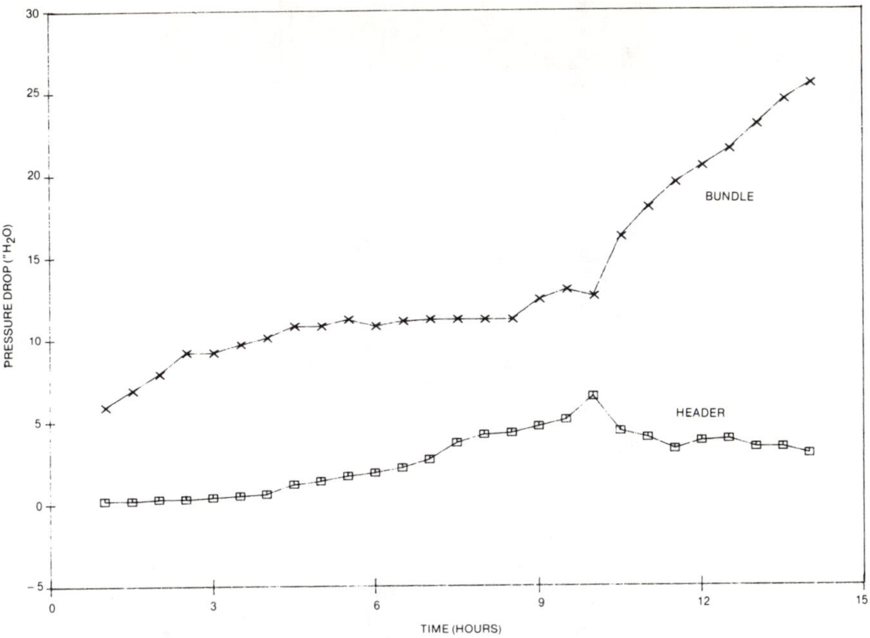


Fig. 8. Pressure-drop traces from first fouling test.

As observed with the bare tube test, heat-exchanger fouling was limited to the upper region of the rig. Fouling occurred primarily on the lower surface of the ceramic header and the upper 46 cm (18 in.) of the tube bundle (Fig. 9). Further down in the rig, the finned tubes showed at most a light dusting of coal ash (Fig. 10). The majority of the deposits were of a powdery nature and were easily removed either by brushing or "soot blowing."

Based on this first test, it appeared that the axial finned tubes were not, in general, more prone to fouling than bare tubes. However, it was clear that fouling in the header wake zone was still severe.

A second, nine-hour fouling test was run to examine the dependence of fouling rate on tube bundle configuration. Six tubes were removed from the bundle, leaving the pattern illustrated in Fig. 11. The pattern was expected to significantly reduce the tube/tube and tube/wall bridging by the coal ash that was observed in testing the full tube bundle.

Pressure-drop traces for the header and tube bundle for both fouling tests are presented in Figs. 12 and 13. These data show that fouling rates can be effectively slowed through increases in tube bundle spacing. As yet undetermined is whether periodic ash removal through soot blowing is necessary for long-duration operation or whether an acceptable steady state fouled condition may eventually be attained.

An examination of the rig after the second test showed that the header was only slightly fouled (Fig. 14). The finned ceramic tubes were moderately fouled but no tube-to-tube bridging was observed. As in the previous test, fouling was limited to the upper 46 cm (18 in.) of the finned tubes.

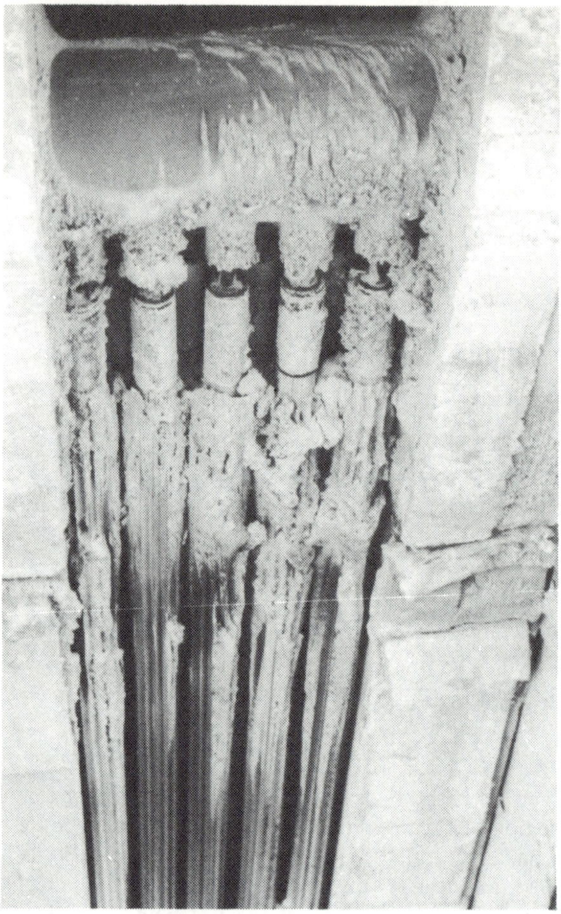

Fig. 9. Fouling of full tube bundle after first fouling test (14 h).

**Performance Tests**

Two natural gas-fired tests were conducted to establish the heat-transfer characteristics of the finned ceramic tube heat-exchanger configuration. Goals for testing were to duplicate previously achieved tube-side operating conditions of $6.8\times10^5$ Pa (100 psia) and 1533 K (2300°F) exhaust temperature. Because of an insufficient number of tubes suitable for high-pressure operation, only eight tubes were used in the tube bundle. An initial test was conducted for approximately five hours (Figs. 15 and 16). Shell-side inlet temperature was set at approximately 1283 K (1850°F). Tube-side air flows were periodically increased, resulting in successively higher tube-side pressures. As air flow increased, tube-side exhaust temperature decreased. Near the end of the test, tube-side air flow was maintained constant but shell-side inlet temperature was raised. Testing was halted when a sudden drop in tube pressure suggested a ceramic component failure.

A post-test inspection of the heat exchanger indicated that one of the eight ceramic tubes had failed. The failure occurred in the nonfinned end

Fig. 10. Test bundle after first fouling test.

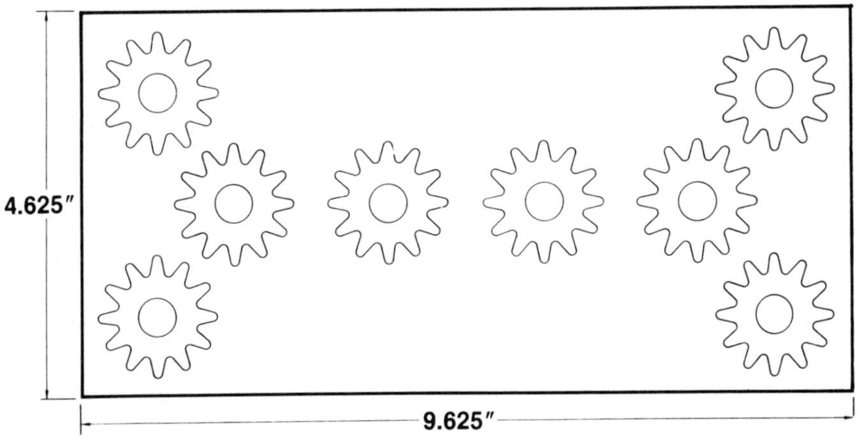


Fig. 11. Reduced tube bundle geometry.

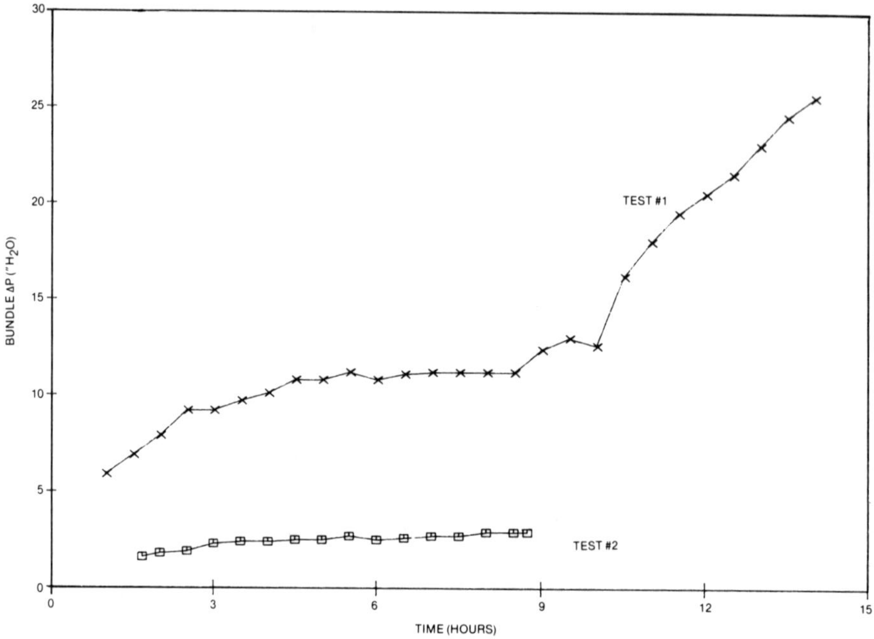


Fig. 12. Tube bundle pressure-drop traces.

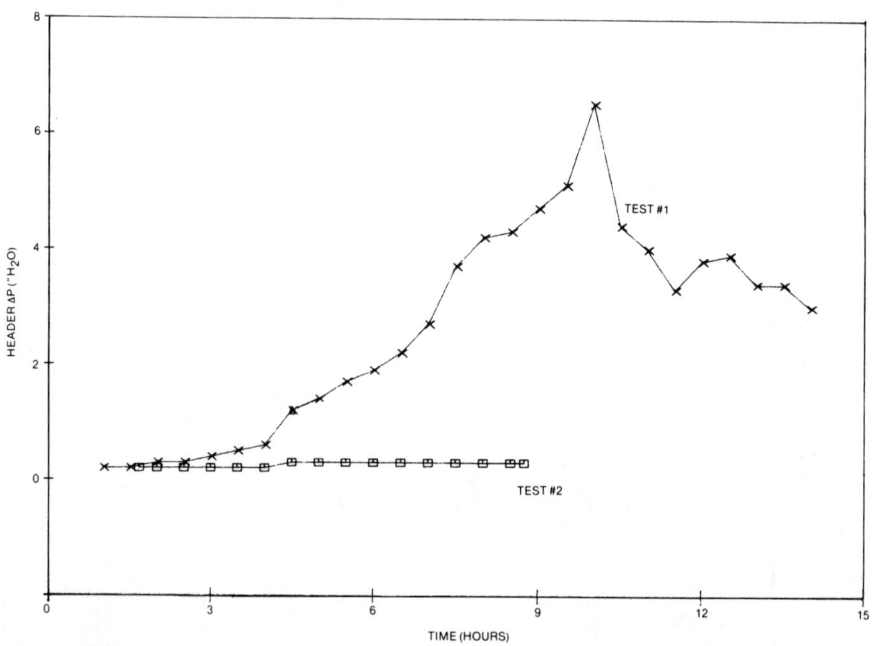


Fig. 13. Ceramic header pressure-drop traces.

Fig. 14. Fouling of reduced tube bundle after second fouling test (9 h).

portion of the tube at the ceramic/metal tube compression joint. Examination of the failed tube revealed wall thicknesses in the fracture area as thin as 1.6 mm (0.0625 in.), well below the nominal 3.2 mm (0.125 in.) thickness anticipated.

In addition to the tube failure, the ceramic header had sustained damage. Contact stresses arising from the clamping configuration acted to fracture the rim of the header exhaust flange. Although significant damage resulted to the flange, an adequate seal was still maintained to allow rig operation at high pressures.

During this initial high-pressure test, tube-side pressures of approximately $4\times10^5$ Pa (60 psia) were achieved. Shell-side inlet temperatures up to 1450 K (2150°F) were reached with tube-side exhaust temperatures ranging from 922 to 1144 K (1200° to 1600°F).

The experimentally observed tube-side temperatures were compared with analytical predictions using a previously developed heat-exchanger design com-

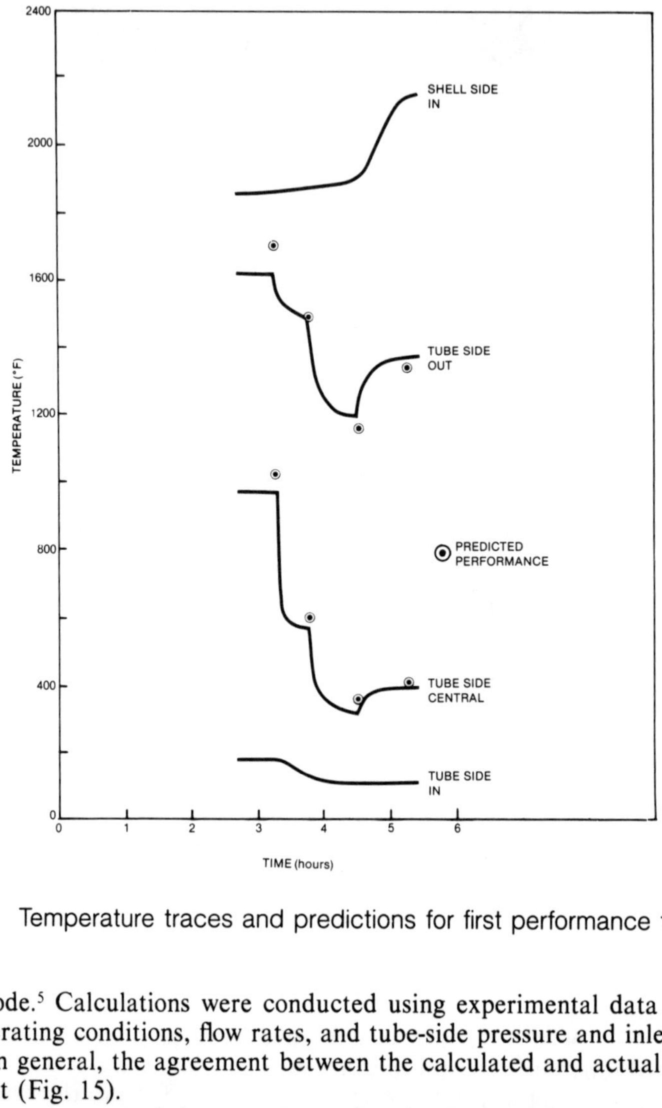


Fig. 15. Temperature traces and predictions for first performance test.

puter code.[5] Calculations were conducted using experimental data for shell-side operating conditions, flow rates, and tube-side pressure and inlet temperature. In general, the agreement between the calculated and actual results is excellent (Fig. 15).

The agreement of the experimental and calculated heat-exchanger performance supports fin design calculations that indicated that the short fin configuration adopted would be nearly 100% efficient in augmenting heat transfer through the fin. For a 100% efficient fin, heat transfer through the fin is directly proportional to the fin surface area, assuming negligible heat-transfer resistance through the tube wall and across the interior tube surface.

For the second performance test, the V-band clamp concept was abandoned, as no backup headers suitable for pressure testing were available. Instead, the backpressure orifice plate was held in position against the header using an externally mounted pneumatic piston. This sealing configuration was used successfully in previous studies with nonflanged ceramic header designs.

Testing was conducted for approximately seven hours. Early in the test it was discovered that a good seal between the header flange and orifice plate

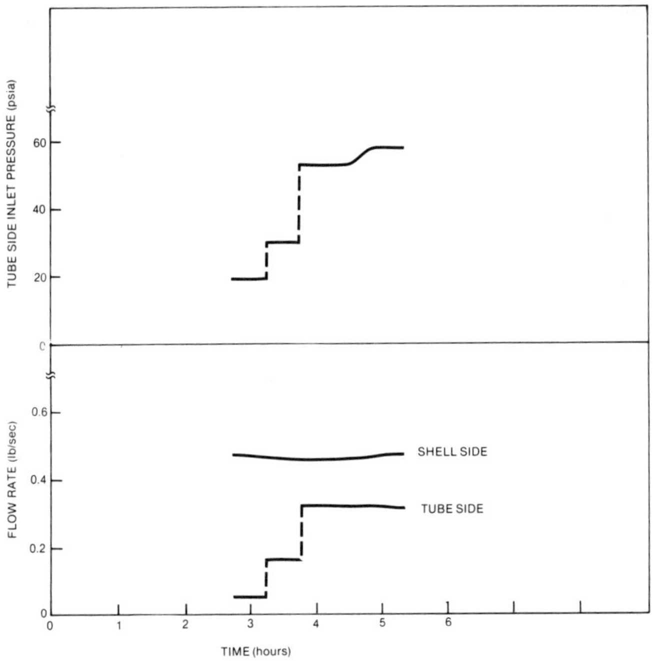


Fig. 16. Flow and pressure traces for first performance test.

could not be achieved. This was attributable to warpage of the orifice plate. Consequently, it was decided that the remainder of the test would be run at relatively low pressures with the intent of increasing temperatures to the highest achievable levels.

Figure 17 presents test data from the second performance test. Shell-side temperatures up to 1533 K (2300°F) were achieved. Tube-side pressures ranged from 1.2 to $2\times10^5$ Pa (3 to 15 psig). Tube-side inlet temperatures were varied over the test duration from ambient levels to 839 K (1050°F).

Figure 17 indicates that measured tube-side exhaust temperatures on the order of 1283 K (1850°F) were achieved. However, several factors suggest that higher temperatures were actually attained. The thermocouples used to measure tube-side exhaust temperature extended through the water-cooled orifice plate. As a result of external ducting modifications associated with the use of the piston sealing configuration, these thermocouples extended only a few centimeters into the ceramic header. Consequently, conduction and radiation losses to the orifice plate were significant. Compounding the radiation loss was the fact that the thermocouples were not shielded to reduce radiation losses because of the need to penetrate the orifice plate.

To investigate potential thermocouple heat-loss magnitudes, the heat-exchanger design computer code was used to calculate tube-side temperatures at the ceramic/metal tube joint and at the header exhaust. Agreement of the test data and predictions at the ceramic/metal tube joint is excellent (Fig. 17). Similar agreement is seen for tube-side exhaust temperatures below 1256 K (1800°F). Above this temperature, the test data and predictions show an

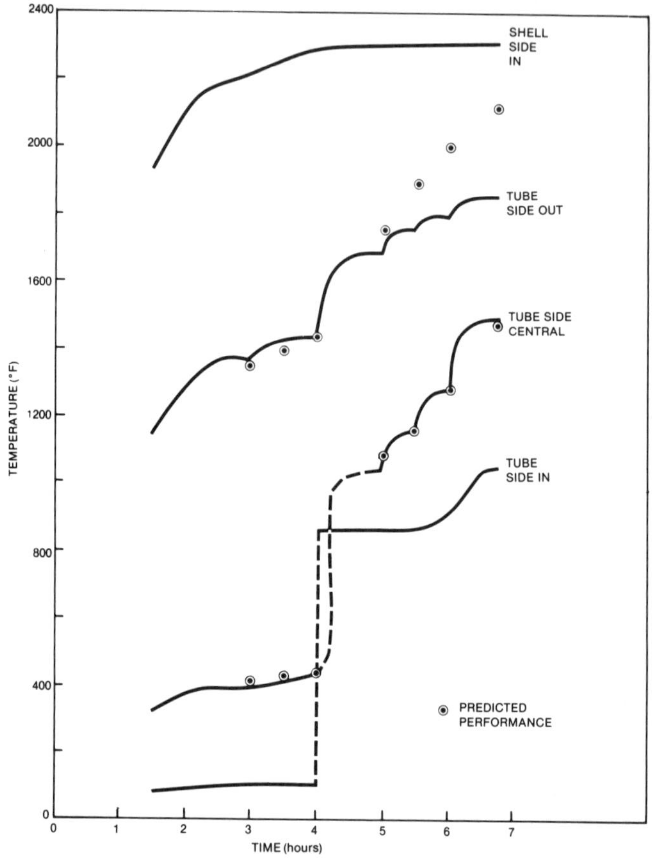


Fig. 17. Temperature traces and predictions for second performance test.

increasing disparity with increasing temperature. This is in line with radiation being proportional to temperature raised to the fourth power. The results of this analysis suggest that exhaust-gas temperatures on the order of 1422 K (2100°F) were achieved with the measured data indicating lower levels due to heat losses to the water-cooled orifice plate.

**Summary**

Earlier programs had demonstrated the viability of the axial flow ceramic heat-exchanger concept. These earlier studies involved the use of 2.54 cm (1 in.) OD and 1.27 cm (0.5 in.) ID, 4.6-m (15 ft) long bare ceramic tubes. The reported effort was conducted to continue the advancement of the high-pressure, axial flow ceramic heat exchanger by assessing the use of: (1) axially finned ceramic heat-exchanger tubes to enhance heat transfer, (2) a hybrid ceramic/metal tube assembly configuration to reduce the ceramic content of the heat exchanger, and (3) a modified hot ceramic header design that used a clamping arrangement for joining the header to downstream metallic piping.

The current program used heat-exchanger tubes composed of 3-m (10 ft) long, axially finned ceramic tubes atop 1.5-m (5 ft) long metallic tubes. The ceramic tubes contained 12 axial fins of 6.4 mm (0.25 in.) height on a tube base of 2.5 cm (1 in.) OD and 1.9 cm (0.75 in.) ID. All ceramic parts employed in the program were fabricated using sintered $\alpha$-SiC.

Two series of heat-exchanger tests were performed. The first, run at low tube-side pressures, investigated shell-side fouling through coal-firing of the heat exchanger. The second test series evaluated heat-exchanger performance at elevated tube-side pressures using natural gas firing. Approximately 65 h of operation were accumulated with shell-side temperatures ranging up to 1533 K (2300°F). Tube-side exhaust temperatures to approximately 1422 K (2100°F) were achieved. Tube-side pressures to $4 \times 10^5$ Pa (60 psia) were achieved, approximately one-half of the original program goal.

Test results demonstrated that axial finning of ceramic tubes is an effective and predictable method of enhancing heat transfer. Heat-exchanger performance agreed well with analytical predictions which implied that the fin efficiency of the relatively short fins employed was near 100%. Although fin durability was not extensively evaluated, no damage to the fins was observed over the test duration. The testing indicated that the hybrid ceramic/metal tube concept is a viable means of reducing the length of ceramic tube required.

Fouling tests were run at a shell-side inlet temperature of 1261 K (1810°F) to allow comparisons with previously run bare tube fouling studies. The fouling tests demonstrated that axial finning does not result in any observable increase in tube fouling relative to bare tubes. Over most of the tube length, very little fouling occurred. However, the critically fouled region in the heat exchanger remained the wake region downstream of the ceramic header, as observed in the earlier studies. High turbulence levels and a crossflow velocity component in this region led to high fouling rates and significant heat-exchanger blockage. With close-packed tubes, severe fouling occurred within 10 to 20 h. The fouling was restricted to a region approximately 46 cm (18 in.) or four to five header diameters downstream of the ceramic header.

Fouling tests of two tube bundle geometries documented the effectiveness of large tube spacing in reducing heat-exchanger blockage rates. Recommendations for future improvements include removing the ceramic tube finning for a distance of four header diameters downstream of the ceramic header to further open up this wake zone and the use of soot blowing for periodic fly ash removal.

The program demonstrated that the fabrication of a flanged ceramic header that can withstand moderate internal pressures was feasible. However, manufacturing yields need to be improved to establish commercial viability. Further attention must be directed to the header sealing requirement. The V-band clamping configuration evaluated in this program was unsuccessful due to excessive contact stresses in the flanged section of the header.

As a result of the fabrication problems experienced with the sintered $\alpha$-SiC finned tubes, it must be concluded that, at the present time, finned ceramic tubes of this material on the order of 3 m (10 ft) long are not economically feasible for commercial applications. High-quality parts can be fabricated but yields are relatively low due to manufacturing flaws and unpredictability in meeting design specifications. In addition, current extrusion technology for sintered $\alpha$-SiC is limited in terms of allowable wall thickness. This wall thickness limitation may also limit the commercial application of finned ceramic tubes due to reduced internal operating pressure capabilities. However, advances in the use of water-based slurries for extrusion may increase manufacturing yields and circumvent existing wall-thickness limitations.


## Acknowledgment

This work was sponsored as part of the Surface Gasification Materials Program directed by the Morgantown Energy Technology Center for the U.S. Department of Energy, Office of Fossil Energy.


## References


[1]M. E. Ward, N. G. Solomon, M. E. Golden, and C. E. Smeltzer, "Development of a Ceramic Tube Heat Exchanger with Relaxing Joint," DOE Report FE–2556–30, June 1980.

[2]M. E. Ward and A. H. Campbell, "Ceramic Heat Exchanger Technology Development," DOE Report DE–4898–F, November 1982.

[3]M. E. Ward and A. J. Kubasco, "Ceramic Heat Exchanger Technology," EPRI Report AP–2883, March 1983.

[4]K. O. Smith and R. B. Poeppel, "Assessment of a Failure Probability Prediction Technique for Ceramic Heat Exchanger Components"; paper No. 85-GT-178 presented at the 30th ASME International Gas Turbine Conference, Houston, Texas, March 18–21, 1985.

[5]K. O. Smith, "Ceramic Heat Exchanger Design Methodology," Solar Turbines Report SR83–R–5051–20, January 1984.

[6]K. O. Smith, "Internal Heat Transfer Characteristics of Silicon Carbide Heat Exchanger Tubes," *ASME J. Heat Transfer*, **106** [8] 672 (1984).

[7]K. O. Smith, "Finned Ceramic Tube Heat Exchanger Technology Development," Final Report, Argonne National Laboratory Contract No. 31–109–38–6796; in preparation.

# Vertical Tube-Style, All-Ceramic, Air-to-Air Heat Exchanger

ROBERT G. GRAHAM

C & H Combustion
Troy, MI 48083

An alternative design is described, in which the waste-heat gases pass through the inside of the tubes while combustion air, being preheated, passes around the outside.

Due to the substantial increase in fuel costs in the last 10 years, a market has developed for all-ceramic, air-to-air heat exchangers. These exchangers are applicable to any process that has high temperature, corrosion, and/or particulate-laden flue gas, such as the exhaust from secondary aluminum melting furnaces, where use of metal recuperators is not practical. C & H Combustion has already installed silicon carbide shell and tube recuperators, in which there are no metals in contact with the flue gas. In the C & H external flue gas (E.F.G.) model, the flue gas passes outside the tubes and the heated air inside.

These units are in-service on investment casting and aluminum remelt furnaces. One is scheduled to start up this fall on a high-temperature steel drop forge furnace, and a test cell is being installed in the flue of a scrap copper melting furnace. They will handle flue gases at temperatures up to 1482°C (2700°F), and return combustion air to the burners from 426° to 871°C (800° to 1600°F). The furnaces range in size from 580 000 to 17 400 W (2 000 000 to 60 000 000 Btu/h) with fuel reduction up to 50%.

The difficulty of working with ceramics, as compared to metal, is threefold—first, the brittle nature of the ceramics; second, the ceramic-to-ceramic seals; and third, the increased material cost.

The brittleness problem was addressed by using heavy (3-in. thick) tube sheets constructed of special silicon carbide shapes set in standard firebrick flues. The exchanger tubes (extruded or slipcast silicon carbide) are set in the sheet with considerable tolerance and allowed to "float." All tube-to-tube–sheet contact is in compression with no shear stresses on either.

The seals are ceramic fiber and are buried in the tube sheet out of the flue-gas stream. As the tubes expand, they grow into the seal. This, coupled with air pressure on the tube sheet, increases sealing efficiency as the recuperator comes to temperature.

The cost of all the ceramics has been decreasing rapidly because of the number of orders placed and the improvement of manufacturing technology by the ceramic companies. Zone selection of ceramics, to meet lower temperatures and less attack, also has reduced exchanger cost.

As an extension of this program, C & H is now testing and starting to market a vertical exchanger (the I.F.G. model) where the flue gas flows inside the tube and heated air outside. This style recuperator can be used on smaller furnaces and will take up less space than the E.F.G. model described above.

This design also addresses the problems industry is experiencing with cube- and plate-type ceramic recuperators. These recuperators have a relatively high pressure drop on the flue-gas side, and can be used only with clean flue gas because they are susceptible to plugging and fires. Also, cube style cannot be cleaned or soot blown, nor can individual ceramic components be replaced.

The vertical exchanger, with silicon carbide internals, is shown in Fig. 1. The air-side access panel is removed. The tubes are carried by a jack, or sprung arch, in the bottom of the exchanger. This arch is composed of special silicon carbide shapes that allow the tube to move in the seal, similar to the E.F.G. model. Since each tube is vertical, the seal is kept in compression, regardless of either the tube sheet or tube movement. The flue gas flows up the tubes and directly out of the stack. The tube sheet has C & H's bell-shaped inlet to each tube. This substantially reduces pressure drop through the exchanger.

The top tube sheet is ceramic, carried by a stainless steel support plate that is air-cooled on the clean air side. Like the bottom support, each tube is buried in the sheet and has an individual ceramic fiber seal. As the tube expands upward, it grows into the seal and leakage is reduced. Since the top plate is in the coolest zone, the seal is kept at low temperatures, which increases the seal's resilient life.

The clean air is zoned across the outside of the tubes with steel baffles. The air will be pressurized, usually 20 to 24 ounces, and any air-side leakage is through baffle tube penetrations that have sufficient clearance to allow the tubes to move freely.

If any tubes need replacing, the top section of the exchanger can be unbolted and removed and the tubes easily replaced.

Wire brush cleaning is possible on the interior of the tubes through an access port built in either the bottom of the transition or the top stack, depending on customer preference. Like the E.F.G. style, this machine allows for individual ceramic maintenance, changing of tube seals, and cleaning, and it will take fires, acids, high temperature, and process upset.

The test program is underway and final results are due by July 1985. The exchanger shown in Fig. 2 is being fired with flue-gas temperatures up to 1315°C (2400°F) and preheating air to temperatures varying from 426° to 760°C (800° to 1400°F). Results to date show that there are distinct differences when comparing the I.F.G. to the E.F.G. and cube-style exchangers. Some are as follows:

(1) *Leakage*: The vertical recuperator has a single pass on the flue-gas side with only one tube sheet at high temperature. With the constant compression between tube and tube sheet, the leakage is about one-quarter that of a comparable external flue-gas model (0.5% of the clean air flow). This small amount of leakage is from the air side to the flue-gas side.

(2) *Heat transfer*: Because the flue gas is inside the tubes and at a relatively low velocity to keep down pressure drop, the heat transfer is less than comparable shell and tube exchangers. The next test module will have tubes with internal and external fins, the combination of which will increase heat transfer and reduce the physical size of each machine.

(3) *Size*: The compactness of the tube bundle, coupled with the single-pass vertical arrangement, results in a unit that is one-third smaller in volume than an equal-in-heat-transfer E.F.G. model. It is still larger than a cube, but, because of its configuration, can be installed easily in most existing furnace stacks.

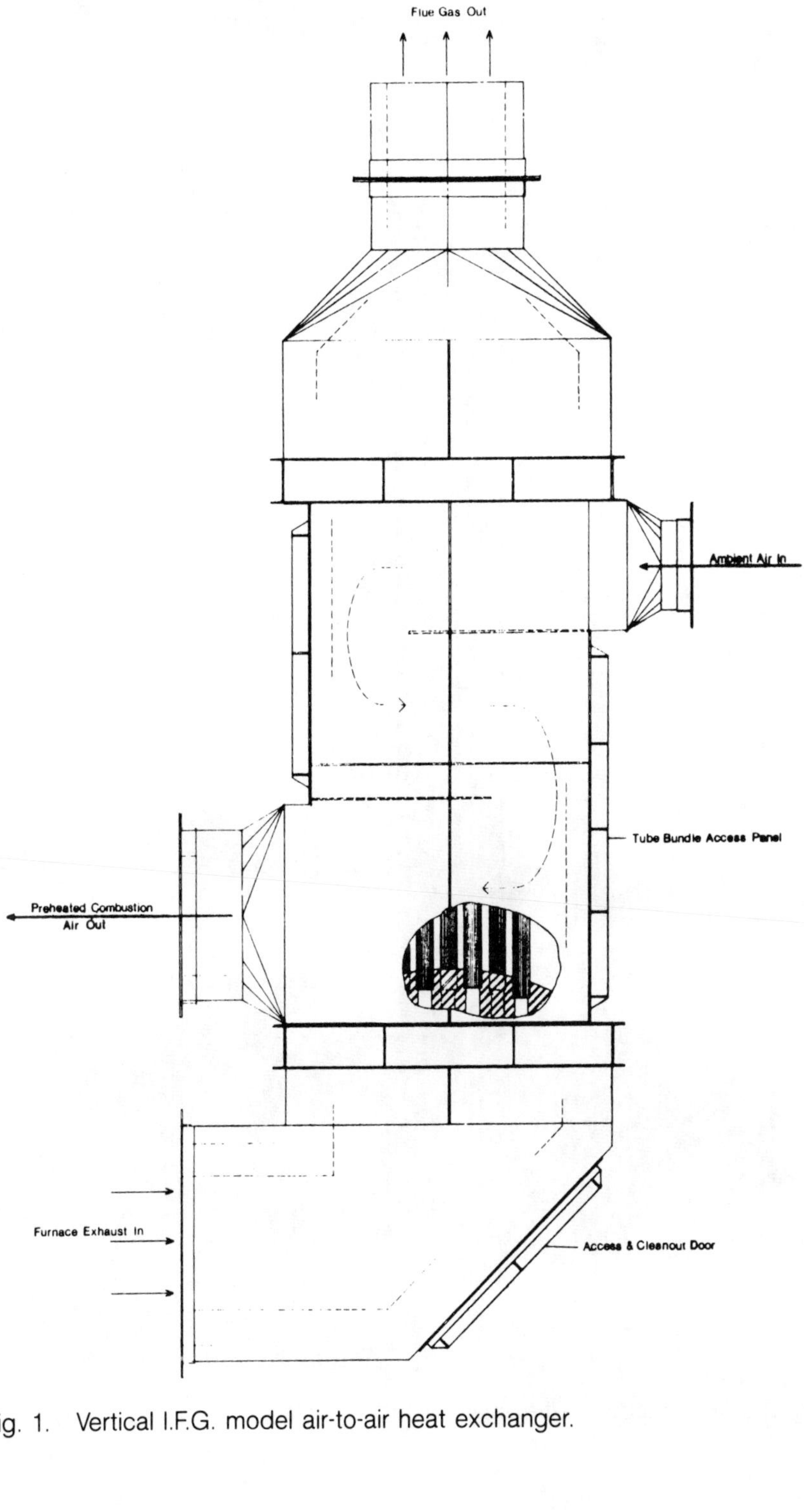


Fig. 1. Vertical I.F.G. model air-to-air heat exchanger.

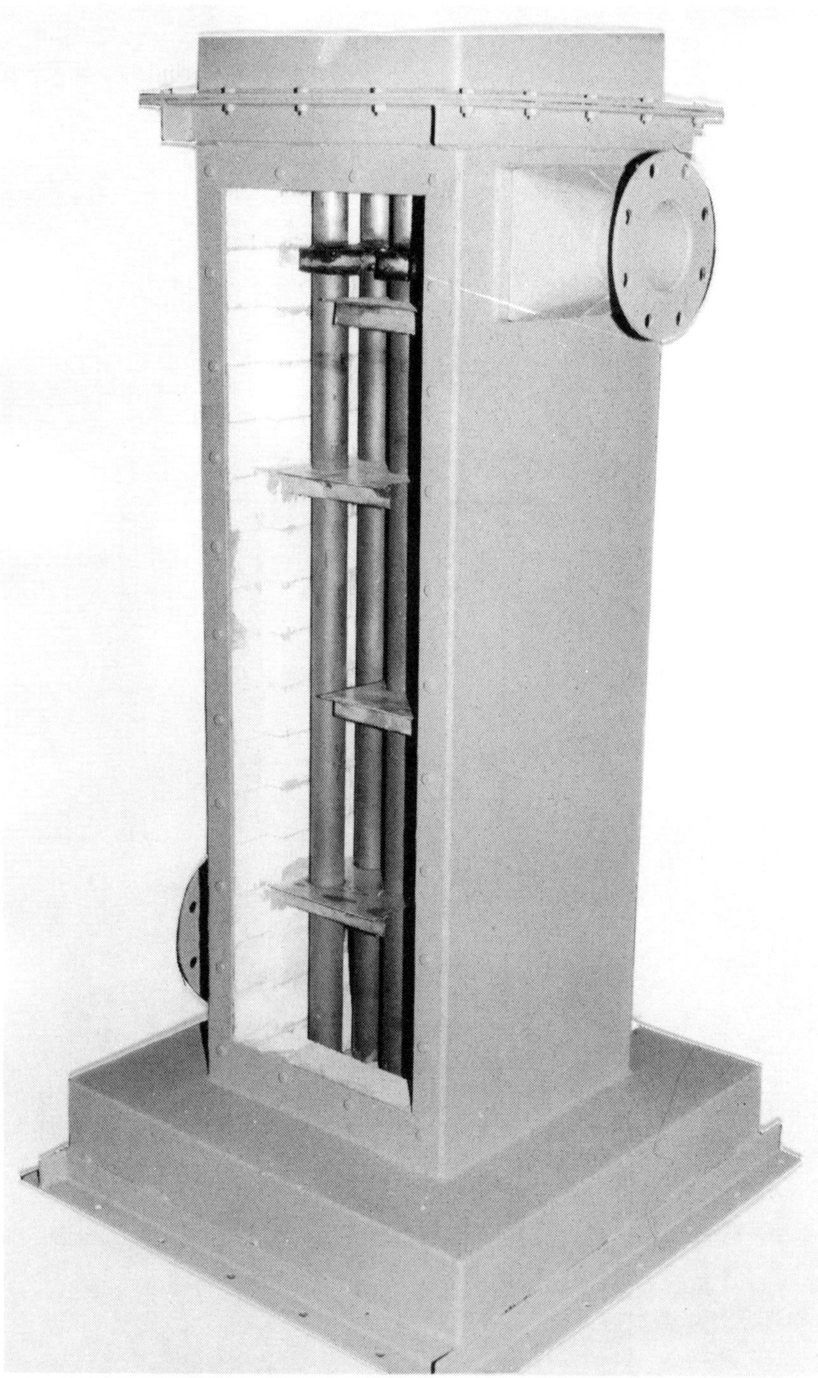

Fig. 2. I.F.G. model heat-exchanger test module.

(4) *Cost*: The tubes are the most expensive item. The balance of the exchanger is standard high-temperature furnace construction. In some applications, it will be possible to use less expensive silicon carbide tubes in conjunction with high-alumina tube sheets. It is then anticipated that the exchanger will be even more competitive.

In summary, the ceramic external flue-gas model is still the most versatile, is easy to maintain, and has low pressure drops on both the air and flue-gas sides. However, there is a market need for the vertical I.F.G. exchanger. It has a higher flue-gas pressure drop, yet this drop is still lower than that of the cube exchanger. The vertical I.F.G. recuperator is easy to install on smaller furnaces. Unlike the cube, it can be brush-cleaned, maintained, and is not susceptible to plugging or fires.

Further testing will supply additional data needed to predict accurately ceramic and seal life. An order has been received for the first field unit, scheduled for fall 1985 installation. Test results from the field and laboratory will be published by early spring 1986.

# Selected Case Histories of the GTE Ceramic Recuperator

JOHN L. FERRI AND J. M. GONZALEZ

GTE Products Corp.
Towanda, PA 18848

A compact, cross-flow, ceramic recuperator is being used successfully in a variety of high-temperature waste-heat recovery applications. In a recent DOE-sponsored program, 175 recuperators were installed in 38 plant sites to evaluate performance and fuel savings. This paper details three of those installations. The selected furnaces operated at temperatures from 871° to 1371°C (1600° to 2500°F), and the resulting preheated air temperatures ranged from 371° to 815°C (700° to 1500°F). The fuel savings ranged from 12 to 61% and averaged 38%. A majority of the installations recorded payback periods of less than 2.5 years.

The recuperators used were manufactured in three sizes: a 10 by 10 by 10 in. ceramic core with a recommended capacity of 0.2 mm W (0.6 mm Btu/h), a 12 by 12 by 12 in. core with a capacity of 0.4 mm W (1.5 mm Btu/h) and a 12 by 12 by 18 in. core with a recommended capacity of 0.4 mm W. The ceramic cores are constructed from cordierite, a magnesium aluminum silicate (MAS) selected because of ease of fabrication, relatively low thermal expansion, good thermal shock resistance, and certain corrosion-resistance characteristics. The ceramic core is manufactured by bonding MAS sheets to ribs of the same materials. The bonding agent is also a cordierite ceramic material. The sheets and ribs are arranged to form a cross-flow heat exchanger (see Fig. 1). The recuperator consists of the ceramic core, a resilient fibrous refractory seal, and a metallic housing (Fig. 2). The unit is under a slight compression from a spring assembly. A key feature of present GTE recuperators is that the preheat air makes three passes in the recuperator while the exhaust gas passes through once. The air side passes are made possible by baffles in the metal housing (Figs. 3 and 4).

The areas of each pass are selected to optimize the heat transfer for a given pressure drop. The triple-pass recuperator is a distinct improvement over the earlier single-pass model because the air preheat temperatures are considerably higher. An added advantage of the triple-pass recuperator is that the maximum temperature difference between the exhaust gasses and the preheat air is reduced. Temperature-induced stresses are therefore lower, thus increasing the reliability of the recuperator.

### Limitations

The test sites were selected primarily for compatibility with the GTE recuperator. Conditions known to seriously degrade the recuperators' performance were avoided for obvious reasons and are listed below.

#### Temperature

The upper operating limit of the recuperator is approximately 1371°C (2500°F). However, to provide a safety margin, the maximum recommended

Fig. 1. GTE ceramic matrix.

temperature of the exhaust entering the recuperator is 1287°C (2350°F). This does not eliminate applications at temperatures over 1287°F since fluing design can reduce the inlet temperatures.

**Corrosives**

Products of combustion containing boron or alkali metals (Li, Na, K) will react with and shorten the useful life of the recuperator. Any reactions will proceed faster as temperatures and concentrations increase. Recommended applications are those with relatively clean flue products.

**Condensables**

Products of combustion containing material with dew points higher than the exhaust outlet temperature of the recuperator will condense and eventually plug the exhaust passages. An example is the Mo in Hastalloy. After oxidation, $MoO_3$ is formed, which melts at 795°C (1463°F) and sublimes at 1155°C (2111°F). Any vapors then condense and plug the recuperators. This material is easily removed with compressed air. However, other materials may collect and form glass phases which could be difficult, if not impossible, to remove.

**Combustibles**

Excess fuel operation is not recommended. Any combustibles in the flue gases will combine with the small amounts of air from leakage around seals. Combustion will then occur in the recuperator and dramatically shorten its life.

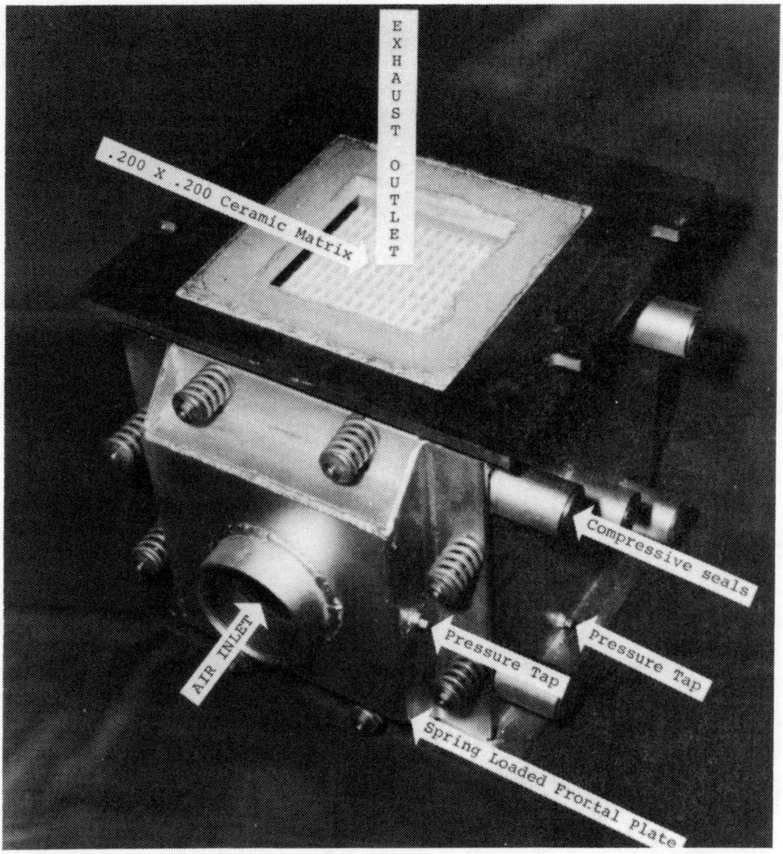


Fig. 2. GTE model R-1000 TPX ceramic recuperator in refractory-lined metal housing.

**Advantages**

Compact units can be used where space is limited. Ceramic units can withstand much higher temperatures than metallic recuperators without the need for cooling air. Ceramic cores can be replaced quickly and easily. Depending on conditions, operating life will vary and replacement of the ceramic core is required when leakage increases to a point where the Btu/h input to the burners is inadequate to maintain the required productivity. Replacing the core involves removing the old core and installing a new one along with gaskets and baffles as required. This can be done on-site, or the recuperator(s) can be returned to GTE for rebuilding. Under ideal conditions, core replacement can take as little as 1 h/recuperator.

**Applications**

Specific applications are described in detail for three companies.

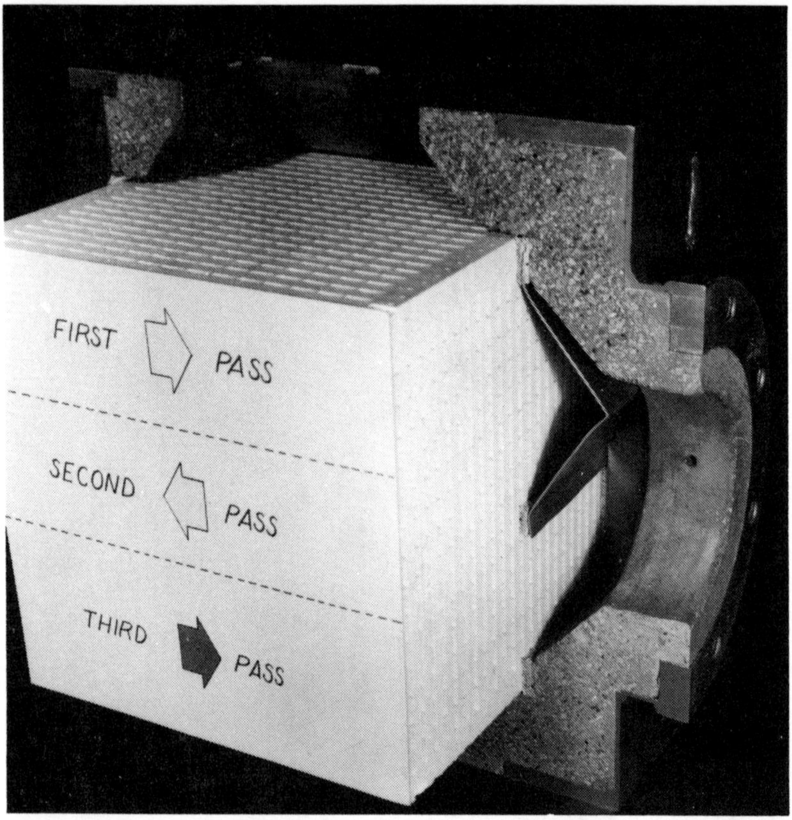


Fig. 3. GTE super recuperator cross section (right view).

**XTEK, Inc., Cincinnati, Ohio**

*Operation:* This plant anneals, carburizes, and/or heat treats 26 000 000 pounds of product annually and consumes 76 600 000 cubic feet/year of natural gas. The plant operates 6 500 h/year. Furnaces used are 2 retort atmosphere control-type pit furnaces, 10 roll furnaces, and 13 draw furnaces. Natural gas is the primary fuel with propane as a backup.

*Furnace Description:* Furnace No. 1026 is a cylindrical retort atmosphere control-type pit furnace which is approximately 20 feet high, 10½ ft. OD, and the retort ID is 6 ft. Annual production rate is 354 000 pounds of carburized product and 1 400 000 pounds of heat-treated product. Product size varies from 3 000 to 60 000 pounds each. The furnace operates 24 h/day and 6 500 h/year. Annual fuel consumption is 12 000 000 cubic feet of natural gas. Operating temperature is 954°C (1750°F) for carburizing and 843°C (1550°F) for heat-treating. The furnace was originally equipped with (12) North American No. 4442-4 burners with a nominal rating of 156 600 W (540 000 Btu/h) each. Three zones are used: top (zone 1) with two burners, center (zone 2) with six burners, and bottom (zone 3) with four burners. The burners fire

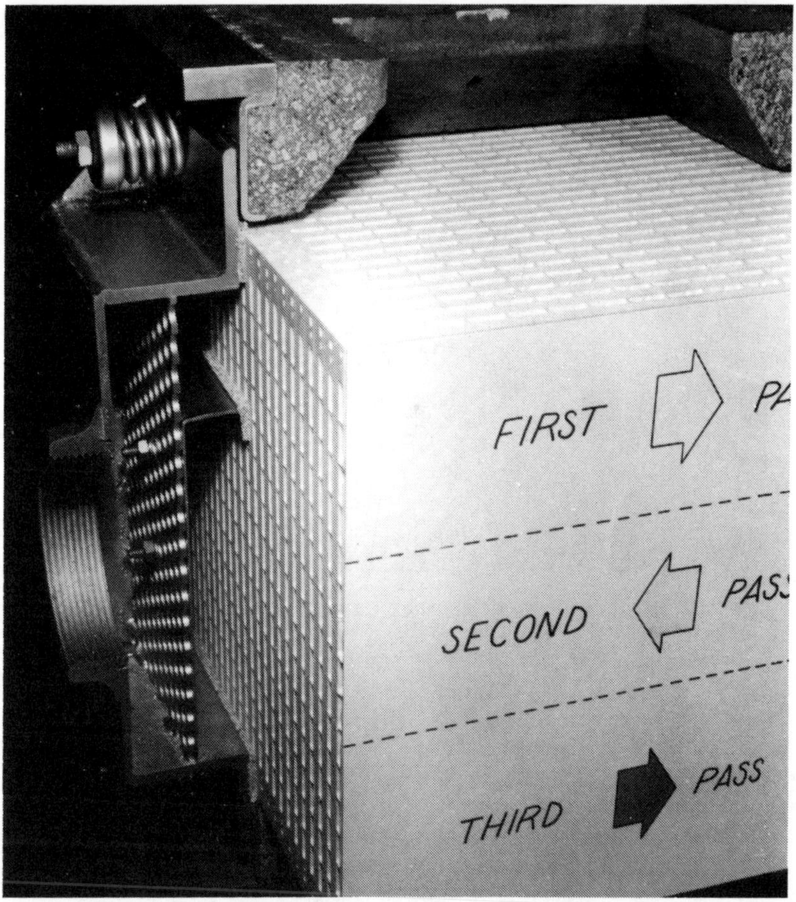


Fig. 4. GTE super recuperator cross section (left view).

radially in the space between the refractory wall and retort. Air flow is held constant and set can be on ratio at high fire. Gas flow is throttled to control temperature. The nominal firing rate is estimated to be 30 to 40% maximum.

*System Design:* The retrofit (Fig. 5) calls for (12) R0600 TPX (0.125 in. × 0.200 in.) recuperators. Each will be mounted to an Eclipse No 163MVTA-PCA burner. Ratio control will use a North American No. 7266 multiplying regulator per zone impulsing No. 7218 pressure-balancing regulators. Zone configuration will remain as before, i.e., zone 1 (top), two burners; zone 2 (center), six burners; zone 3 (bottom), four burners. Furnace pressure will be controlled by linking the eductor air valve of each zone with the zone's main air valve. Thus, exhaust flow through the recuperators will vary simultaneously and proportionally with main input flow.

The furnace is used for carburizing at 954°C (1750°F) and heat-treating at 843°C (1550°F). The predicted performance of each recuperator is as follows:

- RETORT PIT HEAT TREATING FURNACE (10-1/2 FT. O.D. X APPROX. 20 FT. DEEP)
  - TANGENTIAL FIRED WITH (12) RECUPALOT (BURNER/RECUPERATOR/EDUCTOR) PACKAGES
  - THREE ZONES OF CONTROL
- OPERATING TEMPERATURE: 1800 °F (1800°-1550 °F)
- PREHEAT TEMPERATURE: 1075 °F
- MEASURED FUEL SAVINGS: 60.7%
- 5-1/2 MM BTU/HR. INPUT USING (12) BURNERS
- (12) R0600 TPX (.125 IN. X .200 IN.) RECUPERATORS
- COLD SIDE PNEUMATIC MASS FLOW RATIO CONTROL
- DRAFT INDUCEMENT BY (12) EDUCTORS WITH MECHANICALLY LINKED IMPLIED FURNACE PRESSURE CONTROL
- FUEL: NATURAL GAS

Fig. 5. Site #04, XTEK, Cincinnati, OH.

a. *Input Conditions*
Fuel: natural gas
Recuperator air inlet: 38°C (100°F)
Recuperator exhaust inlet: 954°C (1750°F)
Excess air rate: 20%
Exhaust flow rate: 4 800 scf/h
Air flow rate: 4 400 scf/h

b. *Predicted Performance—R0600 TPX (0.125 × 0.200)*
Preheated air outlet: 554°C (1030°F)
Exhaust outlet: 548°C (1020°F)
$\Delta P$ (air side): 13.0 in. wc
$\Delta P$ (exhaust side): 0.38 in. wc
Theoretical fuel savings: 33%

*System Operation:* The system became operational during September 1982. Calibrations were done, and the excess air rate was set to 15% at higher fire (174 000 W (600 000 Btu/h)/burner). The system is operating well and preheats approach 579°C (1075°F) during idling.

*System Economics:* The total retrofit installed costs were $116 830. XTEK reported savings of 60.7%, which amounted to $42 111 annually and resulted in a simple payback of 2.8 years.

*Update:* As of April 1984, the furnace has operated over 10 000 h without any recuperator-related problems.

**Hammond and Irving, Auburn, New York**

*Operation and Furnace Description:* The furnace chosen for the recuperator demonstration is used to heat steel billets prior to shaping by a hydraulic press. Operating temperature is 1148° to 1204°C (2100° to 2200°F). The furnace is batch-loaded through one door. Material is brought to temperature and individually removed for shaping. The interior is 8 ft wide, 7 ft deep, and 6 ft high and is lined with insulating fire brick. A flue is located at the top. The system uses four (4) North American 6442-7B burners with a total input of 13.4 kJ (12.8 mm Btu) gas or No. 2 oil. Fuel input is controlled proportional to air pressure with a pressure-balanced system. Furnace pressure is controlled by a North American electronic pressure indicated controller (EPIC) operating an air jet damper at the roof flue.

*Recuperated System Design (Fig. 6):* Modifications for recuperation include replacing the existing burners with four North American 6825-7A burners each close-coupled to a R1500 TPX (0.200 in. × 0.200 in.) recuperator. Total input will be 7.4 kJ (7.08 mm Btu), assuming 55 MPa (8 psi) and 537°C (1000°F) preheated air to the burner. The recuperators will be sidewall-mounted at hearth level using individual flue gas eductors to convey the flue gases through the recuperators. Furnace pressure will be controlled with the existing EPIC operating a motorized valve supplying air to the eductors. Fuel/air ratio control will use a North American No. 7266 multiplying regulator impulsing No. 7218 pressure-balancing regulators. An orifice in the main air line will impulse the multiplying regulator, causing fuel pressure to vary proportional to air flow.

The predicted performance of each recuperator is as follows:

a. *Input Conditions*
   Fuel: natural gas
   Recuperator air inlet: 38°C (100°F)
   Recuperator exhaust inlet: 1204°C (2200°F)
   Excess air rate: 10%
   Exhaust flow rate: (high fire) 14 960 scf/h
   Air flow rate: 13 600 scf/h

b. *Predicted Performance—R1500 TPX (0.200×0.200)*
   Preheated air outlet: 593°C (1100°F)
   Exhaust outlet: 621°C (1150°F)
   $\Delta P$ (air side): 6 in. wc
   $\Delta P$ (exhaust side): 0.7 in. wc
   Theoretical fuel savings: 36%

- BOX FORGE REHEAT FURNACE
  - EXHAUST AT HEARTH LEVEL
  - BURNERS MOUNTED HIGH IN OPPOSING WALLS/2 PER SIDE
- OPERATING TEMPERATURE: 2200 °F
- PREHEAT TEMPERATURE: 1100 °F
- MEASURED FUEL SAVINGS: 37.4%
- 7 MM BTU/HR. INPUT USING (4) BURNERS
- (4) R-1500 TPX .200 IN. X .200 IN. RECUPERATORS
- COLD SIDE PNEUMATIC MASS FLOW RATIO CONTROL
- DRAFT INDUCEMENT BY EDUCTORS WITH AN ELECTRONIC FURNACE PRESSURE CONTROLLER
- FUEL: NATURAL GAS OR #2 FUEL OIL

Fig. 6. Site #07, Hammond & Irving, Auburn, NY.

*System Operation:* The recuperated furnace was put into operation in mid-April 1982. The multiplying regulator malfunctioned, causing the furnace to operate with an excess fuel condition for an undetermined period of up to 1 week. Replacement of the regulator restored normal operation. Preheated air temperatures are approximately 593°C (1100°F) when the system is operating at high fire.

*System Economics:* The installed cost of the retrofit was $66 643. The reported savings were 37.4%, amounting to $48 386 annually. The simple payback was 1.4 years.

*Update:* As of April 1984, the furnace has operated approximately 8 000 h, and a new furnace has been installed with GTE recuperators.

**Pfizer Company, Minerals, Pigments and Metals Division, Canaan, Connecticut**

*Operation:* Production of calcium metal and lime/limestone products for the steel and lead industries. The plant works on a round-the-clock basis 365 days/year. All five operating furnaces are recuperated, with one employing a metallic recuperator and the other four using ceramic recuperators.

*Furnace Description:* The calcium metal vacuum retort is an indirect-fired furnace. The load is charged into a large-diameter tube which is then sealed and evacuated. Five such tubes are heated in a five-zone furnace (one in each zone) by five oil-fired Hauck burners each rated at 232 000 W (800 000 Btu/h). The burners were located in the back wall of the furnace and fired into the furnace chamber which is maintained at about 1204°C (2200°F). The exhaust gases exit through flues located in the back wall about 3 ft. from the burners. The furnace is charged once a day and brought up to the operating temperature in two hours and maintained at the operating temperature for the rest of the day.

*System Design:* The retrofit (Fig. 7) design calls for five Model R0600 TPX 0.125 in.×0.200 in. recuperators mounted directly to the exhaust flue. The exhaust gas passes through each recuperator horizontally and exits through a vertical duct fitted with an eductor. The transition section is equipped with a clean-out door. The preheated air from each recuperator is ducted through a 3 ft. length of stainless steel (flex connector) to a North American Model 6825-7 hot-air, oil-fired burner rated at 174 000 W (600 000 Btu/h). The burners are controlled with a flow-balanced mass flow regulator connected to an air/oil ratio regulator. The furnace pressure is automatically controlled and the exhaust flow through each recuperator is maintained by the eductors. Each burner is independently controlled.

The predicted performance of each recuperator is as follows:

a. *Input conditions*
Fuel: No. 2 fuel oil
Recuperator air inlet: 38°C (100°F)
Recuperator exhaust inlet: 1204°C (2200°F)
Excess air rate: 10%
Exhaust flow rate: 4 646 scf/h
Air flow rate: 4 400 scf/h

b. *Predicted Performance—R0600 TPX (0.125×0.200)*
Preheated air outlet: 635°C (1175°F)
Exhaust outlet: 690°C (1275°F)
$\Delta P$ (air side): 15.0 in. wc
$\Delta P$ (exhaust side): 0.46 in. wc
Theoretical fuel savings: 38%

*System Operation:* The vacuum retort furnace was started up in July 1981 and operated at 1204°C (2200°F) in a continuous mode. Preheat air temperatures of 621° to 648°C (1150° to 1200°F) have been reported. The furnace is operating very satisfactorily.

*System Economics:* Total expenditures on this retrofit were $123 547. Costs directly attributed to the recuperator retrofit including equipment, design,

- VACUUM RETORT FURNACE
  - BURNERS MOUNTED ON ONE SIDE WALL OF FURNACE - FIRING ABOVE THE RETORT
  - RECUPERATORS MOUNTED NEAR BURNERS LOW ON SIDE WALL
  - FIVE ZONES OF CONTROL
- OPERATING TEMPERATURE: 2200 °F
- PREHEAT TEMPERATURE: 1175 °F
- MEASURED FUEL SAVINGS: 29.7%
- 3 MM BTU/HR. INPUT USING (5) BURNERS
- (5) R-0600 TPX .125 IN. X .200 IN. RECUPERATORS
- INDEPENDENT COLD SIDE PNEUMATIC MASS FLOW RATIO CONTROL
- DRAFT INDUCEMENT BY EDUCTORS WITH IMPLIED FURNACE PRESSURE CONTROL
- FUEL: #2 FUEL OIL

Fig. 7. Site #21, Pfizer Co., Canaan, CT.

materials, and labor amounted to $76 547. The unrecuperated furnace over a six-month period averaged 20.83 gallons of No. 2 fuel oil/h. The recuperator furnace averages 14.58 gallons/h for a 29.7% fuel savings. The annual savings amount to $65 027 based on fuel oil costs of $1.20/gallon. The simple payback is 1.2 years. It should be noted that this furnace was retrofitted with larger retort tubes, and processes more load than the comparison furnace. These data (and the magnitude of the change) were not transmitted to GTE until shortly before this report was begun. The result would be to increase the savings and shorten the payback period.

*Problem Areas:* Pfizer purchases scrap oil for their furnaces. During operation, the plant engineers have inadvertently recalibrated the burners to operate fuel-rich due to large viscosity variations of the oil. GTE believes that operation at, or near, 0% excess air accelerates the deposition of sulfur in the exhaust passages of the matrix. Furnace pressure and preheat temperatures are monitored and a change in either one by more than 15% indicates that the exhaust channels require cleaning. The matrix is cleaned (at temperature) with an air lance, and a blade-type metallic scrapper through the clean-out door in the exhaust flue transition section. No other problems have been reported.

*Update:* As of April 1984, Pfizer is very satisfied with the performance of the GTE recuperators and has recuperated all the furnaces in the plant. Cleaning is required after every 10 months of operation because of plugging from product and fuel-related particulates.

### Conclusion

In these three case histories and in the other test sites, the recuperators lived up to expectations, and results of a phone survey in April 1984 showed that, of the 38 sites that participated in the program, 40% have already installed GTE recuperators on additional furnaces and 48% intend to install GTE recuperators when funding is approved. The DOE goal of accelerating industrial acceptance of this technology appears to have been achieved.

### References

[1]"Technology Acceleration Program for the GTE Ceramic High-Temperature Recuperator," Draft Final Report, U.S. Department of Energy Contract No. DE–FC01–80CS 40330, January 1983.

[2]"The GTE Ceramic Recuperator for High Temperature Waste Heat Recovery," R. E. Dorazio et al., 1984 Industrial Energy Conservation Technology Conference.

[3]M. R. Ally et al., "Optimization of Multipass Cross-Flow Heat Exchangers for Waste Heat Recovery Applications," 1984 National Heat Transfer Conference, Niagara Falls, New York, August 5–8, 1984.

# Design of a Durable, Compact Ceramic Heat Exchanger

R. N. KLEINER, L. E. COUBROUGH, AND L. R. STRASBAUGH

Coors Porcelain Co.
Golden, CO 80401

The methodology used in the design of a compact ceramic heat exchanger is described. It consists of developing a thermal-stress-resistant ceramic material, selecting matrix dimensions, and applying finite element computer modeling techniques to derive the parameters resulting in the minimum thermal stress for maximum heat-exchanger durability.

Metal heat exchangers have been used in industrial heat-recovery applications for many years. Ceramic heat exchangers are relatively new, but are gaining acceptance as more field experience becomes available. The advantage of the ceramic units over metal units is the higher-temperature capabilities of the ceramic which preclude the need for costly over-temperature controls and dilution air systems. The problems encountered with ceramic heat exchangers to date have been variable leakage and lack of durability. These problems are associated with the high thermal stresses in the compact designs and with the seals required to connect the ceramic to the metal ductwork.

A new ceramic heat-recovery product has been developed for the metal heat-treating industry. The product is a compact ceramic recuperator designed to overcome the leakage and durability limitations which arise from the thermal gradients inherent in the use of this type of product.

At present, 9000 hours durability has been achieved on three recuperators at one commercial facility, 8000 hours on three recuperators at another installation, and 4500 hours on four heat exchangers in a third facility.

The heart of the recuperator is a honeycomb matrix made from a new, highly thermal stress-resistant, mixed-oxide ceramic developed especially for this application. The housing and seals are designed to provide uniform compression in the air and gas stream directions. An integral, ceramic diffuser and radiation shield is included as part of the recuperator.

### Technical Approach

Successful development of a heat-exchanger product requires a multidisciplinary approach. It requires technology in the fields of heat transfer, combustion engineering, thermophysics, mechanics, computer science, materials engineering, and ceramic processing. A team effort is required. Coors expertise is in the areas of materials engineering and ceramic fabrication. Input in the other areas was acquired by working closely with other companies and consultants specializing in the other fields.

The methodology used for the heat-exchanger design consisted in developing a thermal stress-resistant ceramic material, selecting crossflow design

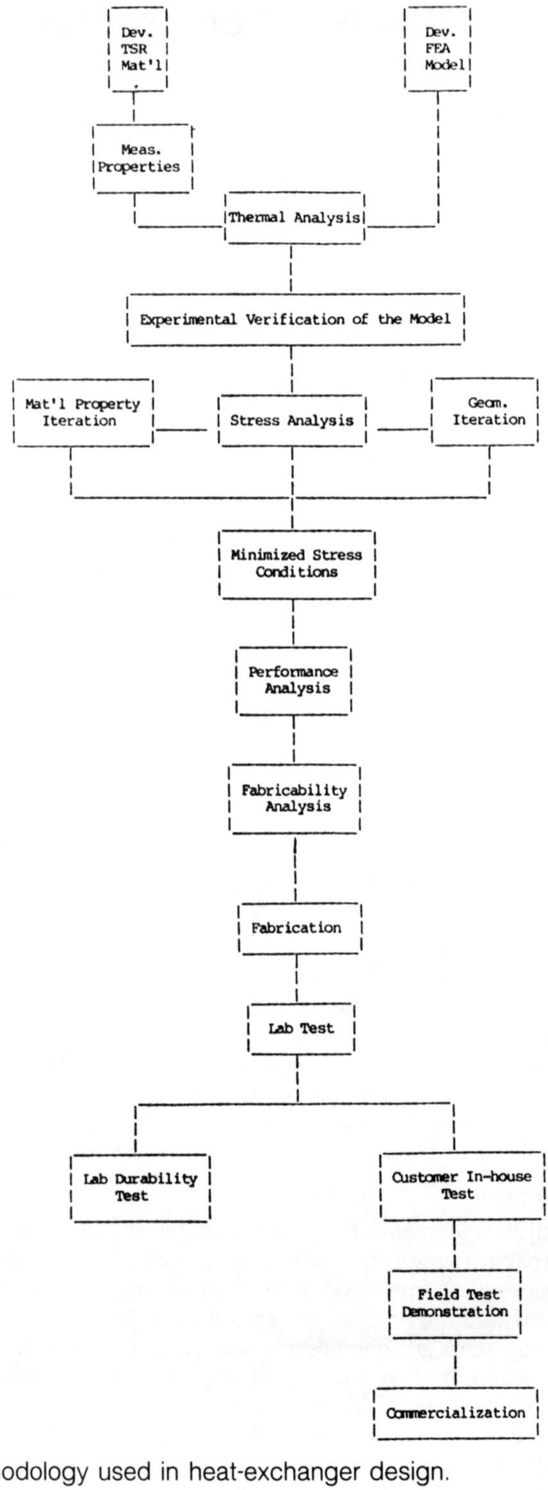


Fig. 1. Methodology used in heat-exchanger design.

dimensions, and using finite element computer modeling to derive the minimum stress (maximum durability) conditions. The conditions leading to the lowest thermal stress were then considered with respect to heat-exchanger performance and fabricability. This approach is outlined in Fig. 1.

Cordierite was considered as a heat-exchanger material but was found to lack sufficient thermal stress resistance in gas turbine heat-exchanger applications. A new ceramic material was designed with emphasis on low thermal expansion, low modulus of elasticity, and high surface fracture energy to resist crack propagation. The mixed-oxide ceramic material developed does not have the strength of the dense, brittle, high-alumina ceramics, but it is not required in these industrial applications where there are no significant mechanical loads. The properties of this material are shown in Table I.

Table I. Properties of Mixed-Oxide System

| | |
|---|---|
| Thermal expansion (in./in.$\times 10^{-6}$ at 1000°C) | 675 |
| Modulus of rupture (MPa) | 24 150 |
| Elastic modulus ($\times 10^6$ MPa) | 5.5 |
| Surface fracture energy | |
| $\gamma_{WOF}$* (ergs/cm$^2 \times 10^4$) | 4.75 |
| $\gamma_{NBT}$† (erg/cm$^2 \times 10^4$) | 1.90 |
| Thermal conductivity (cal/cm/s/°C) | 0.003 |
| Specific heat (cal/g°C) | 0.228 |
| Melting point (°C) | >1700 |

*Work of fracture technique.
†Notch beam technique.

The finite element analysis model used consists of a separator plate, ribs on the exhaust side, and ribs on the air side, as shown in Fig. 2. This plate can be mathematically moved throughout the stack. The ANSYS computer program was used. The physical and thermal properties and the geometric dimensions of the mixed oxide were entered into the model. When the exhaust

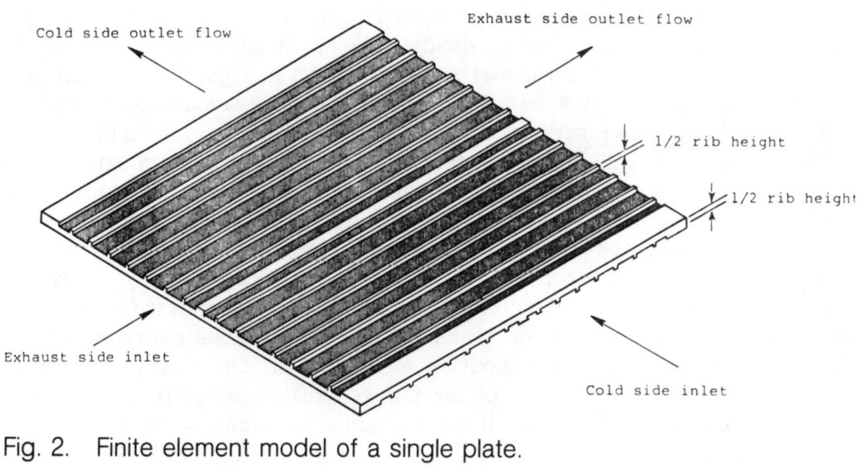


Fig. 2. Finite element model of a single plate.

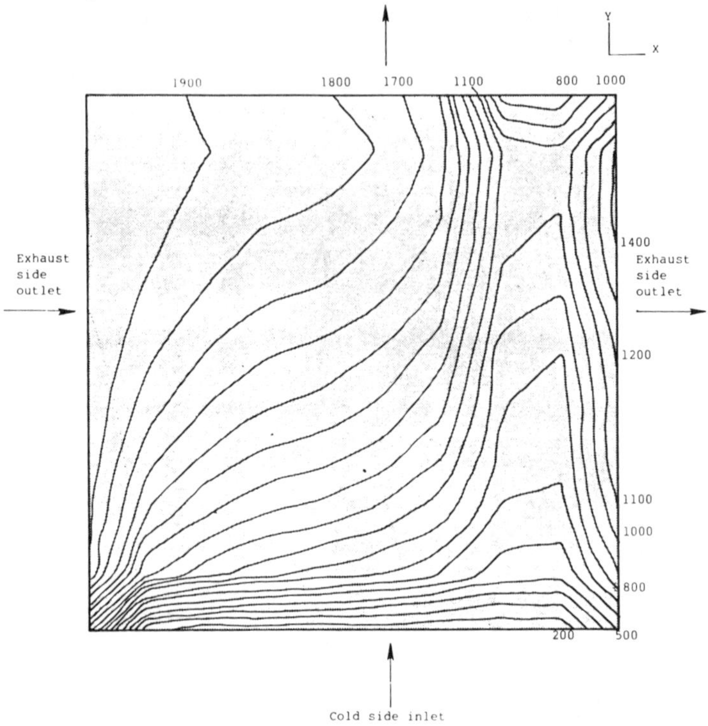


Fig. 3. Temperature distribution in separator plate. 1093°C (2000°F) exhaust inlet temperature.

and air inlet temperatures and flows were entered into the program, temperature distributions were derived and these temperatures were confirmed experimentally, indicating validity of the model. Isotherms on the separator plate at 1093°C are shown in Fig. 3. Thermal stresses were then calculated from the program. A stress distribution in the separator plate of our first-generation mixed-oxide material at 1093°C steady-state operation is shown in Fig. 4. Stress contours around local model areas are shown in Figs. 5 and 6. Heat-up and cool-down transients were also analyzed. The stresses developed during cool-down of this first-generation material are shown in Fig. 7. The maximum stress developed was 20 700 (3000 psi), which is close to the strength of the material. A simple elastic model was considered, that is, if the modulus of rupture was greater than the stress developed, the matrix was considered strong enough to prevent crack initiation. The computer model was verified by testing recuperators from this early formulation under these conditions; a crack developed. The location of the crack agreed with the prediction from the computer model. A modification was made to the first-generation material and the thermal expansion and elastic modulus were lowered. The steady-state stress distribution was similar to that of the first-generation material. The stresses developed during cool-down conditions were quite different, however, as shown in Fig. 8. The maximum stress was 7245 MPa in contrast to 20 700 MPa in

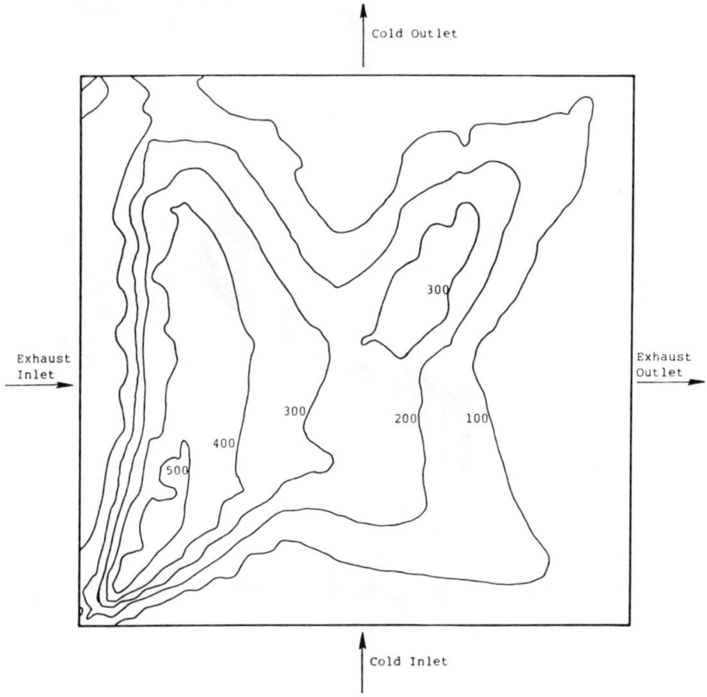


Fig. 4. Stress distribution at 1093°C (2000°F) in first-generation mixed-oxide material (stresses in psi).

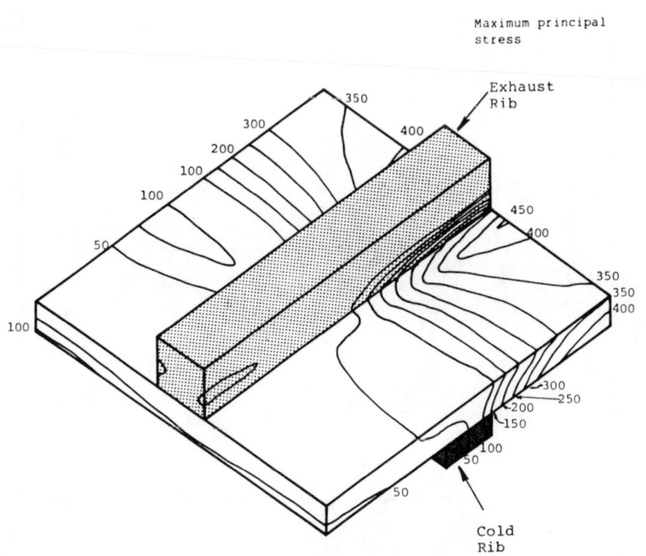


Fig. 5. Stress distribution in local model at 1093°C (2000°F) under steady-state conditions. Exhaust rib (stresses in psi).

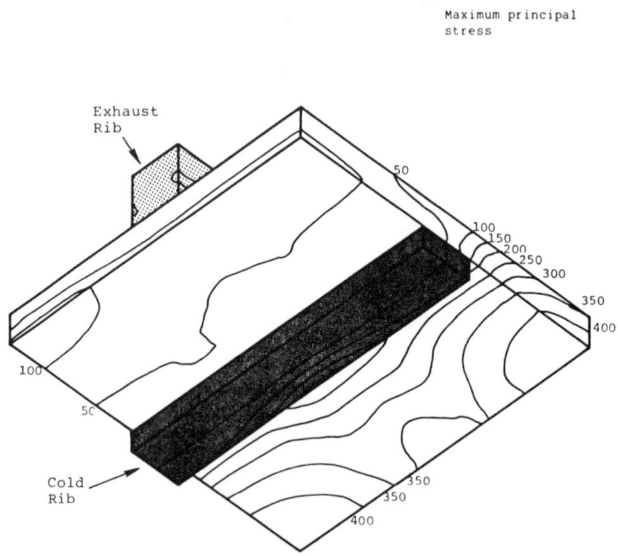


Fig. 6. Stress distribution in local model at 1093°C (2000°F) under steady-state conditions. Air rib (stresses in psi).

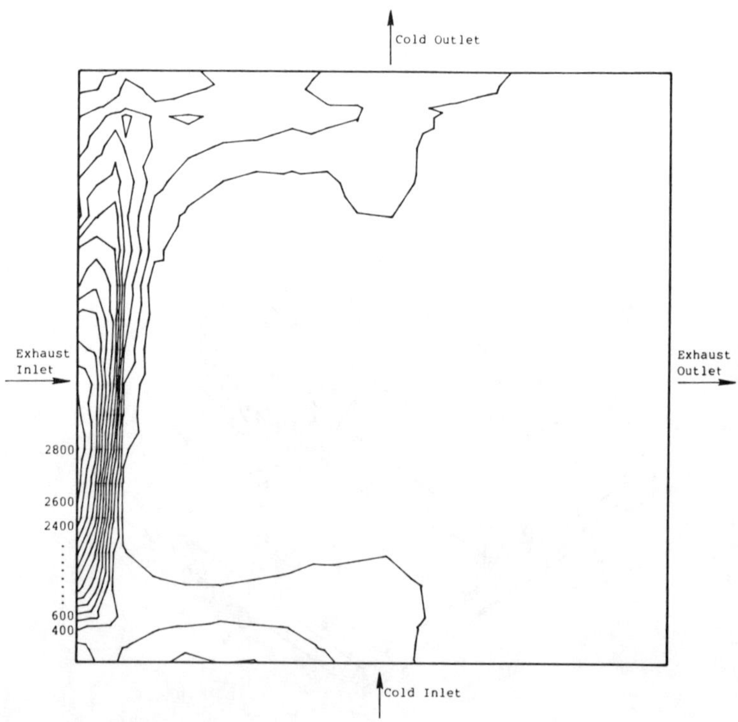


Fig. 7. Maximum principal stress developed in first-generation mixed-oxide material under transient cool-down conditions from 1093°C (2000°F) (stresses in psi).

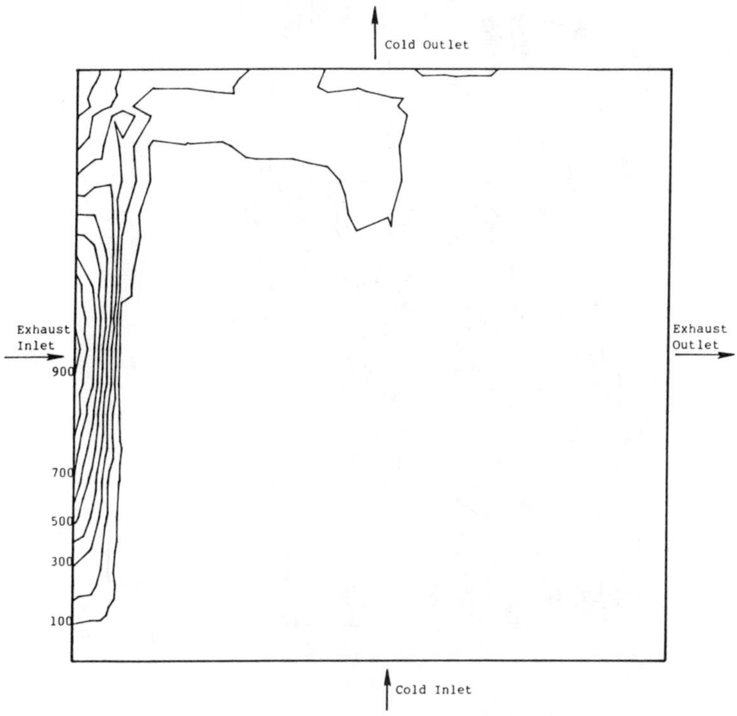


Fig. 8. Maximum principal stress developed in second-generation mixed-oxide material under transient cool-down conditions from 1093°C (2000°F) (stresses in psi).

the first-generation material and recuperators made from this material did not develop cracks under these test conditions.

The effect of a blocked channel was also studied. The temperature distribution is shown in Fig. 9. The stress in the air ribs is shown in Fig. 10. There was an insignificant change in the stress developed due to the blocked cells.

## Heat-Exchanger Product

### Design

The design of the heat-exchanger matrix currently being marketed is shown in Fig. 11. The dimensions of the air passages are 0.093 by 0.582 in. and the exhaust passages are 0.239 by 0.582 in. The backweb is 0.070 in. The housed unit is shown in Fig. 12. The gasket material is a ceramic felt. The housing is stainless steel and is spring-loaded in the air and exhaust directions. Transitions and flanges can be made available in many sizes. A refractory base plate, which serves as a diffuser and radiation shield, is supplied with the heat exchanger.

### Performance

The performance of a 12 by 12 by 12 in. matrix is shown in Figs. 13 and 14. The durability of the recuperator at high temperatures is currently being

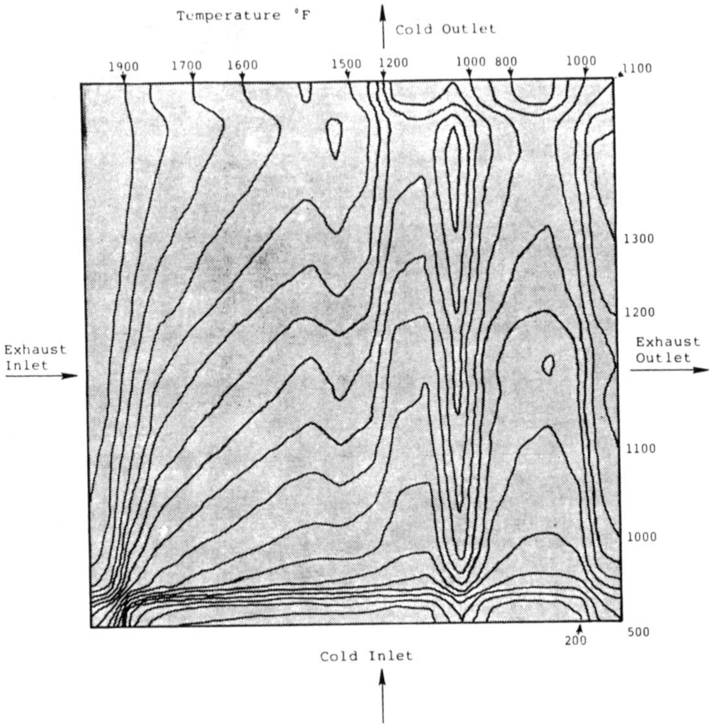


Fig. 9. Temperature distribution due to blocked cold-air channel.

determined. The material has shown the potential in laboratory tests for operation at 1371°C (2500°F), which will extend its application to the forging industry where furnaces are presently being retrofitted. It is initially being used in less severe, lower temperature heat-treating applications 1093°C (2000°F) to gain product performance and lifetime data in production environments.

**Demonstrations**

In the field test demonstrations at three commercial heat-treating companies, fuel savings in the 35 to 50% range have been realized since conversion to these recuperators. Some of the savings is attributed to excess air adjustment made when the burners were tuned for the recuperators. Productivity improvements were also realized. Monitoring of the durability is continuing.

**Conclusions**

A compact ceramic recuperator has been designed for industrial heat-recovery applications.

Finite element analyses were used for evaluating effects of material properties, varying heat-up and cool-down conditions, and different heat-exchanger designs. The analyses permitted a very rapid assessment of these effects on the thermal stresses developed in a ceramic recuperator to facilitate design optimization. This approach is very cost-effective and time-effective compared

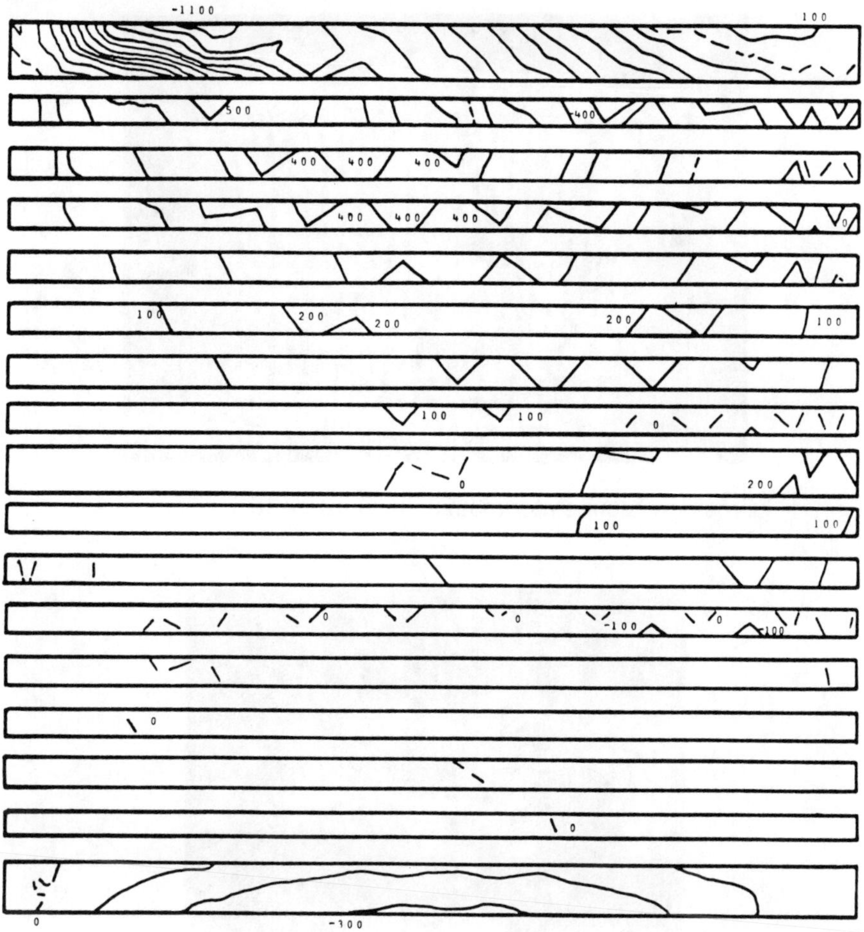


Fig. 10. Stress distribution in cold-air ribs of first-generation mixed-oxide material due to blocked cold-air channel (stresses in psi).

Fig. 11. Crossflow recuperator matrix (12 by 12 by 18 in.).

Fig. 12. Housed crossflow matrix (12 by 12 by 12 in.).

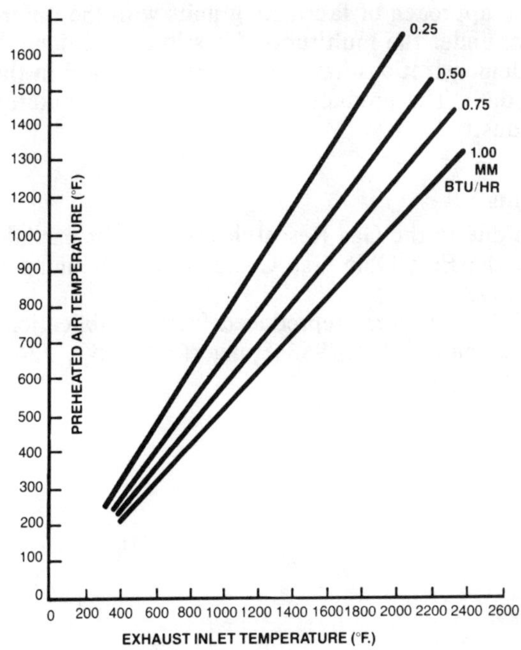


Fig. 13. Preheated-air performance curve for Model 1212 industrial heat exchanger.

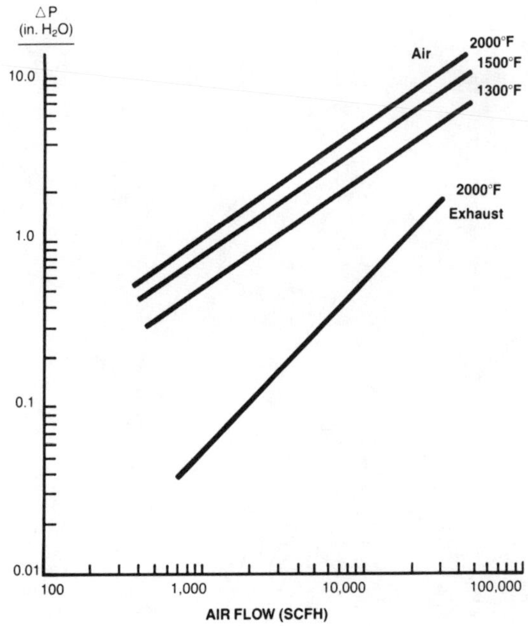


Fig. 14. Pressure-drop curves for Model 1212 industrial heat exchanger.

to the alternative approach of fabricating units with the different parameters and testing them under the multitude of possible conditions.

Field test demonstrations have been very successful in the 12-month operating time to date. The product is presently being marketed for the metal heat-treating industry.

**Acknowledgments**

Thanks are due to the Gas Research Institute for partial support of this program and to the East Ohio Gas Company for obtaining the commercial demonstration sites.

Portions of this text were reproduced from a publication available from Government Distributors, Inc., 966 Hungerford Drive, #24 Rockville, MD 20850.

# A Ceramic Finned-Plate Recuperator for Industrial Application

MURRAY COOMBS, HAL STRUMPF, AND DAVID KOTCHICK

AiResearch Manufacturing Co.
Torrance, CA 90509

High-level recuperation of high-temperature industrial furnaces offers an economically effective means for improving both process efficiency and fuel utilization. A ceramic recuperator capable of operating in flue gas temperatures of up to 1350°C and providing a combustion air preheat temperature of 1100°C can provide in excess of 50% savings in fuel consumption over an unrecuperated furnace. This recuperator consists of an array of cast-ceramic finned plates. The fin geometries are such that when the plates are stacked together, they form the heat transfer flow passages for both the flue gas and combustion air streams. A reference design for an industrial recuperator system was created. The current development efforts conducted on this recuperator concept, as well as plans for future activities, are described.

There is an increasing concern in the United States, directed to improving the efficiency of natural gas usage, particularly in the industrial sector. As a result, the Gas Research Institute (GRI) has undertaken several programs to develop equipment to enhance the efficiency of natural gas utilization. One such program is the GRI "Industrial Compact Ceramic Finned-Plate Recuperator." The objective of this program is to develop a ceramic recuperator capable of operating in a variety of high-temperature industrial furnace flues that can provide preheated combustion air at temperatures up to 1100°C to the furnace burner(s). The energy savings of such a system are illustrated in Fig. 1. In this figure, the percent fuel savings is plotted vs the flue gas temperature and the air preheat temperature for systems operating at 20% excess air. Thus, for a flue gas temperature of 1350°C and an air preheat temperature of 1100°C, the theoretical fuel savings achievable is approximately 65% of fuel consumption of an unrecuperated furnace operating at the same temperature levels. While other considerations may limit the degree of recuperation acceptable with a given furnace (e.g., process control, $NO_x$ formation, etc.), high-level recuperation can offer a dramatic improvement in process efficiency for high-temperature industrial furnaces.

### Concept Description

The recuperator core consists of a series of ceramic finned plates stacked to form alternate air- and gas-side flow passages. The plates have fins cast integrally on both sides, as illustrated in Fig. 2. Two finned plates stacked together form a flow passage equal in height to twice the height of the individual fins. Although the plates are not thermally connected, this unique configuration

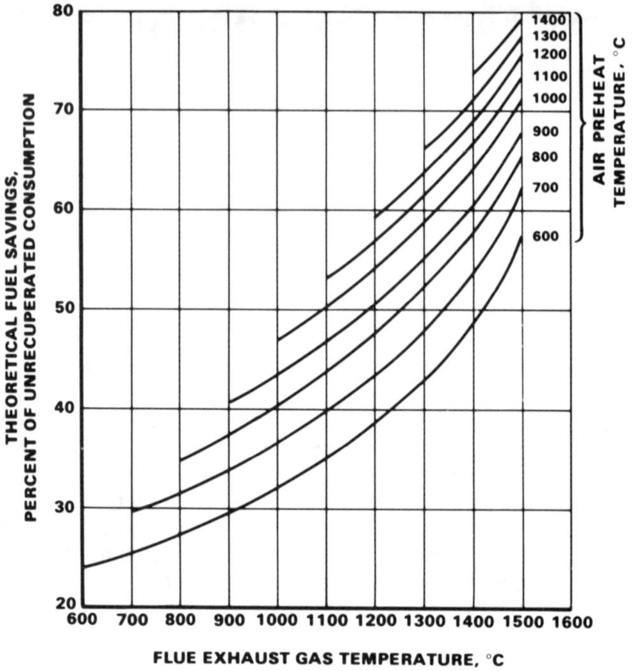


Fig. 1. Potential fuel savings with recuperation.

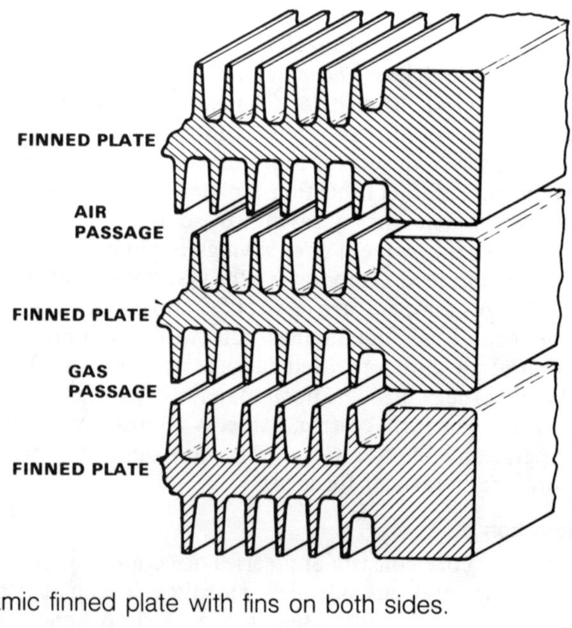


Fig. 2. Ceramic finned plate with fins on both sides.

allows mechanical sealing of the plates with no reduction in the heat-transfer capability. This configuration is possible since, after the plates are stacked, the nonbonded fin tips are arranged in adiabatic planes in the center of the flow passages; even in thermally bonded heat exchangers, heat is not conducted across these planes. Because the fin tips and plates are free to grow as necessary with temperature differentials, the absence of thermal and structural bonding has the advantage of reducing stress between the fins and plates.

The plates have straight counterflow fins on one side and mixed counterflow/crossflow fins on the other side. The plates are stacked so that adjacent plates have identical gas-side and mirror-image air-side fin patterns. The plates are sealed around the edges with a ceramic-fiber insulation that confines the two fluid streams within their respective flow passages. On the air side the triangular-shaped crossflow sections are used to conduct the air into and out of the pure counterflow section of the core.

**Reference System Design**

A reference system design was created to establish the requirements for a finned-plate recuperator operating in a typical industrial application. The application selected was a steel soaking pit. This furnace is currently fired with a single 23 GJ/h cold-air burner and is operated around the clock, seven days a week, with flue gas temperatures ranging from 930° to 1350°C. A typical heating curve for an 11-h cycle is shown in Fig. 3. This heating curve is the same as that employed for the design of the GRI tubular ceramic recuperator described in Ref. 1; therefore, a direct comparison can be made between two different design approaches for a recuperator capable of providing 1100°C preheated combustion air to the furnace burner.

The design requirements for the soaking pit recuperator are summarized in Table I. Several plate geometry configurations using various fin heights and

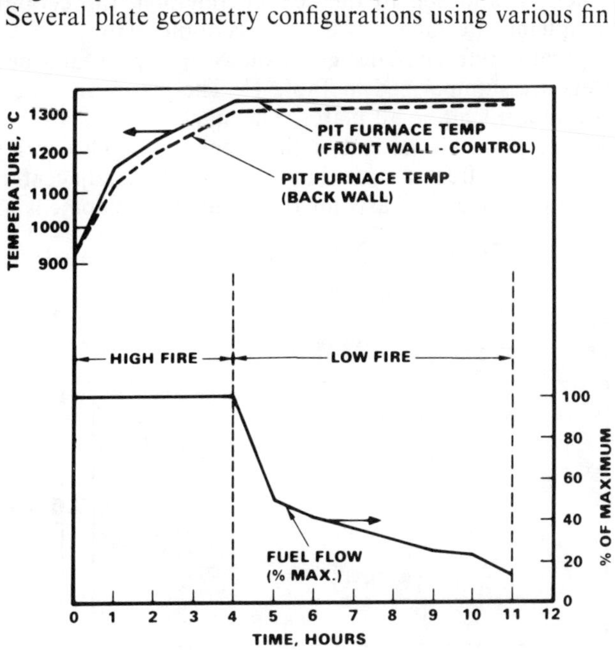


Fig. 3. Typical heating curves for steel soaking pit.

Table I. Steel Soaking Pit Recuperator System Requirements

| | |
|---|---|
| Air inlet temperature, °C | 38 |
| Air outlet temperature, °C. | 1093 |
| Flue gas inlet temperature, °C. | 1343 |
| Overall effectiveness | 0.809 |
| Fuel-to-air weight ratio | 0.0528 |
| Airflow rate, kg/s | 1.09 |
| Flue gas flow rate, kg/s | 1.15 |

fin densities were investigated, and some configurations resulted in relatively high stress levels. The stress level is related to the gas flow length of the heat exchanger—the greater the flow length, the lower the stress. This is due to the fact that a longer flow length reduces the thermal gradient in the flow length direction and, hence, reduces the stresses. The stress situation must be balanced against manufacturing considerations, which indicate the desirability of small plates. Very large plates might result in bowing and sealing difficulties as well as process and handling problems.

The plate design selected is shown in Fig. 4 (gas-side surface shown). The overall plate dimensions are 0.61 by 1.22 m. The maximum stress induced by the thermal gradients in the plate was calculated to be on the order of 24 MPa, which should yield a high component reliability.

On both sides of each plate is an extended lip around the periphery. When the plates are stacked, these lips form channels around the edges of the air and gas passages. A compliant ceramic material is applied to seal the passages from each other. In addition to the sealing function, the ceramic material provides a compliant interface between the ceramic plates.

The recuperator reference design is shown in Fig. 5, and details of the core parameters are summarized in Table II. The plates are stacked upright to maintain a uniform load on all plates. The stacked plates are held together by a spring-loaded plate at one end of the recuperator. This spring load also provides the force to effect an air-tight seal between adjacent plates. The reference design provided the basis for the ensuing development phase of the program.

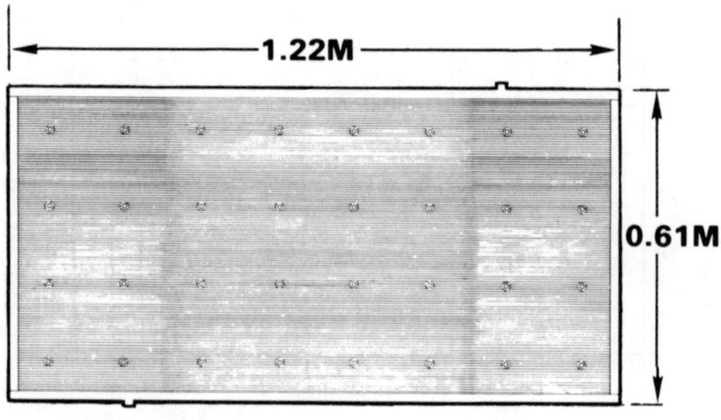


Fig. 4. Reference design finned plate (gas side).

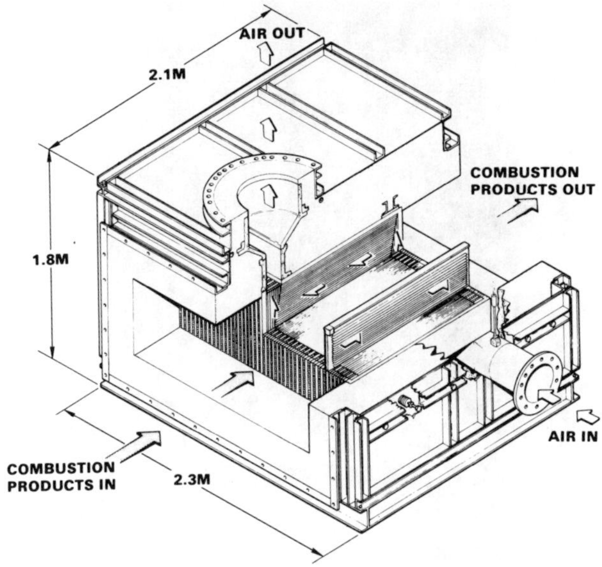


Fig. 5. Recuperator reference design configuration.

Table II. Reference Recuperator Design Summary

| | |
|---|---|
| Gas flow length | 121.9 cm |
| Width | 61.0 cm |
| Stack height | 144.3 cm |
| Number of plates | 55 |
| Core weight | 1500 kg |
| Core volume | 1.07 $m^3$ |
| Air pressure drop | 3.59 kPa |
| Gas pressure drop | 1.00 kPa |

**Fabrication Development**

The fabrication development task, subcontracted to Norton Company of Worcester, Massachusetts, was undertaken to demonstrate the required fabrication capability with increasingly larger components of the necessary complexity. The plates were fabricated from siliconized silicon carbide (NC-430). The first small finned plates fabricated were used to demonstrate the required shape capability and seal technology, and to measure mechanical properties. Subsequent components were used to demonstrate larger-sized fabrication capability.

The small laboratory test components were 0.28 by 0.28 m and incorporated most of the features attendant in the full-size components; precision/slip casting was used to form the components. Figure 6 shows the two sides (air and gas) of the finned-plate laboratory component. The fins, bosses, and header bars were within the specified tolerances. Additional plates without fins were fabricated for seal formation and thermal shock testing.

Fig. 6. Laboratory-size finned plate.

Fabrication of larger (0.61 by 0.61 m) finned plates was completed after the laboratory size demonstration. The size of the component created handling problems; however, achieving the required complexity of shape was not a problem. Figure 7 shows an as-cast finned plate. Fabrication of the finned plates required for a prototype demonstration unit (discussed below) is currently in progress.

### Component Development

Experiments were performed to develop the compliant edge seal, predict and measure the limiting thermal gradients on individual plates, and measure the strength of as-processed plates. All experiments yielded satisfactory results.

Fig. 7. As-cast prototype finned plate (edge view).

Laboratory test plates were used to evaluate various edge-seal configurations and materials. High-density ceramic-fiber insulation products were tested over the range of anticipated operating temperatures for compatibility with the silicon carbide components. The best candidate materials were introduced between two laboratory test plates in the high-temperature test furnace shown in Fig. 8. Loads and temperatures were varied for each seal configuration and candidate material. Figure 9 shows the performance of the selected seal system as a function of seal pressure, temperature, and number of insulation paper

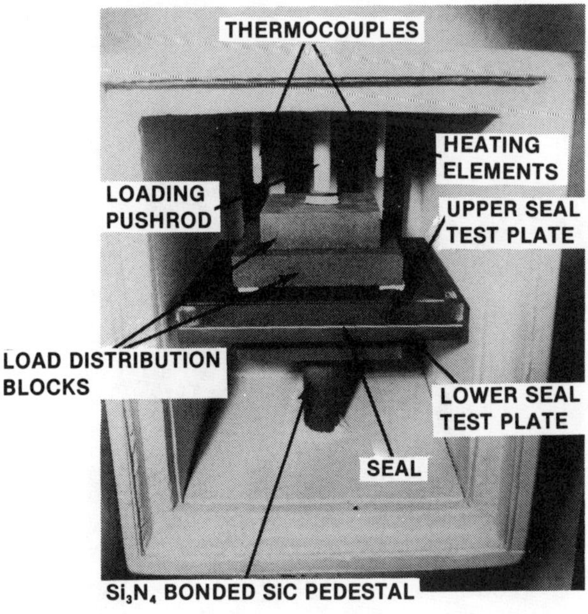


Fig. 8. Ceramic finned-plate laboratory seal-test furnace.

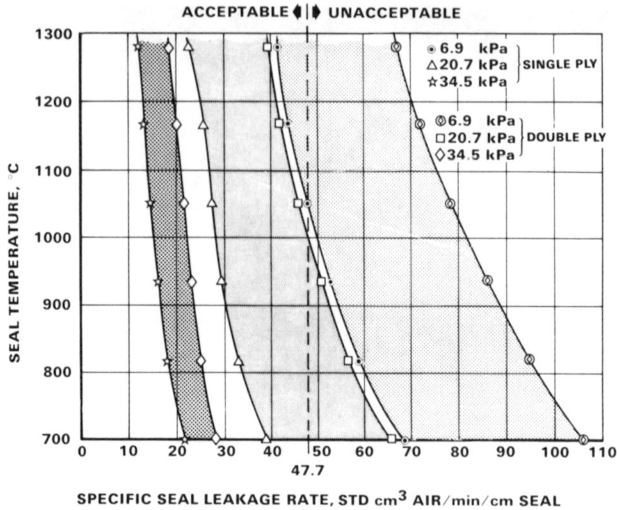


Fig. 9. Leakage rates for selected seal design.

plies. Acceptable seal performance is indicated in Fig. 9 and is equivalent to a leakage of 1% of the total airflow for the reference recuperator design.

Thermal gradients and thermal stress within the finned plates were predicted using finite element techniques. Experiments were developed to impose thermal gradients on laboratory-sized plates. An instrumented test plate (unfinned) is shown in Fig. 10. Gradients imposed over the plates (to induce

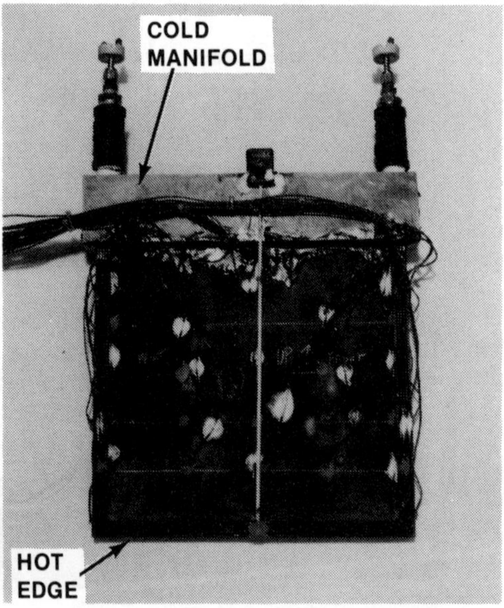


Fig. 10. Instrumented plate for thermal shock testing.

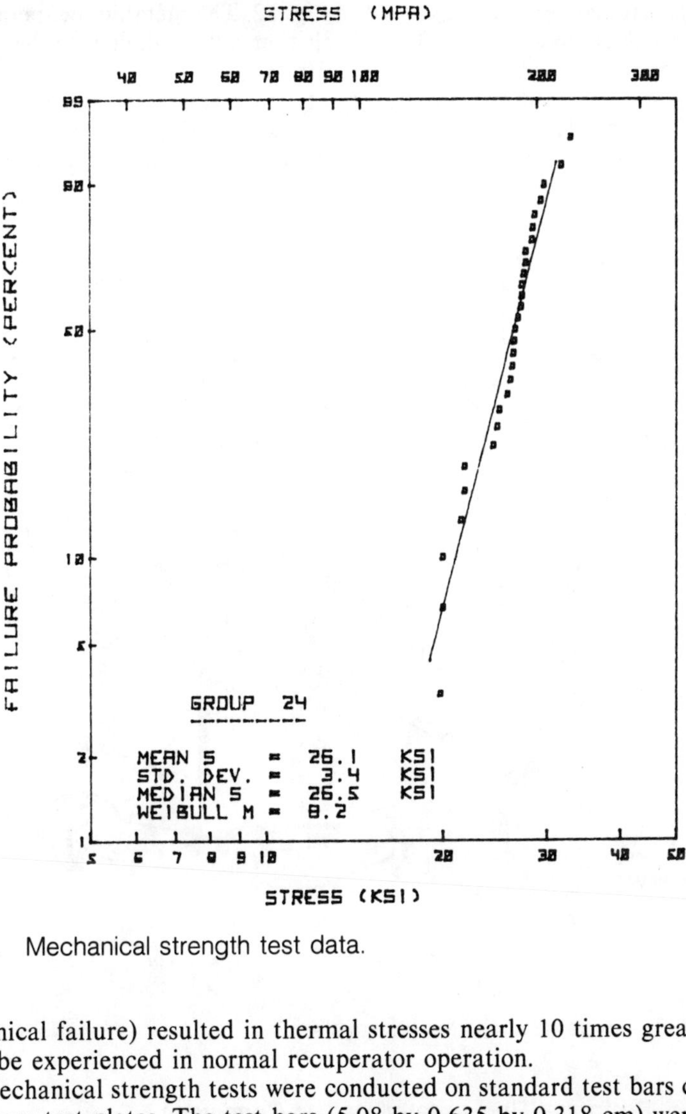


Fig. 11. Mechanical strength test data.

mechanical failure) resulted in thermal stresses nearly 10 times greater than would be experienced in normal recuperator operation.

Mechanical strength tests were conducted on standard test bars cut from laboratory test plates. The test bars (5.08 by 0.635 by 0.318 cm) were tested in four-point bending with quarter-point loading (outer span=3.81 cm, displacement rate=0.051 cm/min) at 1260°C and plotted on a failure probability diagram to determine the Weibull modulus and characteristic strength. These data are shown in Fig. 11.

**Prototype Design**

To demonstrate the fabrication feasibility and performance of the finned-plate recuperator, a scaled-down version of the reference design will be built and tested. A plate size of 0.61 by 0.61 m has been selected. With these smaller plates it is not possible to heat the combustion air from ambient temperature to 1100°C without incurring excessive thermal stress. Thus the prototype recuperator will employ a conventional metallic plate-fin recuperator in a series arrangement, as illustrated in Fig. 12. The metallic recuperator will heat the combustion air to 700°C. This air then will be ducted to the ceramic recuperator where it will be heated to 1100°C.

The design conditions for the prototype ceramic recuperator are summarized in Table III. The prototype unit requires 21 plates. The plates will have the same fin geometries and details as the full-size reference system design.

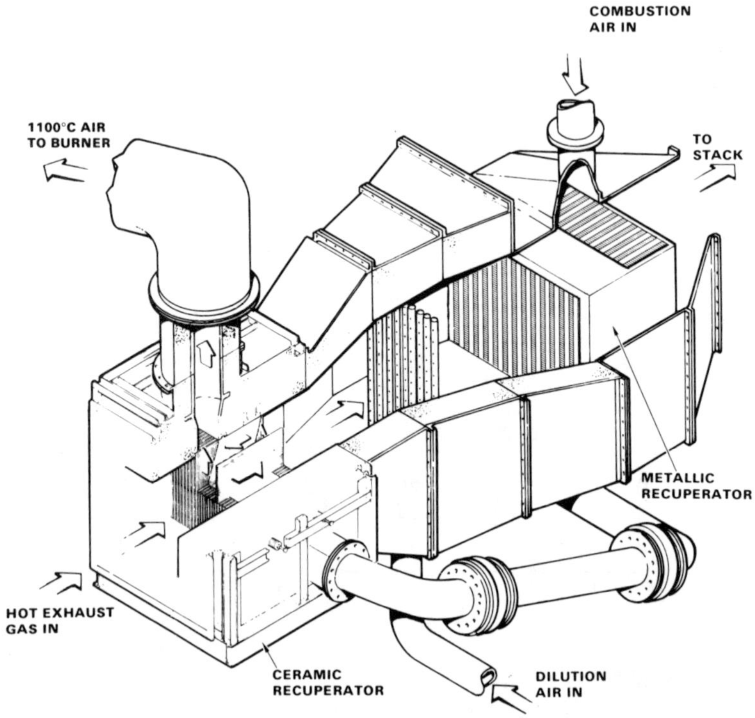


Fig. 12. Prototype recuperator test system.

Table III. Prototype Recuperator Design Conditions

| Parameter | Value |
|---|---|
| Equivalent burner rating, GJ/h | 5.3 |
| Airflow rate, kg/s | 0.50 |
| Gas flow rate, kg/s | 0.53 |
| Air inlet temperature, °C | 706 |
| Air outlet temperature, °C | 1093 |
| Gas inlet temperature, °C | 1343 |
| Gas outlet temperature, °C | 1032 |
| Effectiveness | 0.608 |
| Air pressure drop, kPa | 2.99 |
| Gas pressure drop, kPa | 1.25 |

Testing of the prototype unit is scheduled for late 1985 Subsequent to the prototype tests, a full-scale unit will be used to gain operational performance and economic data in an industrial host site.

## References

[1]H. Strumpf, D. Kotchick, and M. Coombs, High-Temperature Ceramic Recuperator and Combustion Air Burner Programs, Final Report GRI–83/0039, 1984.

---

This paper was also presented at the 1984 International Gas Research Conference.

# Ceramics in Fluid Bed Heat-Recovery Boilers

HOWARD W. WILLIAMS AND LAWRENCE C. HOFFMAN

Aerojet Energy Conversion Co.
Sacramento, CA 95814

JOHN B. PATTON

U.S. Department of Energy
Idaho Operations Office
Idaho Falls, ID 63401

New engineering ceramics applications are being explored to increase fluid bed recovery boilers operating temperatures. High-performance compact heat-exchange concepts are being researched for glass, metals, and other high-temperature industrial furnace applications. Completed engineering work and plans are described, as well as specific needs and projected applications.

### Fluid Bed Waste-Heat Recovery Boilers History

The fluid bed heat-recovery boiler was developed in England during the early 1970s to overcome problems of corrosion and fouling. Known as Fluidfire®, it has been applied successfully in the United Kingdom, Europe, and Japan. It has been developed to only a fraction of its full potential.

In 1981, Aerojet expanded its fluid bed product line to include waste-heat recovery by licensing the English Fluidfire® technology for the United States. Also, in 1981, Aerojet and the U.S. Department of Energy (DOE) Office of Industrial Programs signed a cooperative agreement to evaluate and further develop the fluid bed waste-heat boiler.

The DOE initiative was based on technical and economic studies combined with R&D work which showed a long-term payoff for industrial energy conservation. In August 1983 the Department of Energy advertised an additional project to develop a high-temperature ceramic distributor plate for fluid bed waste-heat recovery equipment. The distributor plate must have numerous evenly spaced orifices to evenly distribute hot furnace exhaust gases flowing into the bed. The DOE realized early on that the distributor plate is the only high-temperature component in the fluid bed system. The fluid bed and boiler tubes are usually only 204° to 315°C (400° to 600°F). Aerojet was selected as one of the researchers to develop high-temperature ceramic distributor plate concepts. Aerojet and its associates in fluid bed equipment have used ceramics for distributor plates in many arrangements, including perforated blocks and porous plates. Some untested prototypes of perforated plates have also been built.

### Fluid Bed Waste-Heat Boiler Description

The design approach to using the fluid bed is quite simple:

(1) Arrange the heat exchange tubes in a flat, horizontal, side-by-side assembly. (Fig. 1).

(2) Surround the tubes with a 2-in. thick bed of 0.016-in. diameter spherical alumina "sand" which expands to about 3 in. when the hot exhaust gases pass upward through it (bed shown pushed to the side in Fig. 1).

(3) Finned tubes can be used to minimize tube requirements since the bed keeps the tubes clean.

(4) Place a thin steel plate, with many small perforations, below the bed. This supports the bed and distributes the hot gas over the full bed area (example of a perforated distributor plate is shown in Fig. 2).

The result is a very effective, compact, self-cleaning heat-recovery unit for heating steam or hot water. Its configuration is a horizontal shelf only 3 in. thick and set in a casing.

### Ceramic Distributor Plate Project

The project to develop the ceramic distributor plate was started on August 31, 1984. It is jointly sponsored by DOE-Idaho Operations and the Aerojet Energy Conversion Company.

The first phase is ongoing and will be completed in May 1985. At that time research and engineering studies on the selected concept will be completed, along with plans for development testing in the final phase.

During the study, Aerojet will be soliciting suppliers for data and suggestions on engineering ceramic materials. Of special interest are configurations and materials which meet the stringent engineering requirements of the fluid bed and can be manufactured at reasonable cost.

Special requirements for the distributor plate are outlined in the following paragraphs. Also, the main thrust of the development plans, which will ultimately make the ceramic plate available for use by industry in fluid bed waste-heat recovery, is described.

To understand the importance of the project and the requirements of the high-temperature distributor plate, a brief review of the fluid bed heat-recovery boiler uses and DOE projects is in order.

### Fluid Bed Heat-Recovery Boiler Applications and Developments

#### Characteristics

The Fluidfire® fluid bed heat-recovery system has several important characteristics:

(1) The surface area of all the "sand" bed material particles is very large, sometimes equal to 1 or 2 football fields.

(2) High surface area, combined with rapid bed agitation from fluidization, keep the bed temperature even within several degrees throughout.

(3) Heat-transfer coefficients in the film between the hot gas and the steel tubing are often 5 to 10 times higher than without the bed. Therefore: (*a*) Tubing quantity is greatly reduced, since this film coefficient is by far the dominant factor in setting tube size; (*b*) gas temperature need not be high in the bed to get good heat exchange (260°C (500°F) is usually adequate); (*c*) the hot gas gives up its heat to the bed material at the instant it enters the bed.

These characteristics have been shown to be very remarkable in the dozens of applications in the last decade. They are precisely what is needed to solve the main waste-heat recovery problems.

To see how those problems are solved, consider the following conditions:

(1) The heat-exchange tubes and casing are not heated above about 260°C.

Fig. 1. Fluid bed heat-recovery unit heat-transfer tubing with bed material pushed aside.

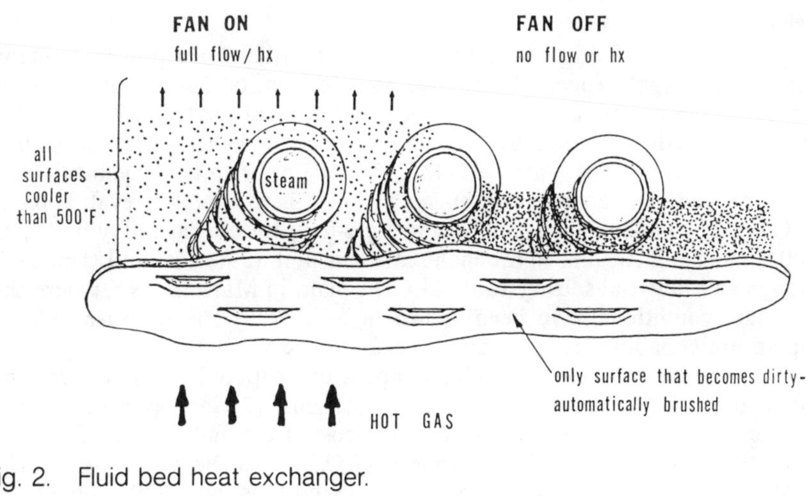


Fig. 2. Fluid bed heat exchanger.

(2) Agitation action of the fluidized bed continuously cleans the heat exchanger.

(3) The unit can be operated on a bypass with an induced draft fan. When the fan is off, the bed is its own shutoff damper.

The DOE expects the resulting benefits to the user will likely be attractive to the point of increasing the amount of heat recovery in the United States.

**Benefits**

The main user benefits of interest are:

(1) Maintenance problem sources essentially eliminated:

(*a*) Fouling eliminated by self-cleaning.

(*b*) Soot blowing and other cleaning maintenance not required.

(*c*) Corrosion is drastically reduced by the low temperatures.

(*d*) Thermal shock problems are practically eliminated by the low temperatures and the heat-storage capacity of the bed.

(*e*) Mechanical distortions and resulting maintenance problems are suppressed by the low and even temperatures.

(2) Operation is simple and independent of the process:

(*a*) The fluid bed and its induced draft fan are on a bypass.

(*b*) No change in exhaust pressure when unit is on or off.

(*c*) Steam or hot water is fed into the existing plant system, with a check valve giving independent operation.

(3) Retrofit is made easier by the compactness and by the flexibility of shape selection.

The beds can be made in a wide range of sizes and with length-to-width ratios up to about 1.5. Beds can be stacked one over the other. Hot gas can flow through beds:

(1) In parallel, to reduce the base area or footprint.

(2) In series, to increase effectiveness.

(3) Combination of both.

The heat-exchanger shape can therefore be made to suit the space available, including tall and thin or low and wide.

**Development Status and Plans**

The Fluidfire® fluid bed heat-recovery system has been proven in many industries and applications. There are, however, limitations in the gas inlet temperature. For some applications, the bottom of the perforated distributor plate is automatically wiped by a brush on wheels. It has not been developed for inlet gas temperatures above 371°C (700°F) prior to the ongoing DOE/Aerojet project. The goal of that ongoing project is to complete evaluation at 593°C (1100°F) and then develop the inlet gas-temperature capability to 872°C (1600°F) and higher, first in a bench-scale test unit at Aerojet and then evaluation in an industrial setting at an ALCOA plant in Massena, NY, where the particulate conditions have been shown to be severe. The increase in inlet temperature capability is important to heat recovery.

An increase in Fluidfire® inlet temperature capability (1) reduces the diluent air flow for high-temperature applications, (2) increases the energy recovered, and (3) reduces size, weight, and cost of the unit. These effects can be lumped into an overall figure-of-merit (F.O.M.), as shown on Fig. 3. The F.O.M. is doubled when inlet temperature capability is increased from 371°C to 593°C and doubled again when it is increased to 982°C (1800°F). The ceramic distributor plate project, described further below, will increase the capability to 1315°C (2400°F) and greater.

**Bench Scale Test Unit**

Located at the Aerojet plant in Sacramento, California, this unit is operated as the development test rig for R&D at higher gas inlet temperatures.

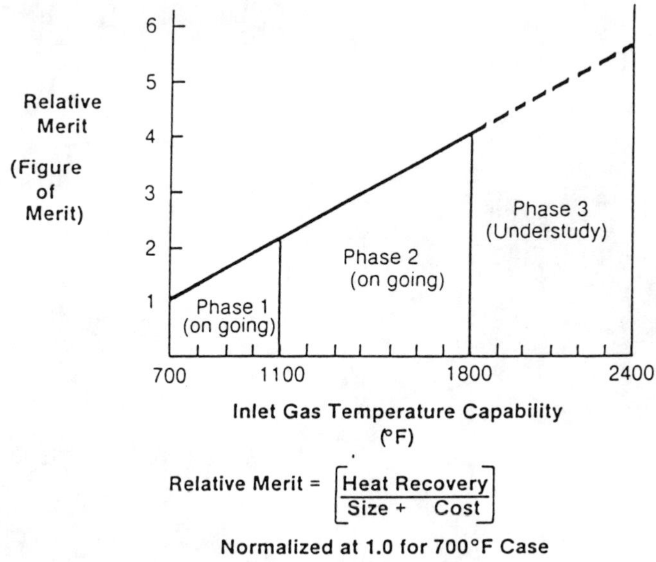


Fig. 3. Fluid bed waste-heat recovery system; relative merit vs inlet temperature capability.

It is shown on Fig. 4. The bench test unit is identical to 1 segment of the 30-segment ALCOA test unit. All key dimensions, except for bed area, are full scale. Tubes, plates, bed material, orifices, and brushes are full size for maximum data applicability.

The heat exchanger was fitted with a dual fuel burner. This will provide the fluid bed with hot dusty gas under conditions that can be varied to cover the range of potential experience at ALCOA and other industrial sites. For example, inlet temperature can be varied from 371° to 982°C, both No. 2 and No. 6 oil have been used, and additional particulate can also be introduced in the gas stream.

The unit is small enough for advanced component features to be tested under rough simulations of operating conditions, allowing quick evaluation and development. Check tests have been run to 760°C (1400°F) gas inlet temperature and some development results have been incorporated into the evaluation unit at ALCOA.

A separate cooperative study has recently been initiated by DOE and Aerojet. The study will investigate use of a ceramic distributor plate to further raise the gas inlet temperature to 1315°C so that performance can be increased even further.

#### Evaluation Test Unit

The evaluation unit was installed by ALCOA on the flue of an aluminum melting furnace (see Fig. 5). The furnace has an 83-ton melting capacity and firing rate of 9 860 000 W (34 million Btu/h). It is charged with molten aluminum and solid scrap aluminum and is fired with No. 6 residual oil.

The Fluidfire® fluid bed heat-recovery unit has its own induced draft fan on the outlet. Hot flue gas at temperatures to 1148°C (2100°F) is drawn from

Fig. 4. Bench scale fluid bed heat-recovery unit with oil burner heat source.

Fig. 5. Fluid bed heat-recovery unit (right) in operation on aluminum melting furnace at ALCOA plant.

a tee in the furnace flue (upper left of Fig. 5) and mixed with diluent air to limit the temperature to 593°C (1100°F). The mixture passes through the Fluidfire® unit (upper right) to extract the heat and generate saturated steam at 175 psig. The cooled gases then pass through the ID fan and out a small stack. Steam generated from waste heat is piped into the steam line of the existing fuel-fired boiler. The existing boiler fuel consumption will be reduced by as much as 4 060 000 W (14 million Btu/h) (see Table I). Testing is being conducted on the unit and results will be published in DOE reports.

#### Conservation Potential

The potential impact on U.S. industrial energy conservation from the fluid bed waste-heat recovery technology has been estimated. Several independent estimates show about a potential $1 billion/year impact. The benefits to industry go beyond the economics, since there will also be a reduced dependency on nonrenewable energy supplies. The economic value of energy saved using fluid bed waste-heat recovery is generally higher than most persons expect. An estimate can be obtained from Fig. 6 for specific applications.

#### Ceramic Distributor Plate Requirements

The primary function of the distributor plate is to support the bed while introducing a uniform stream of gas into it. The distributor is usually designed to provide (1) uniform and stable fluidization, (2) little or no down-sifting of bed material during operation or during shutdown, and (3) low velocities to minimize bed attrition and heat-exchanger tube erosion.

Table I. Fluid Bed Heat-Recovery Test Units

| | Aerojet Bench Scale Unit | Alcoa Evaluation Unit |
|---|---|---|
| Case dimensions (ft) | 4.4×3.6×7 high | 13.8×10.8×14 high |
| Number of beds | 1 | 2 |
| Bed gas flow arrangement | | Parallel flow |
| Bed size (ft) | 2.2×2.8 | 10×12.8 |
| Gas flow (scfm) | | |
| at 593°C | 437 | 14 300 |
| at 871°C | 437 | 14 300 |
| Operating bed depth (in.) | 2.7 | 2.7 |
| Boiler tube fin diameter (in.) | 2.12 | 2.12 |
| Energy saved (mm W)* | | |
| at 593°C | † | 4 |
| at 871°C | † | 5 |
| Furnace firing rate (mm W) | | 9.9 |
| Heat saved÷firing rate | | 41% at 593°C<br>50% at 871°C |

*Expected results with fired boiler and 70% efficient.
†Bench scale unit for test only—not practical to use recovered energy.

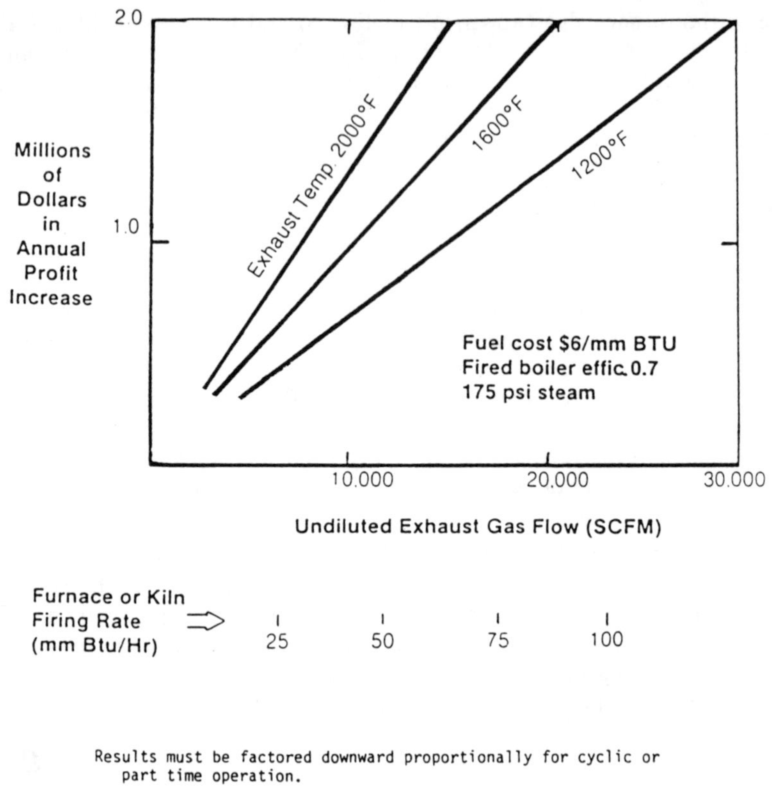


Fig. 6. Fluid bed waste-heat recovery economic benefits.

The uniformity and stability of a fluidized bed depends on the distributor pressure drop. In general, the greater the flow resistance through the distributor, the more uniform and stable the bed will be. The logical conclusion is that the pressure drop across the distributor should be as high as possible. However, a high pressure drop distributor, in addition to requiring a costly gas pumping input, may result in heat-exchanger tube erosion, particle attrition, and reduced heat transfer because of gas "rat-holing" (severe channeling). To balance the needs of bed stability and the potential problems with high pressure drop distributors, it is necessary to understand the need for the pressure drop to preserve bed stability.

The solid particles in a gas fluidized bed, unlike particles in a liquid fluidized bed, are not uniformly buoyed up by the gas. Instead, instabilities develop and the gas tends to collect in particle-free cavities. These cavities are similar in appearance and action to gas bubbles in a liquid. These bubbles tend to coalesce and increase in size as they rise through the particulate bed. The nonhomogeneous character of the bed results in pressure fluctuations. There have been no rigorous efforts to establish the pressure fluctuation in fluidized beds, but the general practice is to assume that the fluctuation is proportional to the bed depth. To accommodate the pressure fluctuations, some commercial fluidized beds have been designed with distributor plate pressure drops as high as 200% of the average press drop across the bed. However, the general approach is to use a design value closer to 30%.

Aerojet's experience with shallow fluidized beds has shown that a somewhat higher pressure drop is required for these systems. The Aerojet/Fluidfire® fluidized bed waste-heat recovery boiler has a distributor plate pressure drop/bed pressure drop ratio of about one-third.

The pressure drop across the distributor is primarily a function of the flow area through the plate. The velocity of the gas streams into the fluidized bed may be calculated based on the simple orifice equation:

$$V = C_f \sqrt{2g\Delta P/\rho} \qquad (1)$$

where $V$ is velocity of gas from distributor, $C_f$ flow coefficient, $g$ gravity constant, $\Delta P$ pressure drop across the distributor, and $\rho$ the density plate of gas.

Assuming a maximum distributor pressure drop of 1.3 in. of water, the velocity at 1315°C (2400°F) would be $V=134$ ft/s. For a superficial velocity of 3 ft/s at 510°C (950°F), the nozzle-to-plate area ratio required would be about 4.5%. This is more than the 2–4% usually recommended for typical fluidized beds. The difference is due to the high temperature drop in the bed. The high temperature at the distributor results in a very low-density gas, and a relatively high flow area is required. Early in the proposed program a range of conditions will be established for the various types of fluid beds.

In addition to the ratio of the distributor throughflow area to the bed area, the number and size of openings in the distributor are also important in the fluidized bed design. In general, aerodynamically, if the openings are too large and too far apart, the bed will not have uniform flow; if the openings are too close together, the adjacent bubbles tend to merge prematurely, resulting in poor solids mixing.

A minimum distributor hole spacing criterion is:

$$\log 3.52 \sqrt{L_B \rho_B} = 1.3 + 0.0192\ (X/D_o) \qquad (2)$$

where $L_B$ is bed depth, $\rho_B$ bed density, $D_o$ distributor orifice diameter, and $X$ distributor orifice spacing.

For a 4-in.-deep bed and a bed density of 100 lb/ft$^3$, the minimum ratio of orifice spacing to orifice diameter would be less than 0.5. This is a very low ratio and it suggests that the use of sintered plates would be aerodynamically feasible. Although it has been shown that a plate with as few as 100 holes/ft$^2$ will yield the same results in deep beds, shallow beds require more holes. The current Aerojet/Fluidfire® design has slots on 0.3-in. centers.

For distributors with simple orifices, bed particles can sift down the holes even when the average velocity is higher than the nominal terminal velocity of the particles (flow imbalance will result in some orifices with lower velocities). For minimum downsifting, the recommended orifice diameter is no more than three times the bed particle diameter. Bed particle sizes range from 0.010- to 0.040-in. diameter. The use of slots, orifices with downflow blockage valves, or orifices with caps will also minimize sieving. Beds with heat-exchange tubes will have smaller orifices because of smaller particle diameters needed for heat exchange. More advanced tube designs will tend to have even smaller bed particles as improved elutriation suppression allows the smaller particles which have high film coefficients of heat transfer.

The basic problems in using ceramics for the distributor plate for a high-temperature waste-heat application consist of (1) structurally supporting the distributor and the bed material, (2) providing suitable openings for the hot gas to go through, and (3) keeping the distributor from being fouled when there are particulates in the waste gas stream. The structural design must consider the modest strength of the ceramics (in tension) and the fabrication of large structures with small elements. The problems with providing suitable gas flow openings are with the extreme hardness of the ceramics that makes them hard to machine, and the high shrinkage which makes controlling the flow passage size very difficult. Potential fouling of the distributor is a universal problem; it is not unique to ceramic distributor design, although the extremely high temperatures present unique problems.

It does not appear feasible to construct a 0.010 slot in a 0.03-in. thick ceramic plate which has the nonsifting characteristics needed. Also, the orifices and bottom of the plate must be cleaned of particulate which accumulates there. This is a unique problem for fluid beds that are used for waste-heat recovery. Waste-heat flows are often laden with particulate, whereas fluid bed combustors are fluidized, with clean air passing through the distributor plate.

Fluidfire® heat-recovery units in operation have used a stainless steel wire brush which periodically cleans both the slot orifices and the underside of the plate automatically.

The problem of keeping the distributor plate from fouling has two facets. In gas streams with only vaporous impurities, one approach is to keep the distributor hot so that no fouling material can condense and foul the critical flow areas. When there are solid particulates in the stream, mechanical removal is required. This necessitates changing the design to accommodate the mechanical cleaning. For example, if particulates are to be recirculated through the distributor for cleaning the openings, they may have to be enlarged to allow the particles to flow through. In another case, if the bottom of the distributor has to be brushed periodically, it may be necessary to design a distributor with the plenum free of supports and obstructions.

The requirements shown in Table II are the major factors which must be considered. The following critical design factors are:

(1) Ceramic plate size (maximum) and hole size (minimum) fabrication limitations.

(2) Very low tensile strength and ductility values of ceramic materials.

(3) Fouling depositions and cleaning methods.

(4) Bed down-sifting through plate perforations during shutdown.

(5) Metal limitations: Limited strength and corrosion resistance in high-temperature furnace gases.

The ceramic distributor plate must obviously be a self-cleaning system comprised of a plate, support structure, end sealing, and cleaning system.

### Selected Concepts

Aerojet established and studied many alternative concepts. Starting with previous experience of the Aerojet team with over a dozen ceramic distributor fluid beds, the concepts were formulated, analyzed, and reviewed. They included plate types, segmented types, and extruded element concepts.

#### Plate Types

For this type, thin plates perforated by various means are used. This most closely resembles the current metal plate state of the art, but has many draw-

Table II. FBWHR Unit Requirements

| Requirement | Range | Nominal Reference |
|---|---|---|
| 1. Bed size, $ft^2$ | 20–120 | 120 (10×12) |
| 2. Temperatures | | |
| Inlet side (max), °C | 815–1315 | 2400 |
| At idle (min), °C | 537 | 1000 |
| Bed side, °C | 204–982 | 950 |
| 3. Cycle time, h | 1–12 | 4 |
| 4. Total pressure drop, in w/c | 3–10 | 6 |
| 5. Unfluidized bed load, $lb/ft^2$ | 20–50 | 50 |
| 6. Fouling particulate loading g/scf | 0.2–2 | 0.4 |
| 7. Plate perforations | Adaptable to 0.020 in. diam. hole, 0.010 in. slot | Provide adequate distribution to simulate existing pattern |
| 8. Plate orifice-to-overall area ratio | 0.02–0.04 | 0.04 |
| 9. Plate cleaning, order of desirability | (1) Self-cleaning (2) Automatic (3) Easily cleaned manually | Continuously self-cleaning |
| 10. Bed sifting | Prevent or handle, recirculate automatically | |
| 11. Cost effective | Including cost of plates, installation, replacement, and maintenance | Including cost of plates, installation, replacement, and maintenance |

backs. Present technology appears to be limited to about 1 ft by 2 ft plate size. This means 50 to 60 plates in one heat exchanger with much assembly work on the 480 linear feet of sealing. Also, the plates require a separate support structure and must be individually replaceable. One of the major problems is the orifice design and fabrication. Straight holes have been cast and laser-cut in various shapes of holes and slots.

The problem is how to design orifices that (1) can be fabricated economically, since many thousands are needed in each heat exchanger to give proper fluidization quality; (2) can be cleaned periodically and readily from below; and (3) will not let bed material sieve when gas flow is stopped (bed material is spherical particles of uniform size and as small as 0.01-in. diameter). The DOE is pursuing resolution of these problems in a separate project.

Aerojet has investigated porous ceramic plate approaches. Aerojet's partners in fluid bed heat-exchange products, Stone International Ltd., has operated fluid bed units with porous plates for heat recovery. To date, no practical methods for cleaning particulate have been established, although several concepts have been tested.

### Segmented Types

These include arch construction with reduced stress loading. Aerojet has experience with this type used with deeper beds. Compared to plate types (1) orifice problems are even worse; (2) assembly does not require hardware, but the high skill level needed is a drawback; and (3) maintenance is difficult. This type of plate construction, especially for shallow beds, does not seem practical for most heat-recovery applications.

### Extruded Element Concepts

This class of concepts is comprised of side-by-side elements that are identical or interlocking. They can be manufactured by extrusion, casting, weaving, or any other method.

All surfaces of an element can be formed by parallel lines in this class. Possibilities include configurations which could be called planks, channels, tubes, tiles, beams, etc. The rationale for this approach is summarized as follows. It may be possible to:

(1) Construct a plate out of a moderate number of elements (10 to 100).

(2) Make the elements identical for economy.

(3) Have elements as long as the full width of the heat exchanger to minimize sealing and joining.

(4) Replace elements singly or as an assembly.

(5) Utilize the interfaces between elements to satisfy requirements of gas distribution orifices, and perhaps include bed sieving prevention design. For example, grooving and overhanging lips may be designed to provide that function.

(6) Use low-cost manufacturing techniques such as extrusion, weaving plus impregnation, extrusion plus green state forming, etc.

This general concept is believed to have the most potential for advanced ceramic distributor plates. The DOE/Aerojet project is concentrating on the best configurations in this class.

# Fluidized Bed Waste-Heat Recovery System

W. E. COLE, R. DE SARO, AND K. D. PATCH

Thermo Electron Corp.
Waltham, MA 02254

J. B. PATTON

U.S. Department of Energy
Idaho Operations Office
Idaho Falls, ID 63401

The fluidized bed waste-heat recovery system is being developed to preheat combustion air for industrial furnaces, using the heat available in dirty flue gas streams. Described in this paper are the concept, laboratory performance test results on a subscale system, and field-test results of a subscale hot test module.

Stack gases in many industries, including aluminum smelting, are discharged directly to the atmosphere at temperatures of over 1093°C (2000°F)—representing a loss of more than 50% of the fuel thermal energy. These large inefficiencies clearly demonstrate the need for recuperation development in the Energy Conservation branch of DOE. One of the contractors in this program is Thermo Electron Corporation.

Thermo Electron is developing a direct contact recuperation system based on fluidized bed technology, as shown in Fig. 1. The fluidized bed waste-heat recovery (FBWHR) system uses an indirect heat-transfer medium of small alumina particles heated by furnace exhaust gases which pass through the distributor plate of a fluidized bed. These heated particles are then removed and placed on a distributor plate in a second fluidized bed, where the ambient combustion air is heated. The cooled particles are then returned to the first bed for reheating.

This concept has a number of advantages compared to alternative waste-heat recovery systems. In particular, the FBWHR system features: (1) wide applicability, (2) low cost, (3) large energy savings, (4) short payback period, (5) high resistance to corrosion and fouling, and (6) high reliability.

The initial development of this concept is for the aluminum smelting industry. In aluminum smelting, waste aluminum feedstock is melted in an open hearth reverberatory furnace. Flue gas temperatures often exceed 1093°C, resulting in a very low efficiency for the smelting process. The aluminum feed consists of a number of different alloys and contaminant materials. Conversely, the product of the smelter is a precisely defined alloy. To effect this change, it is necessary to alter the composition of the melt and remove the contaminants, of which magnesium is the most serious. The magnesium is removed

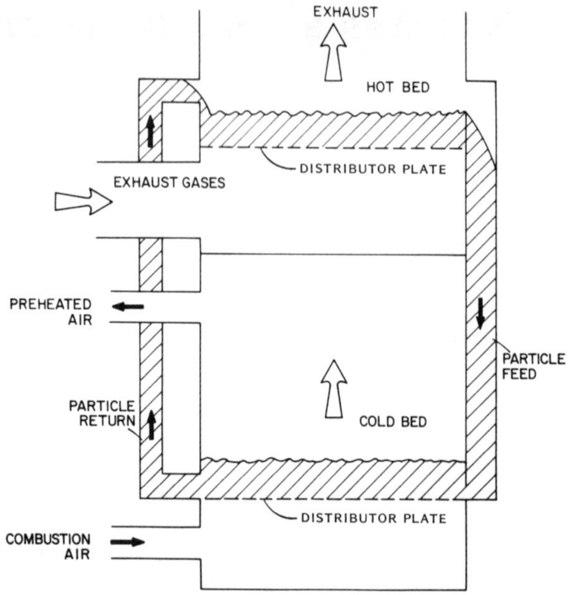


Fig. 1. Fluidized bed waste-heat recovery system.

by bubbling chlorine through the melt. Some of the chlorine, along with sodium and potassium additives, escapes from the furnace with the flue gases, creating an extremely corrosive gas. No reliable waste-heat recovery system is commercially available to recover heat from such flue gases.

The FBWHR system can be fabricated from materials compatible with this corrosive environment. In contrast to other concepts, there is direct contact between the gases and the heat-transfer media in the FBWHR system, and no intermediate heat-transfer surfaces are needed.

Research efforts of the FBWHR system can be classified into two major topics. The first is development of the FBWHR system technology and the concurrent documentation of its performance. The second topic is development of a method for keeping the hot distributor plate, a critical component in the FBWHR system, free of deposits.

**Laboratory Testing**

An experimental program was conducted to develop and adapt fluidized bed heat-transfer technology to the FBWHR system. An isometric drawing of the laboratory test unit is shown in Fig. 2. The unit consists of a hot-side fluidized bed (outside dimensions 8 by 6 by 2 ft) mounted above a cold-side fluidized bed (6 by 4 by 1½ ft), a particle circulating system, a furnace simulator, a combustion air blower, and an exhaust system. The furnace simulator produces clean, high-temperature gas that simulates the temperature, flow rate, and pressure of an aluminum smelting furnace flue. The hot gas passes through the upper fluidized bed, heats the particles, and is then discharged to the atmosphere by the exhaust system. In the lower fluidized bed, a combustion air blower pushes ambient air through the lower distributor plate. The air is heated by the hot particles moving across the lower bed.

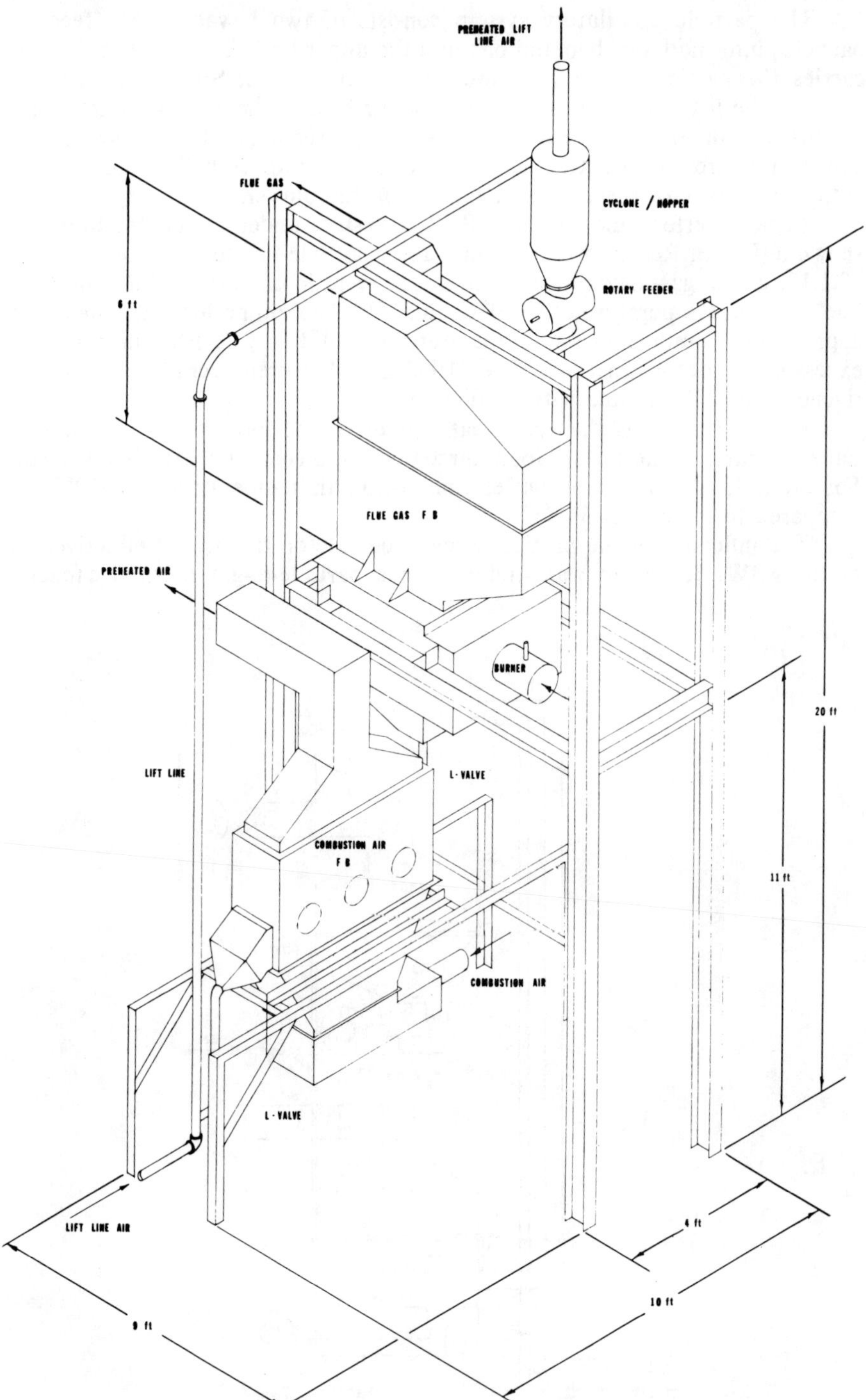


Fig. 2. Laboratory test facility.

The particle circulatory system consists of two L valves, one feeding particles into the lower bed and one into the upper bed. A pneumatic lift line carries the particles from the lower to the upper bed. Since the particles entering the lift line from the combustion air heat exchanger are hotter than the lift line air, the lift line also acts as a heat exchanger. This preheated air can then be brought into the plenum of the lower fluidized bed where it mixes with the balance of the combustion air from the blower.

Typical performance of the FBWHR system under design conditions is shown in Fig. 3. Results were obtained for a particle-to-flue gas capacity ratio of 1.13, a flue gas-to-combustion air capacity ratio of 1.31, and a simulated flue gas inlet temperature of 1110°C (2030°F). System performance included a preheated combustion air temperature of 544°C (1012°F), which was in excess of the design goal of 537°C (1000°F). Measured overall system effectiveness was 0.47, in line with predictions.

Figure 4 presents the analytically predicted performance of the system under identical conditions. Good agreement of theory and data is apparent. For example, the predicted outlet combustion air temperature is 1039°F, as compared to the actual 544°C.

The influence of the particle circulation rate on the overall effectiveness of the FBWHR system was studied over a particle-to-gas thermal capacity

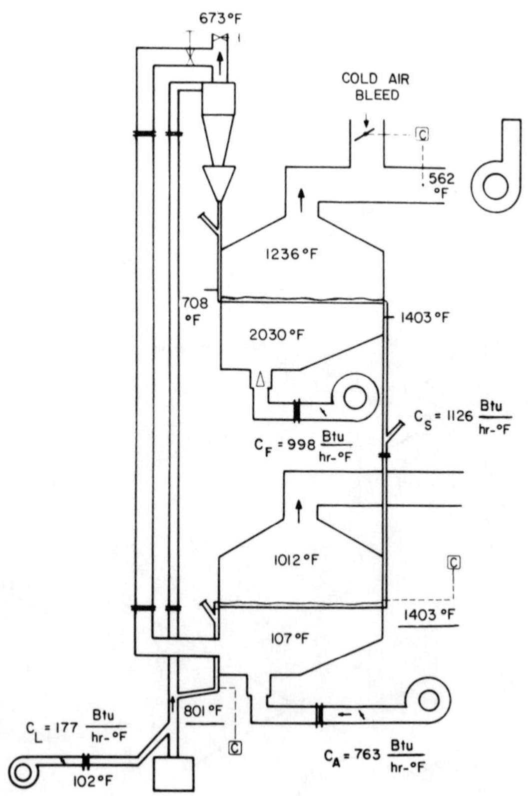


Fig. 3. Performance of FBWHR laboratory system under design conditions.

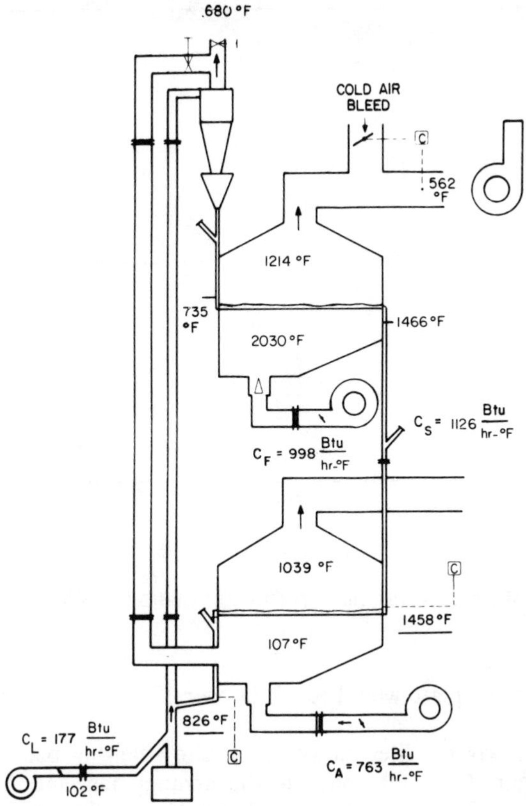


Fig. 4. Performance of FBWHR laboratory system under design conditions (predicted).

ratio of 0.36 to 1.23. The data points from the 12 particle circulation rates are compared to the theoretical predictions in Fig. 5.

Validation of the general trend of increasing the overall effectiveness with an increase in particle-to-flue gas capacity is apparent. As flow of the medium is increaseo at a constant gas flow, more heat can be transferred between the hot and cold fluidized beds. Above particle-to-gas capacity ratios of unity, however, significant increases in particle flow are required to yield a perceptible improvement in effectiveness. For example, increasing the capacity ratio $C_s/C_g$ from 0.92 to 1.09 results in only a 1% increase in theoretical overall effectiveness, from 0.475 to 0.480. Negligible effectiveness increases occur above a capacity ratio of 1.09.

**Distributor Plate Testing**

One concern with the FBWHR system is the potential deposition (or condensing) of particulate matter on the hot distributor plate. This material builds up around the plate holes, increasing the pressure loss and reducing the gas flow rate. Reduction in the flow rate at high firing rates reduces the recuperator effectiveness and, hence, combustion air preheat temperature. A

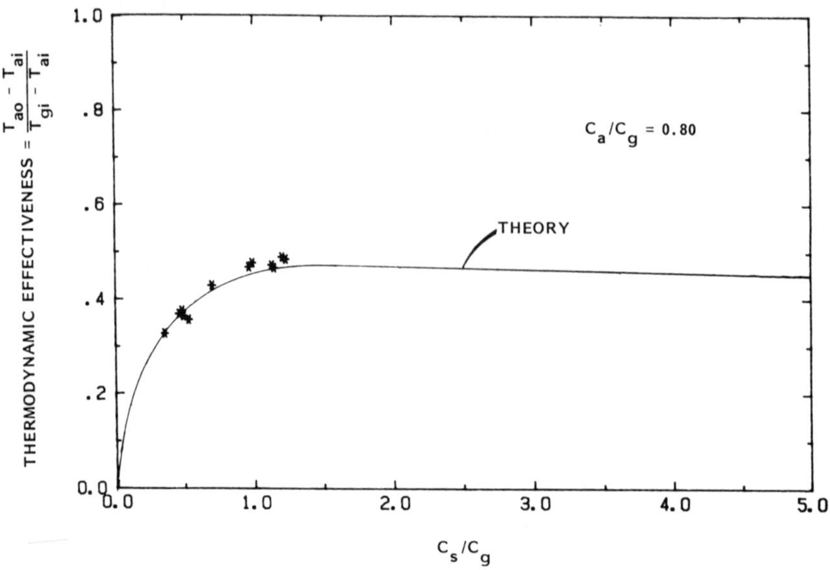


Fig. 5. Effect of particle circulation rate on overall FBWHR system effectiveness.

20% blockage of the plate would result in over 37.7°C (100°F) loss in preheat temperature.

A test unit was installed in an aluminum smelting plant, sidestream to the furnace. Figure 6 is a schematic of this arrangement and the installed unit is shown in Fig. 7. Ten percent of the 3500 scfm of flue gases are pulled by the blower through a 1 $ft^2$ opening in the flue and diluted to 1093°C (2000°F) in the inlet duct; they then travel through the distributor plate. The gases exit the freeboard, are diluted to 260°C (500°F) with ambient air to protect the blower, and are then returned to the top of the flue.

Temperatures, pressures, and flow rates recorded during a typical test run are shown in Fig. 8. The flue gases, extracted at 270 scfm, exit the flue at 1104°C (2020°F) and enter the plenum at 1037°C (1900°F). Heat losses in the insulated plenum and freeboard amount to 3.3% of the total flue gas heat content, resulting in the flue gas exiting the freeboard at 1004°C (1840°F). The distributor plate temperature of 993°C (1820°F) is close to that of the exiting gas temperature. Heat losses in the uninsulated duct are designed to be high, at 21%, to minimize the required dilution air into the fan. The distributor plate pressure loss is 5.5 in. wc and the orifice pressure loss is 0.7 in. wc.

The field-test unit was operated for approximately 3000 h. Initial tests showed that the plate fouled in approximately 125 to 350 h. Fouling was typically indicated in those tests by an increase in the distributor plate pressure loss and a decrease in the gas flow rate, as shown in Fig. 9. The plate pressure loss increased from 5 to 12 inches of water after approximately 75 h. At this point, the fan outlet damper was fully open, corresponding to the full-flow operating point of the blower. As the test continued, the flow rate decreased because the blower could no longer maintain the required flow rate.

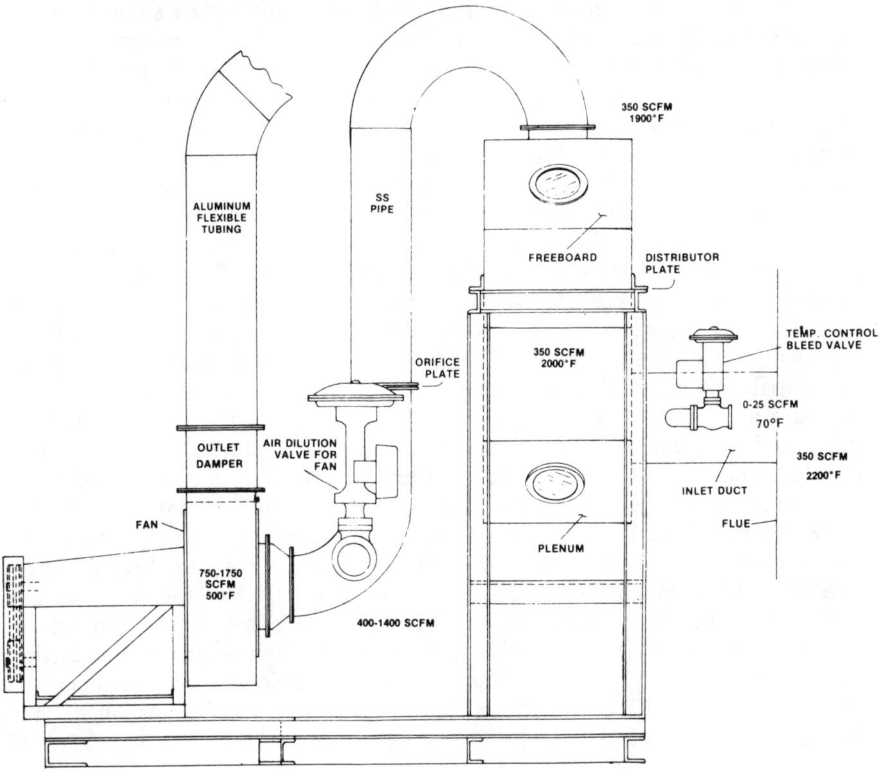


Fig. 6. Fouling test unit.

Fig. 7. One-tenth scale fouling unit installed sidestream to the furnace.

Analysis of the material deposited on the distributor plate showed that it was predominantly aluminum and magnesium. Both of these materials can be traced to the chlorinating process. Aluminum results from chlorine reacting with the aluminum, volatilizing, and then reacting with water to form alumina. Magnesium similarly results from chlorine reacting with the magnesium, becoming entrained in the flue gases, and reacting with water. Other minor ingredients found in the deposit included iron, sodium, calcium, potassium, silicon, chromium, and zirconium. These materials are either fluxes added to the melt or materials of construction for the furnace.

An air lance was used on-line to prevent the plate from fouling during operation. The air lance consists of a 1/16- or ⅛-in.-diameter nozzle at the end of a ¼-in.-diameter stainless steel tube. The tube fits through the side of the plenum wall so that the nozzle is positioned in the plenum directly under the distributor plate. The opposite end of the tubing is connected to a high-pressure air supply. The distributor plate is cleaned by the high-velocity air, generated by the nozzle, as it impacts the distributor plate. The impact force can be controlled by varying the air pressure, altering nozzle geometry, or positioning the nozzle at varying distances from the plate. The entire distributor plate can be cleaned by manually translating and rotating a single air lance. The air lance was used once a day on the test unit for about three minutes. The flow rate through the air lance was less than 0.01% of the total flue gas flow rate through the test unit, resulting in a negligible loss of gas temperature.

The ability of the air lance to keep the plate clean during on-line testing is depicted in Fig. 10. With on-line cleaning, the fluidized bed was operated for about 1000 h with no indication of fouling. This is 3 to 8 times longer than had been possible without on-line cleaning. Since there was no indication of

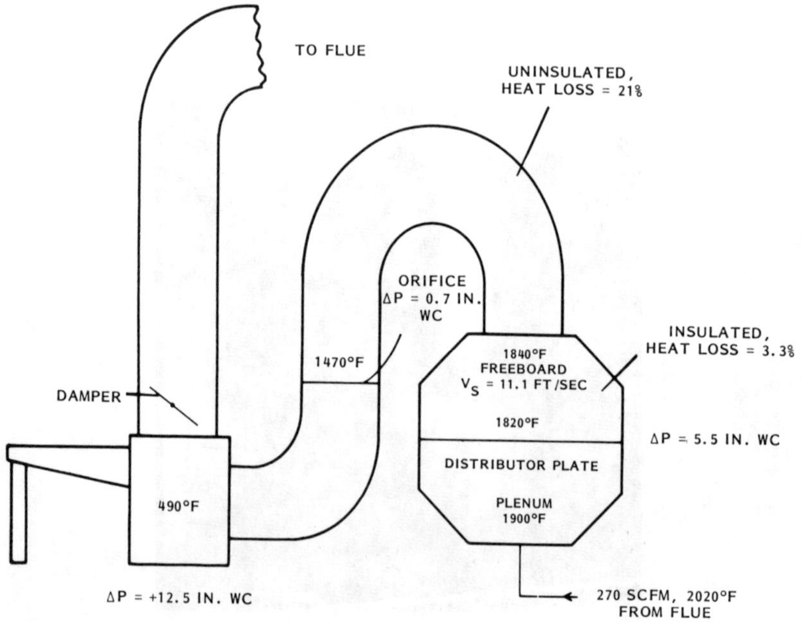


Fig. 8. Typical 1/10 scale operating conditions.

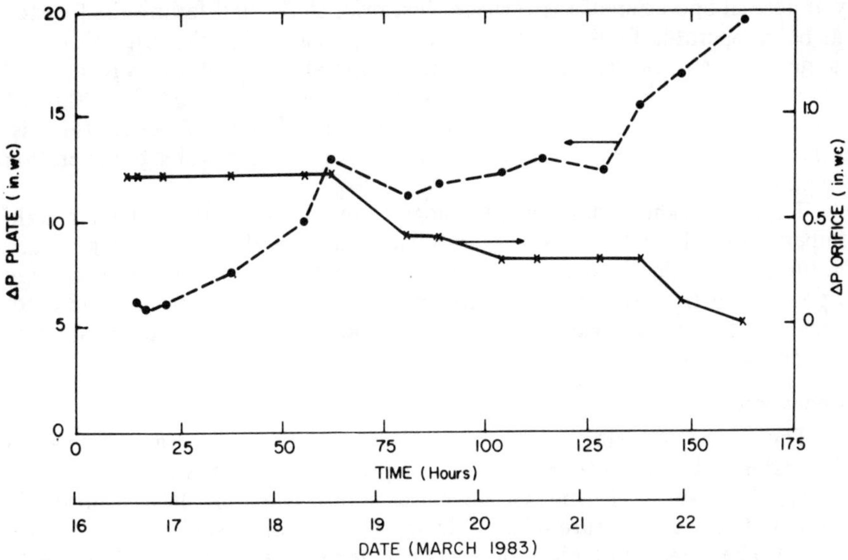


Fig. 9. Distributor plate fouling characteristics without on-line cleaning.

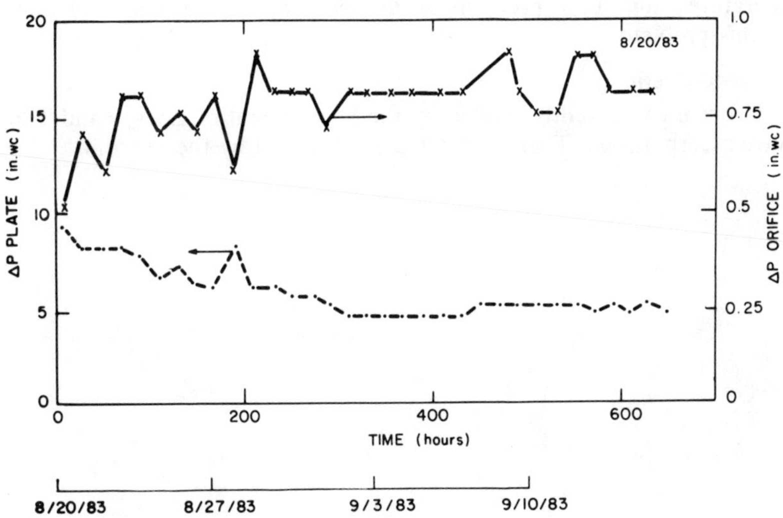


Fig. 10. Distributor plate with on-line cleaning.

fouling prior to shutdown, it is likely that the on-line cleaning test could have continued for much longer than 1000 h.

Two distributor plates failed at this 1000-h point in two separate tests. Both plates warped beyond design limits, resulting in unacceptably high gas leakage past the plate perimeter seal. These failures are believed to be caused by abnormal operation of the furnace. Just prior to the first failure the furnace was being operated fuel-rich. Since dilution air is added to the test unit's inlet flue gases to control the distributor plate temperature, combustion is believed to have occurred. Plate temperature excursions as high as 1648°C (3000°F) were recorded during the 48-h period prior to the failure. Whether this distortion will occur under normal operating conditions after prolonged operation is unknown.

These tests show that the technology to build distributor plates for high-temperature fluidized beds is only marginally capable of meeting design goals. On the basis of this inadequacy, DOE is currently developing the technology to produce distributor plates with higher temperature capabilities. These advanced plates will increase the operating range of fluidized bed waste-heat recovery systems.

## Conclusions

The FBWHR system is a viable concept to recover the waste heat from dirty industrial flue gases. Results of this program have shown:

(1) The overall thermal effectiveness is approximately 48%, resulting in an air preheat temperature of 537° from 1093°C flue gases.

(2) The distributor plate does collect deposits and tends to block with time. However, an air lance easily cleans the plate and prevents deposits from forming.

(3) Deformation of the plate, possibly caused by excessively high flue gas temperatures, may be a problem. Additional work is required to define and solve this problem.

## Acknowledgment

This work was jointly funded by the Department of Energy and Thermo Electron Corporation. Their support is appreciated by the authors.

# High-Temperature Heat Wheel for Industrial Heat Recovery

RODNEY G. WHITBECK, KLAUS H. HEMSATH, AND ARBIND PRASAD

Midland-Ross Corp.
Energy Technology Div.
Toledo, OH 43696

A heat wheel system using a ceramic heat-exchange matrix has been developed which can provide a means of significantly improving overall process thermal efficiency, especially of high-temperature processes. The unique design features of this heat wheel reduce inherent leakage and maintenance problems, and expand temperature and pressure limits associated with existing heat wheel regenerators. A laboratory scale demonstration of the concept, supported by the Gas Research Institute, was completed in March 1984 at the Midland-Ross Energy Technology Division Technical Center in Toledo, Ohio. The demonstration showed that the heat wheel performed at approximately 67% thermal effectiveness over the range of test conditions with acceptable, predictable rates of leakage and trouble-free mechanical performance, and provided a steady supply of high-temperature, high-pressure combustion air.

Recovery of heat from process flue-gas streams is a well-known method of returning heat to the process for the purpose of decreasing fuel consumption and operating costs. It has been shown that heat recovery from high-temperature 1350°C (2500°F) flue-gas streams can reduce fuel consumption by up to 65% when compared to the same equipment without heat recovery.[1] Various types of heat recuperators and heat regenerators have been used for this purpose at more moderate temperatures, but none have been proven on high-temperature installations.

As discussed in this paper, recuperators are heat exchangers in which one fluid is separated from the other by a barrier. Regenerators are heat exchangers in which the heat-transfer medium moves between the hot and cold streams, alternately accepting and rejecting heat.

The objective of this program has been to develop a regenerative heat wheel with the following characteristics:

(1) High-temperature operating capability. Applicable to processes with flue-gas temperatures up to 1350°C (2500°F).

(2) Compact size.

(3) High thermal effectiveness.

(4) Low, predictable, consistent leakage rates.

(5) Excellent economic benefits.

### Existing Equipment and Limitations

Recuperators are normally fabricated from alloy steels and are limited in operating temperature range and thermal effectiveness. New recuperator development programs are investigating the use of ceramic materials in standard recuperator configurations to increase effectiveness, to decrease size, and to reduce cost.

Heat wheel regenerators have been used for low-temperature processes below 500°C (1000°F) at low cross-stream pressure differentials of typically less than 1 kPa (4 in wc), and have been proven to be very efficient, compact, and usable at high volume flows. However, at temperatures above 550°C (1000°F), heat wheels have not been widely used, primarily due to the increasing difficulty of controlling leakage. At higher temperatures, deformation of the monolithic heat wheel matrix becomes severe and sealing against it becomes difficult. High-temperature deformation problems also result in increased maintenance and downtime.

Over the years, heat-recovery devices have found wide applications but advancement of the state-of-the-art to higher temperatures and flows has been slow. In Fig. 1, an attempt has been made to summarize the current market for heat-recovery devices across the entire process flow and temperature range.

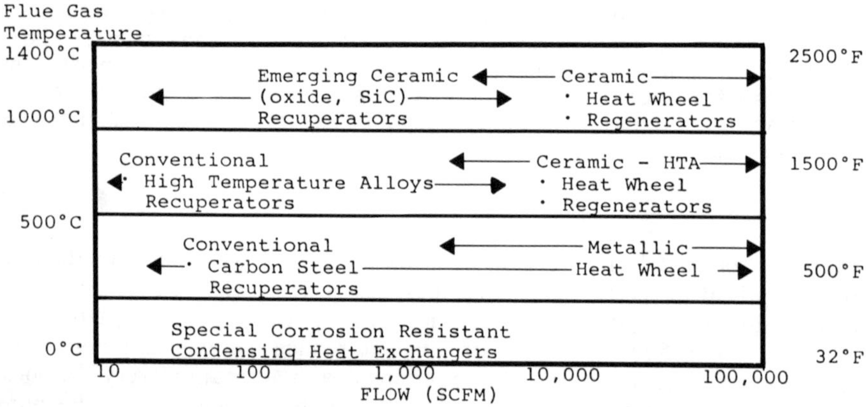


Fig. 1. Heat-recovery devices: overview.

To take advantage of emerging low-expansion ceramic materials, a new-concept heat wheel was designed. The device can take advantage of the best features of the heat wheel and can eliminate or improve on design deficiencies experienced with the Lungstrom-type heat-recovery devices to high-flow and high-temperature applications. After proof of concept, this novel design has a potential to expand the market for heat-recovery devices to high-flow and high-temperature applications.

### New-Concept Design

The new heat wheel is a patented design, protected under U.S. Pat. No. 4 321 961.[2] It is shown schematically in Figs. 2 and 3, and is essentially a segmented annulus rotating about a stationary central duct. The annulus fixtures the ceramic matrix heat-exchange medium which rotates between hot and cold process streams. In this configuration, the matrix can deform without being constrained and is free from stress buildup due to thermal and mechanical loading. These stresses have caused failures in earlier ceramic heat

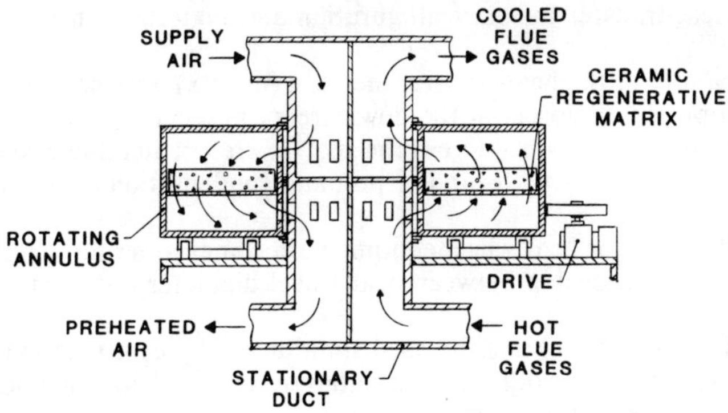


Fig. 2. Schematic: longitudinal section view.

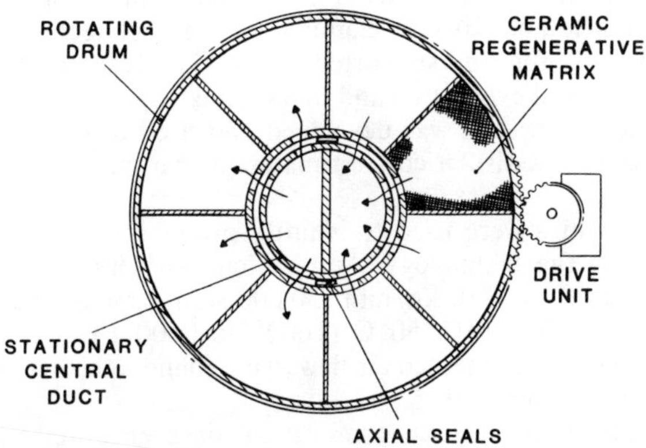


Fig. 3. Schematic: transverse section view.

wheel designs, particularly near the central hub.

The flows enter the heat wheel through a split duct and are distributed into the segment chambers. Heat transfer takes place by forced convection and conduction as the gases pass through the matrix in opposite directions. In the case shown in Fig. 2, the matrix hot face is the bottom face while the top face is cold.

The design concept uses cross-stream (axial) seals which seal between rotating and stationary duct components. These components have machined surfaces and present fewer seal interfacing problems than designs which use the matrix surface for cross-stream sealing. The circumferential seal, which prevents leaking from inside the wheel to the outside, is also achieved by sealing between machined surfaces. The drive is external and operates at low temperatures.

The heat wheel has several design and operating features, listed below, which give it a competitive advantage when compared to conventional recuperators.

(1) Control of process temperatures possible by variation of heat wheel rpm.

(2) Heat-transfer matrix configuration and material can be "application specific."

(3) Failure of the heat-transfer medium (matrix) does not result in failure of the equipment or upsets in the downstream process.

(4) Thermal gradients across the matrix are predictable and permit use of specific materials to withstand or promote condensation of flue-stream constituents.

(5) Payback for heat wheel equipment becomes lower for larger units due to the square relationship between heat wheel diameter and heat-transfer surface area.

(6) The heat wheel may be used simultaneously as a reaction unit and a heat-recovery device. Addition of a catalytic material for chemical removal of flue-stream contaminants is an example.

## Initial Work

The need for high-temperature heat-recovery equipment for the steel industry first led Midland-Ross to examine heat wheel technology. The new-concept heat wheel program, supported by the Gas Research Institute, began with conceptual model evaluation and market analysis. A mathematical model of the heat wheel operation was developed and used to provide design information and sizing to assist in cost estimates for various commercial applications.

The early studies were used to identify prime applications which would most benefit from the technology. Applications with high flow rates, i.e., 50 kg/min (1500 scfm) to 1750 kg/min (50 000 scfm); medium to high flue-gas temperatures, i.e., 500° to 1350°C (1000° to 2500°F); and approximately equivalent flue gas and preheated air flows were found to have the best payback and highest market potential.

Prime applications include combustion air preheating and process air heating in a variety of industries.

To meet the market need and to test the concept on a pilot scale, a 50 kg/min (1500 scfm) heat wheel was designed and built.

## Test Facility

A test facility was built to verify theoretical performance calculations; obtain thermal, mechanical, and wear data; and gain operating experience during process-simulation tests. The facility consists of a hot gas source, a high-pressure cold air source, pressure control dampers, associated ductwork, instrumentation, and controls. Figure 4 shows the pilot scale heat wheel installed in the laboratory facility.

Figure 5 is a schematic of the test setup.

## Test Results

Tests were performed during the operating period to determine thermal effectiveness and leakage rates at varying cross-stream pressure differentials and temperatures. Overall mechanical performance and component wear were also evaluated.

### Thermal Effectiveness

The thermal effectiveness of the heat wheel varied between 66 and 70% and was independent of input temperature and leakage during testing.

Fig. 4. Pilot scale heat wheel at Midland-Ross.

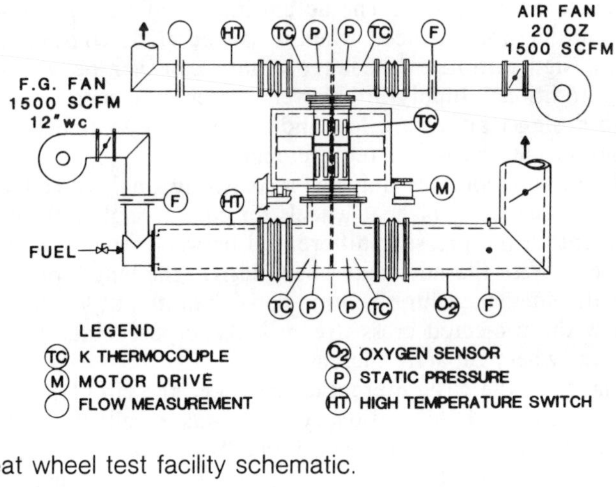


Fig. 5 Heat wheel test facility schematic.

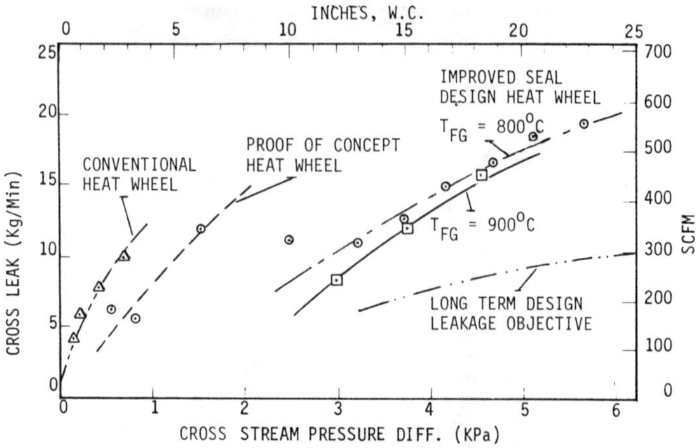


Fig. 6 Cross leakage as a function of cross-stream pressure differential.

#### Seal Performance

The leakage rates varied with cross-stream pressure differential but were consistent and predictable. Cross leakage as a function of pressure differential is shown in Fig. 6 for testing at 800°C (1500°F), 900°C (1650°F), and for heat wheels tested as part of earlier projects.[3] The advanced-design heat wheel leakage performance shows a significant improvement over standard heat wheel designs. In addition, design modifications offer the prospect for further axial seal performance improvement. The actual impact of leakage on design performance is considerably reduced when it can be made constant and can be considered a design parameter. Leakage volume can then be compensated for by providing additional supply air blower capacity. In most cases, costs associated with design leakage are low and are directly related to the costs of additional blower electric power requirements.

The pilot scale laboratory unit has a relatively small capacity compared to the anticipated capacity of heat wheels for commercial applications. Leakage is dependent on the pressure differential between streams and the length of the axial seal. With the new heat wheel design, the length of the axial seal is proportionally smaller at high volume flows than it is at low flows. Figure 7 illustrates how the projected cross-stream leakage percentage of total flow is reduced as heat wheel capacity increases.

Performance of the circumferential seals, i.e., the circular seals, between the rotating annulus and the stationary duct, was excellent. Leakage across the circumferential seal varied between 1 and 4% of total flow. Circumferential seal leakage improved as the sealing components wore together during operation.

#### Mechanical Performance

The performance of the segmented ceramic heat-exchange matrix verifies that the Midland-Ross design can use ceramic elements for high-temperature applications without element failure. Figure 8 illustrates the matrix mounting method in the heat wheel. Expandable packing material surrounds the matrix elements.

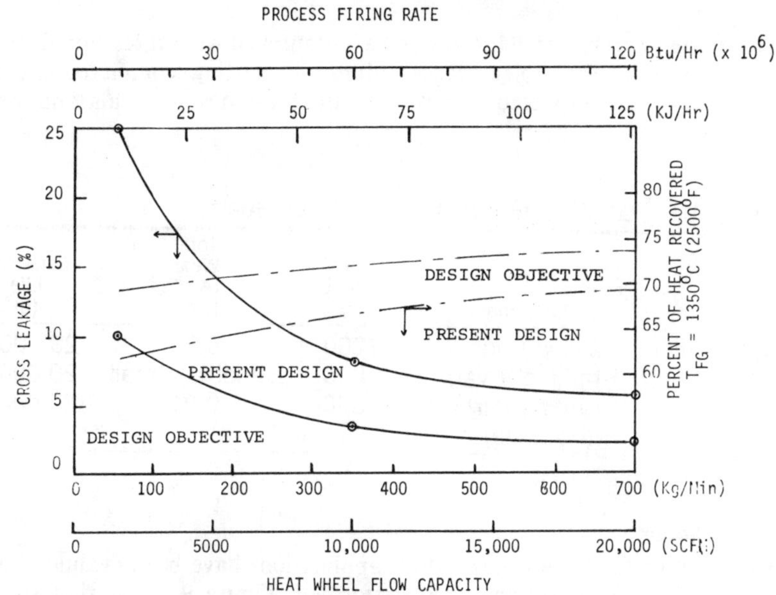


Fig. 7. Cross leakage vs heat wheel capacity.

Fig. 8. Condition of ceramic heat-exchange matrix after 1000 h testing.

**Wear**

Seal design and materials selection have improved wear rates significantly during the course of the project. Recent durability testing, completed in early 1984, produced the expected component life estimates for seals and wear components shown in Table I.

Table I. Wear Data: Circumferential and Axial Seals

| Seal | Component | Duration of Testing (h) | Total Wear (Avg) (mm) | Expected Life (h) |
|---|---|---|---|---|
| Circumferential | Sealing rings | 1200 | 0.38 | 23 000 |
| Axial | Stationary vane | 800 | Variable–Small | 20 000+ |
| | Rotating surface (hot side) | 800 | 0.01 | 20 000 |

**Future Applications**

Cost factors for several heat wheel applications have been evaluated and a simple payback analysis has been performed. Figure 9 shows that simple payback varies from approximately two years at low flow to less than one year at flue gas flows of 700 kg/min (20 000 scfm) for new equipment installations. The prices used to produce the payback curve are "selling prices," which include engineering, materials fabrication, typical installation, and general overhead expenses. The significant payback advantage of larger units is due to the reduced proportional cost of the structure due to the square relationship between flow area and heat wheel diameter.

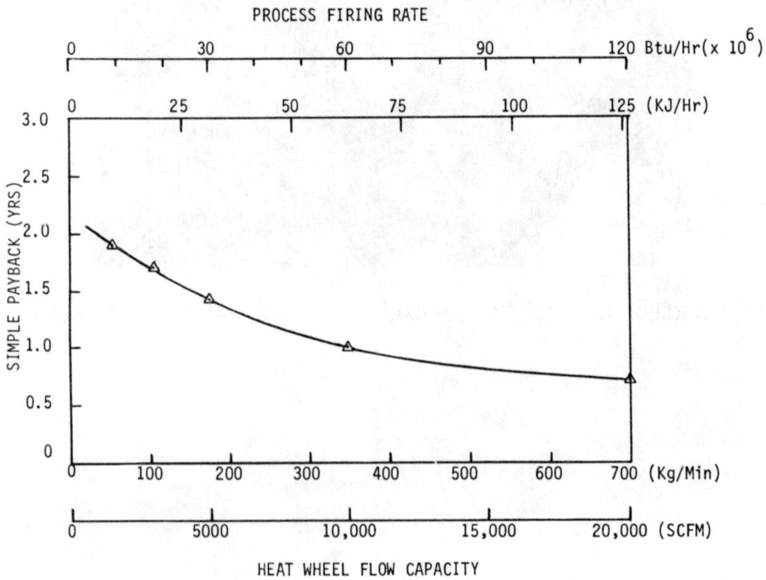


Fig. 9. Heat wheel sizing vs simple payback.

Several target markets for the heat wheel were identified and are listed in Table II. The designations "ceramic" and "metallic" refer to the materials of construction of the internal structural components of the heat wheel.

Table II. Heat Wheel Applications

| Temperature Range | Process Firing Rate | Structure Material | Target Applications |
|---|---|---|---|
| <800°C (1500°F) | <60 mm GJ/h (58 mm Btu/h) | Metallic | Process air heating, heat treating, process temperature control |
| 800°C–1350°C (1500°F–2500°F) | <60 mm GJ/h | Ceramic | Aluminum melters, ceramic processing, forging operations |
| <800°C | >60 mm GJ/h | Metallic | Steel reheat furnaces, heat treating, incinerator heat recovery, fired heaters |
| 800°C–1350°C | >60 mm GJ/h | Ceramic | Steel reheat furnaces, forging operations, glass melting, scale-free steel processing |

### Future Work

Future work on the new-concept heat wheel will be conducted in three phases. The first phase will involve improving the axial seal performance by using a revised seal design and improved material. The second phase will be design and demonstration of a high-temperature, all-ceramic unit. The third phase will be the commercial field demonstration phase of the technology.

### Conclusion

This new-concept heat wheel design offers distinct design advantages over conventional heat wheels. It is compact, has high heat-recovery efficiency, provides consistent leakage performance at low percentage of inlet flow, and is capable of providing high-pressure preheated air for use with preheated air burners.

Laboratory demonstration of the technology has shown that the ceramic heat-transfer matrix provides excellent mechanical and thermal performance and that wear of components is adequate and has potential for significant improvement.

The advances made during this program may expand the use of the highly effective heat wheel regenerator to new industrial applications and processes.

### References

[1]R. K. Shah, C. F. McDonald, and C. P. Howard. "Compact Heat Exchangers—History, Technological Advancement and Mechanical Design Problems," ASME 1980.

[2]K. H. Hemsath, "Thermal Energy Exchanging Device," U.S. Pat. No. 4 321 961.

[3]A. Prasad and J. K. Jasti, "Advanced Regenerative Heat Recovery System"; GRI Annual Report, 1981.

# $Si_3N_4$-BN Composites: Potential Materials for Heat-Exchanger Applications

K. W. Hill, J. D. Lee, and D. R. Petrak

Babcock & Wilcox
Lynchburg, VA 24505

$Si_3N_4$-BN composite materials are excellent candidates for high-temperature heat-exchanger applications. Property data and a feasibility test performed on a heat-exchanger element fabricated from such a composite are described.

Recovery and utilization of waste heat through the use of heat exchangers is a significant part of current energy conservation efforts. Efficient recovery of high-temperature waste heat demands heat exchangers that are constructed of refractory materials, especially for heat exchangers to be utilized above 950°C. The properties of a relatively new ceramic material, a hot-pressed composite of $Si_3N_4$ and BN, make it an interesting candidate for high-temperature heat-exchanger applications.

Particulate composites consisting of $Si_3N_4$ with a dispersed BN phase have recently been examined by several investigators. Mazdiyasni and Ruh[1] showed that the composites have lower dielectric properties and improved thermal shock resistance than $Si_3N_4$ alone. Niihara et al.[2] demonstrated the anisotropic thermal diffusivity of these hot-pressed materials, which is primarily due to the preferred orientation of the BN platelets. Patent literature has suggested that high BN levels in composites improve machinability.[3] Most recently, Lee et al.[4] discussed the results of a systematic characterization of $Si_3N_4$-BN composites* with up to 30 wt% BN. This paper discusses the potential use of $Si_3N_4$-BN material as a heat exchanger, and presents property data and a description of a feasibility test performed on a 30 wt% BN composite.

## Material Fabrication and Properties

Based on a review of prior characterization work done on the system $Si_3N_4$-BN, a 30 wt% BN composite appeared to be an attractive candidate for a high-temperature heat exchanger and was chosen for feasibility testing. The composite was composed of $Si_3N_4$ and 30 wt% BN with several percent oxide densification aid. Powder mixtures were prepared by wet-milling in methanol and dry-milling using alumina media. Billets were uniaxially hot-pressed at 1700°–1750°C and 4000 psi for 2 h in graphite dies coated with BN and lined with graphite foil under a nitrogen atmosphere.

Property data generated are shown in Table I (along with data on a similarly processed $Si_3N_4$ material without BN for comparison) and Fig. 1. The

*Cercomp, Babcock & Wilcox, Lynchburg, VA.

Table I. Properties of $Si_3N_4$ and $Si_3N_4$-30 wt% BN

| Property | $Si_3N_4$ | $Si_3N_4$-30 wt% BN |
|---|---|---|
| Density (g/cm³) | 3.23 | 2.75 |
| XRD phase content | $Si_3N_4$ | $Si_3N_4$, BN |
| Flexural strength (MPa) | | |
| (∥ to HP direction.) | | |
| @ Room temperature | 730 | 290 |
| @ 1000°C | 530 | 260 |
| @ 1200°C | 410 | 190 |
| @ 1400°C | 120 | 80 |
| Young's modulus (GPa) | 280 | 120 |
| Thermal diffusivity ($cm^2 \cdot s^{-1}$)* | | |
| @ RT ⊥ to HP direction | 0.124 | 0.235 |
| @ RT ∥ to HP direction | 0.109 | 0.092 |

*D. P. H. Hasselman, Virginia Polytechnic Institute, Blacksburg, VA.

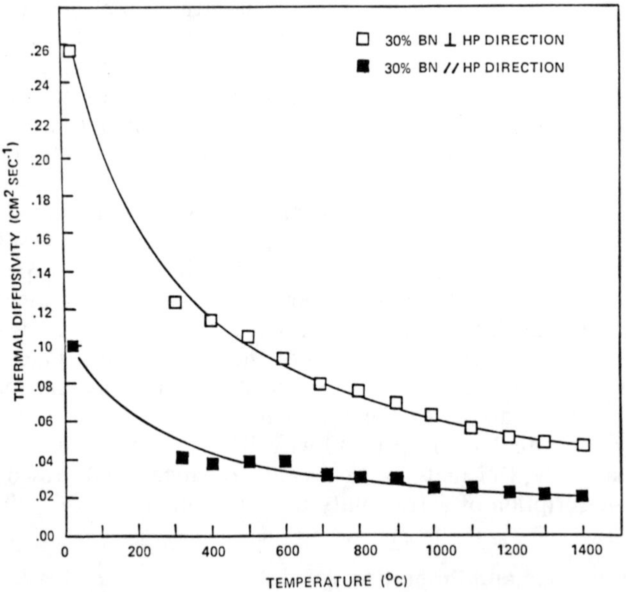


Fig. 1. Thermal diffusivity of $Si_3N_4$-30 wt% BN as a function of temperature.

hot-pressed density of the 30 wt% BN material was 95% of theoretical. Microscopic examination of a ground surface revealed the anisotropic microstructure of the composite, with the BN platelets tending to orient perpendicular to the hot-pressing direction, as expected.

The flexural strength of the 30 wt% BN composite, measured in three-point bend, was roughly one-half that of the $Si_3N_4$ without BN, but did not decrease significantly at 1000°C and dropped only slightly at 1200°C. The Young's modulus of the composite, determined from a load vs deflection plot of the three-point bend test, was much lower than that of the $Si_3N_4$.

Fig. 2. Machined $Si_3N_4$-BN heat-exchanger components.

The anisotropic microstructure of the 30 wt% BN composite was reflected in the thermal diffusivity data. The room-temperature thermal diffusivity of the composite material was much greater in the direction perpendicular to the hot-pressing direction than in the parallel direction, as well as being significantly greater than $Si_3N_4$ in the perpendicular direction. As the test temperature was elevated (Fig. 1), the thermal diffusivity decreased in both directions, although more pronouncedly in the perpendicular direction, resulting in a decrease in the anisotropy of the measurements in the material.

The thermal shock resistance of the composite was not measured quantitatively. However, repeated immersions of a $Si_3N_4$-30 wt% BN material into molten iron did not induce mechanical damage, indicating excellent resistance to thermal shock. As with thermal shock resistance, quantitative measurements of machinability were not made, but a cursory look showed that conventional carbide tooling adequately machined the material.

**Heat-Exchanger Feasibility Testing**

To demonstrate the feasibility of using the 30 wt% BN material as a heat-exchanger element, a two-component recuperator element was designed, fabricated, and tested in a simulated furnace flue at temperatures up to 1204°C.

Two billets (11 cm in diameter by 5.3 cm thick) were hot-pressed from $Si_3N_4$-30 wt% BN composition via the procedure given above. The billets were machined successfully with conventional carbide tooling to provide separate passages for flue gas and preheat air flows when sandwiched together (Fig. 2). The configuration chosen was such that hot flue gas flows straight through the key-shaped holes while the air to be preheated enters one exchanger component through the radial slots, flows through a common center chamber from the first to the second component, and exits radially through the second component's slots (Fig. 3). This design takes advantage of the high diffusivity of the material in the radial direction perpendicular to the hot-pressing direction.

To test the heat-exchanger element, a test fixture assembly was designed and fabricated (Fig. 4). The two recuperator components were sandwiched within several sections of pressed refractory board, which were machined to

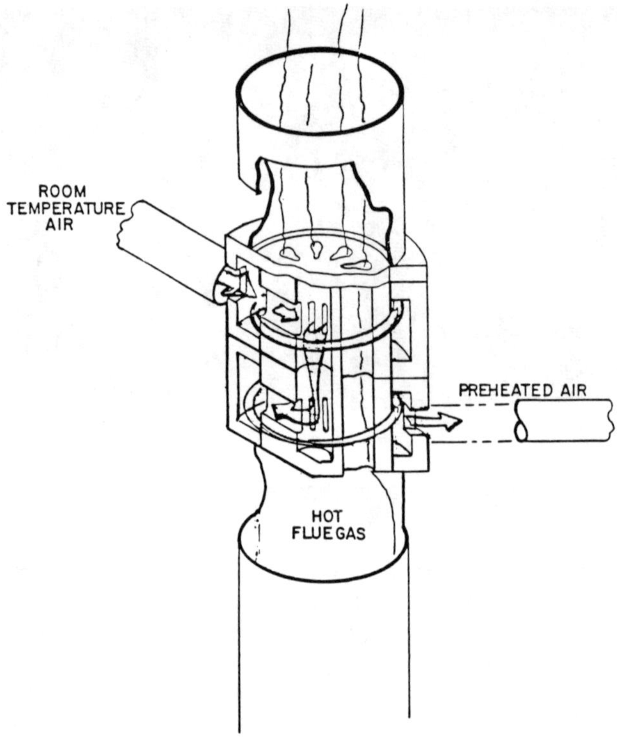


Fig. 3. Ceramic heat-exchanger design.

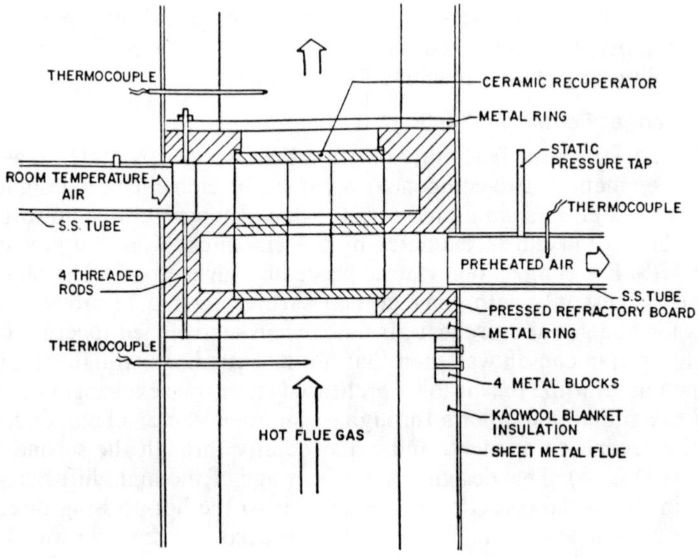


Fig. 4. Heat-exchanger test fixture.

shape. A compression seal between these parts was established using a metallic ring on each end of the subassembly connected by threaded rods. This subassembly was secured within a ceramic fiber-lined sheet metal tube which simulated the flue. Two steel tubes were attached and sealed to the assembly for preheat air entry and exit to complete the fixture.

The heat-exchanger assembly was mounted on a single-burner, propane-tank heat source for testing. Flue gas and preheat air inlet and outlet temperatures and pressure drops were measured. The flow rates of the room-temperature air to be heated were varied from 6.8 up to 38 kg/h at flue gas inlet test temperatures of 816°, 1093°, and 1204°C. Data readings were taken at a particular set of conditions when the preheated air outlet temperature was stable. A second set of 1093°C flue inlet testing was done to include measurement of the flue gas flow rate in the data.

**Results and Discussion**

The feasibility testing of the ceramic heat exchanger was successful; the data are given in Fig. 5 and Table II. Good heat transfer was demonstrated at flue gas inlet temperatures up to 1204°C and preheat air flows up to 38 kg/h. No degradation of the ceramic recuperator element was apparent except some discoloration, probably deposited by the burner exhaust gases. Excellent thermal shock resistance was confirmed since the recuperator was heated to and cooled from testing temperatures in a few minutes.

As plotted in Fig. 5, the increase in preheat air temperatures as a function of flue gas inlet temperatures and air flow rates was fairly consistent. As

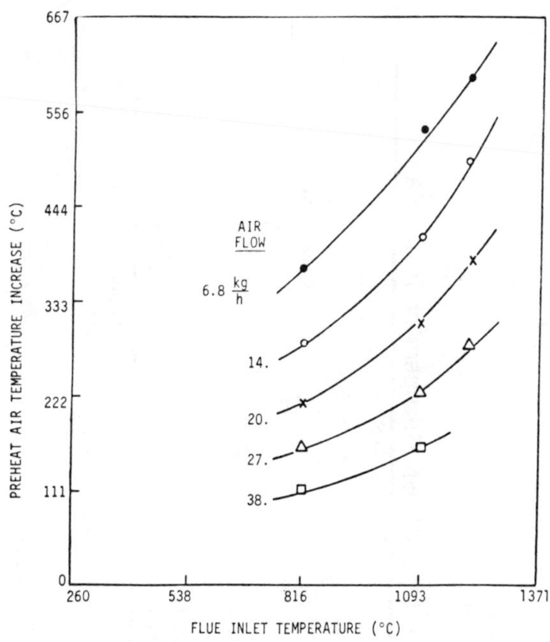


Fig. 5. Air temperature increase as a function of flue inlet temperature and air flow rate.

Table II. $Si_3N_4$-BN Recuperator Test Data

| Flue Side | | | Air Side | | | |
|---|---|---|---|---|---|---|
| Avg. Inlet Temp. (°C) | Maximum Outlet Temp. (°C) | Mass Flow Rate (kg/h) | Avg. Inlet Temp. (°C) | Avg. Outlet Temp. (°C) | Mass Flow Rate (kg/h) | Heat-Transfer Effectiveness* (%) |
| 1093 | 902 | 13 | 21 | 546 | 6.8 | 49 |
| 1093 | 866 | 13 | 21 | 444 | 14 | 39 |
| 1093 | 843 | 13 | 21 | 369 | 20 | 45 |
| 1093 | 829 | 13 | 21 | 311 | 27 | 51 |
| 1093 | 785 | 13 | 21 | 278 | 34 | 56 |

*Calculated using air-side heat gain as actual heat-transfer rate.

expected, the pressure drop across the air side was much greater than across the flue side due to the tortuous path of the preheat air through the recuperator.

From the second set of data taken at a flue gas inlet temperature of 1093°C (Table II), the approximate heat-transfer effectiveness was calculated using the following equation:

$$\text{Effectiveness} = \frac{C_{\text{air}}\,(\Delta T_{\text{air}})}{C_{\text{minimum}}\,(\Delta T_{\text{maximum}})} \qquad (1)$$

where $C_{air}$ is the preheat air-capacity rate (the product of the mass flow rate times heat capacity), $\Delta T_{air}$ the increase in preheat air temperature, $C_{minimum}$ the smaller of the air or flue gas capacity rates, and $\Delta T_{maximum}$ the difference between the flue gas-inlet and air-inlet temperatures. Effectiveness compares the actual heat-transfer rate to the thermodynamically limited, maximum possible heat-transfer rate. The air-side heat-transfer rate was used as the actual rate in the calculations because there was more confidence in these data.

Although the effectiveness values can be considered as approximate only, considering the unbalanced heat-transfer rates which were calculated using the measured data and certain approximations, the range of values from 39 to 56% indicates good potential for this material, especially since the design of the recuperator was not optimized.

## Conclusions

Feasibility testing of a $Si_3N_4$-BN heat-exchanger element successfully demonstrated the material's potential in this type of application. A 30 wt% BN composite was chosen for testing based on its promising properties. This material was machined with conventional carbide tooling into the desired configuration and demonstrated heat-transfer capability and excellent shock resistance with no degradation up to a temperature of 1204°C.

## References


[1]K. S. Mazdiyasni and R. Ruh, "High/Low Modulus $Si_3N_4$-BN Composite for Improved Electrical and Thermal Shock Behavior," *Am. Ceram. Soc. Bull.*, **64** [7] 415–19 (1981).

[2]K. Niihara, L. D. Bentsen, D. P. H. Hasselman, and K. S. Mazdiyasni, "Anisotropy Effects on the Thermal Diffusivity of $Si_3N_4$-BN Composites," *J. Am. Ceram. Soc.*, **64** [9] C-117–C-118 (1981).

[3]L. C. Montgomery, "High Density Refractory Composites and Method of Making," U.S. Pat. No. 4 268 314, May 19, 1981.

[4]J. D. Lee, H. H. Moeller, D. R. Petrak, and P. L. Berneburg, "Properties of $Si_3N_4$-BN Composites"; for abstract see *Am. Ceram. Soc. Bull.*, **63** [3] 422 (1984).

# Development of an Indirect Gas-Fired High-Temperature Heating System

ARVIND C. THEKDI AND STEVEN R. HUEBNER

Midland-Ross Corp.
Energy Technology Division
Box 985
Toledo, OH 43696

MICHAEL A. LUKASIEWICZ

Gas Research Institute
Chicago, IL 60631

An indirectly gas-fired (radiant tube) high-temperature heating system is currently under development at the Energy Technology Division of Midland-Ross Corporation, with the support of the Gas Research Institute. Initial test results of this heating system indicate that significant increases (a factor of 3) in heat flux may be achieved compared to conventional systems. Extrapolation of the results via a numerical model from these preliminary tests to designs with all-ceramic components indicates that increases of a factor of 5 in heat flux may be achievable in practice. These improvements in indirectly fired combustion systems are due to the development of a high-intensity radiant tube burner, advances in the design of high-temperature heating elements, and the availability of high-temperature materials. Currently, where indirect firing would be required at process temperatures above 1100°C (2012°F), electricity is the primary energy source. Commercialization of high-temperature indirect gas-fired heating equipment resulting from this program would permit natural gas to be an economic alternative to electricity at process temperatures above 1100°C (2012°F).

Many industrial heat-treating processes require the use of specially controlled atmosphere gases while the load is being heated to high temperature. In these instances, an indirectly fired combustion system is often used to supply the necessary heat. Indirect heating is achieved by firing a burner into a tube which is located inside the furnace enclosure. The combustion products transfer heat to the tube by radiation and convection. At steady state, the tube transfers this heat to the furnace primarily by radiation. However, sometimes convection may be significant for low-temperature processes. In some cases, electrical heating elements may be used when special atmospheres are required.

At the present time, gas-fired combustion systems are limited to process temperatures less than approximately 1050°C (1922°F). This is primarily due to the fact that metals used in radiant tube construction are limited to 1100°C (2012°F). In addition, the efficiency of gas-fired systems is substantially reduced as the process temperature increases.

Because of the temperature limitations imposed by metallic radiant tubes, many processes cannot use the conventional metal tubes for high-temperature indirect gas firing. Examples of such processes are high-temperature carburizing, sintering and brazing, scale-free billet heating for forging, and vacuum furnace heating. Therefore, these processes may utilize electrical heating elements as substitutes for the radiant tubes. These heating elements are constructed from high-temperature ceramics.

Materials technology is rapidly advancing. Ceramics will soon be available commercially which are capable of operating at temperatures up to 1650°C (3000°F). They may be competitive with or superior to metallic radiant tubes as a source of heat in high-temperature furnaces with the use of proper combustion systems. Coupling these technological developments with a compatible energy recovery system, it may become economically advantageous to apply natural-gas-fired systems to processes where electric heating has dominated. Additionally, because high-temperature capabilities for the tube imply greater rates of heat transfer, fewer radiant tube heating elements would be required and productivity may be increased.

### Objective

The objective of this program is to develop an indirect gas-fired heating system applicable to process temperatures above 1100°C (2012°F). This system is to have an overall fuel efficiency of at least 60% in order to be competitive with electric heating systems. The heat flux of these heating elements is to be in the range of 90 to 270 $kW/m^2$ (28 500 to 85 600 $Btu/h \cdot ft^2$) depending on process temperature. The heat flux capability of existing indirectly gas-fired combustion systems is approximately 25 $kW/m^2$ (7880 $Btu/h \cdot ft^2$). The greater heat flux will reduce the number of heating elements required for a specific application and increase productivity by virtue of the higher heating rates.

### Economic Advantage

There are several economic advantages of an indirectly gas-fired high-temperature heating system. For process temperatures above 1100°C (2012°F), operation of a gas-fired system with an overall energy efficiency of 60% would result in a reduction in energy costs over electricity for energy cost ratios of electricity-to-natural gas greater than 1.67 to 1. Additionally, in this operating range, fewer gas-fired heating elements would be required since they are to have a greater heat flux than electrical elements and generally have a greater surface area than the electrical element.

The indirectly gas-fired high-temperature heating system offers economic advantages for process temperatures below 1100°C (2012°F). The high-temperature ceramic from which the elements are to be constructed is believed to be capable of a longer heating-element life. Also, the higher heat flux capability of the indirectly gas-fired high-temperature heating system should result in increased heating rates and productivity. The energy cost for the high-temperature heating system would be less than the existing gas-fired system since the former is to incorporate a highly effective heat exchanger.

### Existing Equipment and Limitations

The limitations of existing indirectly gas-fired (radiant tube) heating systems were investigated with respect to achieving greater rates of heat transfer and higher energy efficiency. The analysis indicated that the energy flux which can be obtained from a high-temperature heating system is substantially higher than those being used today (see Fig. 1). For example, for a furnace operating at 1000°C (1832°F), it is possible to obtain an energy flux of 200 to 500 $kW/m^2$ (6 to $16 \times 10^4$ $Btu/h \cdot ft^2$), with the use of a proper radiant heating element.

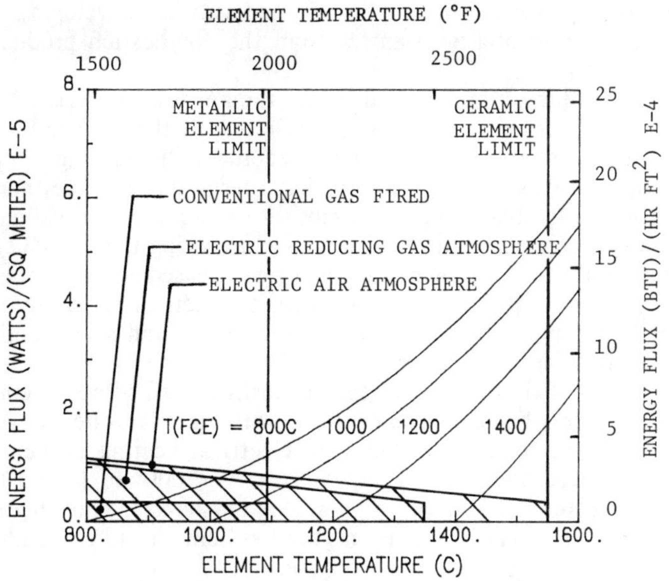


Fig. 1. Energy flux as a function of heating-element temperature for various furnace temperatures.

In the existing gas-fired application, the peak energy flux is 25 kW/m$^2$ (7930 Btu/h·ft$^2$).

The possibilities of achieving the high-energy fluxes, stated above, depend on four major developments which, when used in combination, will allow the furnace designer to obtain all the potential benefits.

These four developments are:

(1) A high-capacity gas-fired burner which can be used to fire at substantially higher rates with complete combustion within the heating element.

(2) An energy transfer device to transfer the released heat from the gases to the heating element or radiant tube surface.

(3) Proper material for the construction of the heating element which can withstand temperatures in the range of 1100° to 1650°C (2000° to 3000°F).

(4) A heat-recovery device which can be used to recover the sensible heat in the flue gases.

At this time, the technology for the heat-recovery device which can be used with the proposed heating system is well established. The other three components need some type of development.

As stated, existing gas-fired combustion systems (excluding immersion applications) deliver a maximum energy flux of approximately 25 kW/m$^2$ (7930 Btu/h·ft$^2$). This limitation is primarily due to the fact that existing radiant tube combustion systems do not have sufficient combustion intensity, i.e., they are not capable of increasing the quantity of fuel which may be burned inside the radiant tube. Therefore, development of a high-intensity combustion system suitable for radiant tubes is necessary.

Second, existing gas-fired systems are not necessarily capable of transferring energy from the gases to the furnace, assuming that greater combustion intensities were possible. This implies the necessity for developing a method to augment the rate of heat transfer from the combustion products to the radiant tube.

Third is the limitation imposed by the radiant tube material. The higher energy transfer by radiation will be possible only if the surface temperature of the heating element or source is increased above the limiting temperatures for the metallic tubes. It may be seen that energy flux may be increased by a factor of 4 or 5 by modifying the combustion system and providing a means of heat-transfer enhancement. However, by eliminating the 1100°C (2000°F) temperature limitation, the energy flux may be increased by substantially greater quantities. This requires the development of practical manufacturing techniques for high-temperature materials in the shapes and sizes applicable for the gas-fired furnaces.

Completion of the three development tasks described above could result in gas-fired systems becoming competitive with electric systems over a wide range of operating conditions. Presently, electrical heating elements are applicable to temperatures in the 1100° to 1650°C (2000° to 3000°F) range at considerably greater energy fluxes than gas-fired systems. Completing the above development tasks would result in gas-fired systems being applicable over a greater operating range than electric systems.

### Innovative Concept

In order that the indirectly gas-fired combustion system satisfies the objectives of this program, the four critical components described earlier must be developed and integrated into a combustion system. The developmental emphasis of this program is related to the high-combustion intensity burner and the high heat flux heating element.

The design for the high-intensity burner is based on increasing the combustion intensity by increasing the turbulence in the mixing region of the reactants and products of combustion. The turbulence in the flame zone is generated by causing the combustion air to flow through a cross section of significantly reduced area with a relatively high ratio of wetted perimeter-to-flow area. Typically, existing gas-fired radiant tube combustion systems achieve uniform temperature profiles along the tube length by generating a long, slow mixing-type flame. A uniform radiant tube temperature profile is needed to maximize the energy flux from the radiant tube. Since the maximum tube temperature is limited, energy flux can be maximized only when the entire tube surface is at this maximum temperature.

A closed-end radiant tube with axial fins was designed to increase the rate of heat transfer from the high-temperature gases to the furnace enclosure (see Fig. 2). The primary concept is that the burner is to fire into the inner tube. Combustion of the fuel is to be nearly complete as the gases make a 180° bend and flow in the annulus with the axial fins. The fins are the primary heat-transfer enhancement device. Their function is to receive energy from the high-temperature gases by radiation and convection and radiate this energy to the outer tube either directly or indirectly.

Where possible, ceramic materials have been used for high-temperature application. The outer tube was manufactured by Norton Company and is a silicon carbide. The inner tube was manufactured by Asahi Glass of Japan and is also a silicon carbide. The axial fins located in the annulus are made of Inconel 601. The next-generation laboratory prototype is to have ceramic fins.

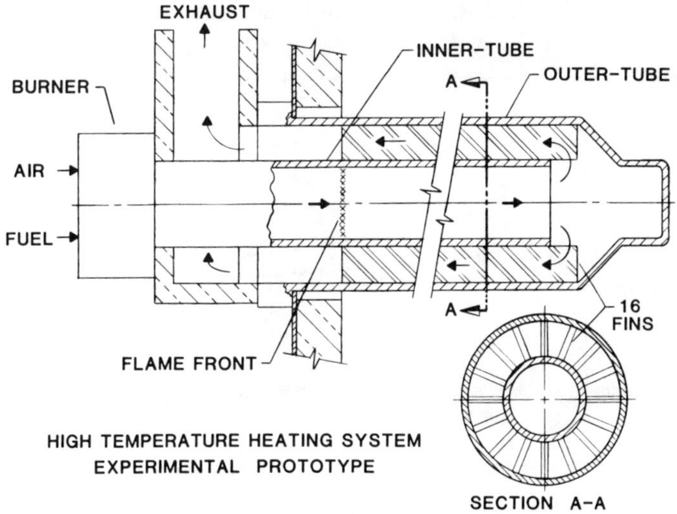


Fig. 2. High-temperature heating system experimental prototype.

### Test Setup

A test furnace has been fabricated for development and testing of the high-combustion intensity burner and the combination of the burner and closed-end radiant tube with axial fins arranged in the annulus. This test furnace is capable of operation at temperatures up to 1650°C (3000°F). An auxiliary heating and cooling system for the furnace makes it possible to maintain furnace temperature, independent of the heat input of the test burner. Therefore, a wide range of firing conditions may be simulated. Inspection ports are located horizontally along the side of the furnace in the plane of the radiant tube. This permits the radiant tube temperature to be obtained by optical pyrometry. Additionally, a combustion-air preheater is available for use with the test facility to simulate the use of a heat-recovery device.

### Test Results

#### High-Intensity Burner Tests

A series of experiments was performed with various burner configurations in order to achieve high combustion intensities. The most noteworthy result of these experiments was the dependence of combustion intensity on the velocity of the fuel entering the combustion zone relative to the combustion air. The experiments were performed in a circular tube which was 11.4 cm (4.5 in.) OD, 10.2 cm (4.0 in.) ID, and 152 cm (60 in.) long. Normally, a conventional radiant tube combustion system would burn $7.87\times10^{-4}$ $m^3/s$ (100 cfh) of natural gas, resulting in heat release of 30 kW in a tube of this geometry. This corresponds to a combustion intensity of approximately 2400 $kW/m^3$ (230 000 $Btu/h \cdot ft^3$).

The criterion for defining complete combustion in these experiments is that the exhaust gases must contain less than 0.001 mole fraction of methane equivalent combustibles. A combustion intensity of 25 000 $kW/m^3$ ($2.4\times10^6$ $Btu/h \cdot ft^3$) has been achieved by properly adjusting the relative velocities of the combustion air and fuel. This represents an order of magnitude improve-

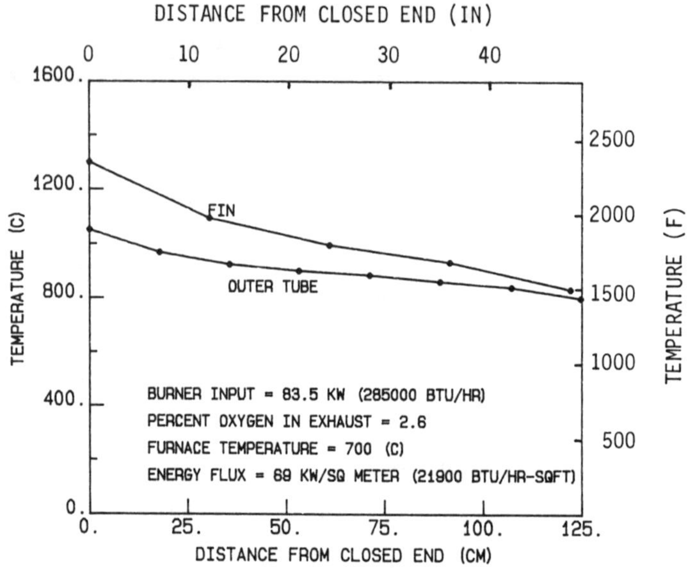


Fig. 3. Radiant tube temperature profile.

ment in the burner performance over the commercially available designs.

**Closed-End Tube with Axial Fins**

A series of experiments was performed with the closed-end radiant tube with axial fins using the high-combustion intensity burner. The purpose of these experiments was to quantify the effect of the increased firing rate and enhanced heat transfer on the heat transfer and thermodynamic efficiency of the system.

The heat flux from the outer tube was calculated by two methods. The first was based on determining heat transfer by computing the difference in total enthalpy between the entering and exiting gas streams. As a check on this calculation, the heat transferred from the surface of the radiant tube by radiation to the furnace was also computed. These two methods agreed within 10% for all experiments. The value for energy flux, shown in Fig. 3, is based on total enthalpy difference since the fuel flow rate as measured by an orifice meter is probably more accurate than the fourth-power temperature differences of the tube temperature (as a function of length) and furnace temperature as measured by a radiation pyrometer.

A significant increase in the average heat flux has been achieved as a result of the application of the high-combustion intensity burner and the axial fins. As stated, conventional systems are capable of a maximum energy flux of 25 $kW/m^2$ (7930 $Btu/h \cdot ft^2$). The greatest heat flux of the closed-end tube with axial fins is 69 $kW/m^2$ (21 900 $Btu/h \cdot ft^2$), representing an increase of a factor of 2.75 (see Fig. 4).

The thermodynamic efficiency of the laboratory prototype is comparable to that of a directly fired combustion system. This is a direct result of the burner construction and the design of the flue-gas exhaust system. It should be noted that no external recuperator was used. The use of recuperators will

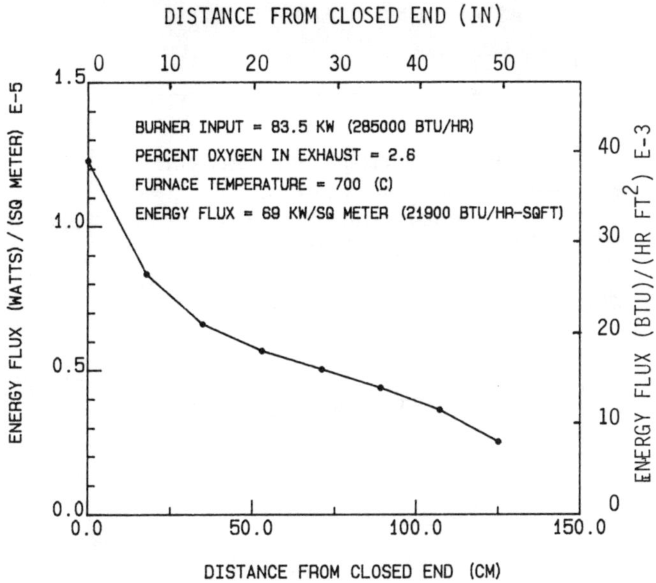


Fig. 4. Local energy flux profile.

further reduce the exhaust temperature and will increase the thermal efficiency.

The temperature and, therefore, the heat flux varies considerably along the length of the radiant tube (see Fig. 4). This is partially due to the relatively low mass flow of gases, resulting from the rate at which the burner was fired without exceeding the material temperature limitation for the metallic fins, and partially to the temperature limitation of the metallic fins. The relatively low mass flow rate causes significantly greater temperature gradients in the gas in the axial directions than would be experienced for higher mass flow rates.

A simplified numerical model has been developed which describes the heat transfer in the experimental prototype. The purposes of formulating this model are to determine a fin configuration which results in a uniform temperature profile along the tube length and to predict performance for systems constructed for high-temperature materials. To establish reasonable confidence in the predictions of the numerical model, it has been fit to the experimental data base.

A series of computer trials was performed using the numerical model. The intention of these trials was to determine the combination of fin lengths which would produce a uniform tube temperature profile. A fin configuration similar to that shown in Fig. 2 was selected, except that the fins were arranged in three lengths. Specifically, 4 fins extend 125 cm (50 in.) from the flame front plane, 4 other fins extend 60 cm (24 in.) from the flame front, and 8 other fins extend 35 cm (14 in.) from the flame front. This configuration was used to produce more uniform temperature profiles, as shown in Fig. 5. Using the fin configuration described above, and increasing the burner firing rate until the same peak temperature was obtained as the experiment with a 700°C (1300°F) furnace temperature, an energy flux of 88 kW/m$^2$ resulted, which

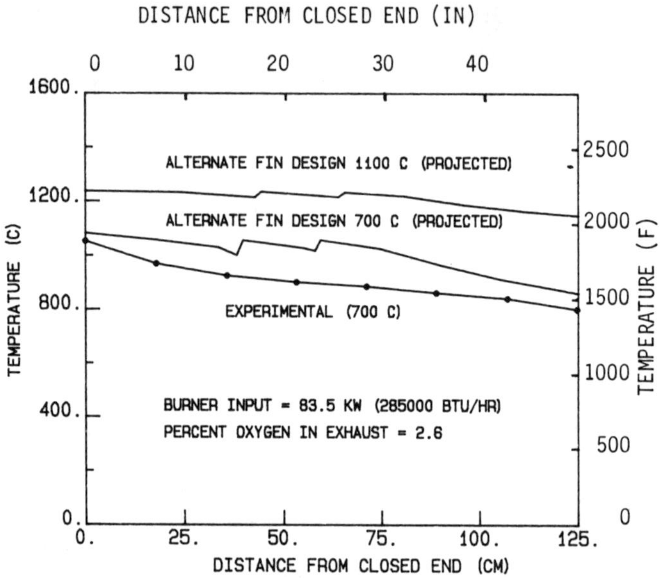


Fig. 5. Radiant tube temperature profile.

is 27% greater than the flux obtained in the experiment with uniform fin lengths. Additionally, it is projected that an energy flux of 72.5 kW/m$^2$ would result in a 1100°C (2000°F) furnace. This is three times the heat flux which can be achieved with conventional designs.

The computer model can be used to analyze a number of other options for heat-transfer enhancement devices. Further analysis, using such options, indicates that, with proper materials and configurations of the enhancement devices, the system can deliver heat fluxes of 500 kW/m$^2$, approximately 10 times higher than conventional units can produce.

**Conclusions**

Considerable progress has been made toward development of the essential components of the indirect gas-fired high-temperature heating system. A radiant tube burner has been developed which is capable of burning 10 times the fuel per unit of volume compared to existing radiant tube burners. This component is quite critical since large increases in heat flux require that more energy be released from the combustion unit volume of the tube. A hybrid (ceramic-metallic) closed-end radiant tube heating element has been tested which produces approximately three times the heat flux of conventional systems. Extrapolation of these results to an all-ceramic heating element with optimum fire design indicates that a heat flux of 72 500 W/m$^2$ (23 000 Btu/h·ft$^2$) may be obtained at furnace temperatures of 1100°C (2012°F).

The indirect gas-fired high-temperature heating system developed at Midland-Ross offers significant potential for improvement over conventional indirect heating systems. Analyses of such a system indicate that operation at high temperature could increase heat flux by at least one order of magnitude. The trend in this direction has been established during the experimental work carried out at Midland-Ross. This may increase productivity and reduce the ratio of radiant tube area to hearth area required for a given furnace. Additionally, the ability to operate at process temperatures above 1050°C (1922°F) with a suitable heat exchanger integrated into the system would allow natural-gas-fired systems to be economically competitive with electric heating systems.

# Ceramic Radiant Tubes

V. JAYARAMAN

Consultant
Livonia, MI 48154

High-temperature processes, which require special atmospheres or simply an air atmosphere free of products of combustion, are currently carried out in electrically heated furnaces. Ceramic radiant tubes offer a fuel-fired alternative to electric heating. They could also permit greater heat dissipation rates. There are many potential applications for ceramic radiant tube heating in the industry. However, successful industrial application demands certain essential properties in these tubes. These properties are listed.

## Heating of Atmosphere Furnaces

In industrial heating and heat-treating practice, it is often necessary to process the work in an air environment free of products of combustion, or in an environment containing specially prepared gases. Some examples are:

(1) Heating of steel with minimal oxidation and decarburization, prior to forging (1371°C (2500°F)).

(2) Sintering of powdered metal preforms (1287°C (2350°F)).

(3) Melting of nonferrous alloys with minimal oxidation, prior to casting (1204°C (2200°F)).

(4) Brazing of automotive, aircraft, and other components (1176°C (2150°F)).

(5) High-temperature carburizing of steel (1093°C (2000°F)).

(6) Clean hardening, annealing, and carburizing of steel (982°C (1800°F)).

The temperatures noted within parentheses above are commonly used for the various processes listed. Such processes can be carried out in furnaces heated by one of two methods.

### Electric Heating

Electric heating can be accomplished through the use of metallic resistors made of nickel-chromium or iron-chromium-aluminum alloys up to about 1093°C. Beyond this temperature, use of nonmetallic resistors like silicon carbide or intermetallic compounds like molybdenum disilicide is required. Special resistors made of graphite or refractory metals like molybdenum and tungsten are suitable for operation in a vacuum only.

Depending on the operating temperature and the environment in which the operation is carried out, metallic resistors can be used at dissipation rates of 6 to 31 $kW/m^2$ (2 000 to 10 000 $Btu/h \cdot ft^2$) of resistor surface. The allowable dissipation rates range from 15 to 93 $kW/m^2$ (5 000 to 30 000 $Btu/h \cdot ft^2$) for silicon carbide, and 15 to 155 $kW/m^2$ (5 000 to 50 000 $Btu/h \cdot ft^2$) for molybdenum disilicide.

Silicon carbide resistors are available in diameters up to 2 in. and effective heating lengths up to 8 ft. They are very fragile and must be handled with extreme care. Molybdenum disilicide elements are available in diameters up to 3/8 in. and effective heating lengths up to 4 ft. They are generally stronger than silicon carbide and require less care in handling.

All electrical resistance elements have certain limitations with regard to the environment in which they can be used. While there are many applications well suited to electric heating, there are many instances where electric heating may not be preferred, economical, or practical. In such cases, indirect fuel firing offers a viable option.

### Indirect Fuel Firing

There are two commonly used methods for separating the work from the products of combustion in a fuel-fired heating system:

(1) The entire work chamber could be a gas-tight muffle and the burners installed outside the muffle. Heat is then transferred to the work by radiation from the muffle.

(2) The burners could be arranged to fire inside gas-tight tubes which are located around the work. Heat is then transferred to the work by radiation from the tubes. Thus, the term "radiant tubes" has come to be applied to these tubes.

Except in the case of small work chambers or special applications, muffles are not used as often as radiant tubes these days.

### Heat Transfer from Radiant Tubes

For heat to be transferred from radiant tubes to a furnace enclosure (and to the work inside the enclosure), the tubes must run hotter than the furnace. From the laws of radiation heat transfer, for any given furnace temperature, the minimum temperatures of the outer surface of the tubes can be calculated for various dissipation rates. Calculated values for a few typical cases are:

| Furnace Temperature, °C (°F) | Dissipation Rate $kW/m^2$ | Outer Tube Temperature, °C (°F) |
|---|---|---|
| 982 (1800) | 15 | 1015 (1860) |
| 982 (1800) | 31 | 1048 (1920) |
| 1204 (2200) | 62 | 1282 (2340) |
| 1204 (2200) | 124 | 1354 (2470) |

It is important to bear in mind that these are the minimum tube temperatures calculated for ideal conditions that do not normally exist in practice. The actual temperatures could be significantly higher.

As the heat dissipated from the outer surface of the tube must come from the flame inside the tube, a further temperature gradient is set up between the outer and inner walls of the tube. The inner wall temperature can be calculated from the basic conduction equation:

$$Q = KA\Delta T/L \qquad (1)$$

where $Q$ is heat transferred through the wall, $K$ thermal conductivity of the wall, $A$ area through which heat is transferred, $\Delta T$ temperature difference between the outer and inner surfaces of the wall, and $L$ thickness of the wall.

### Metallic Radiant Tubes

If we consider the typical case of a metallic radiant tube with a ¼ in. wall and a thermal conductivity of 180 Btu per hour per square foot per °F per inch thickness, operating in a 982°C furnace at a dissipation rate of 15 mW/m$^2$ the $\Delta T$ through the wall of the tube can be calculated from the above equation as $-13.8°C$. This is not a significant gradient, and is to be expected in the case of metallic tubes with high thermal conductivity. As the outer tube temperature, in this case, was 1015°C, the inner wall temperature would be 1019°C (1867°F). As the products of combustion leaving the tube must be hotter than the inner wall (to be able to transfer any heat to the tube), they could run from 1037° to 1093°C.

As the strength of metallic radiant tubes drops off significantly after about 1037°C, the dissipation rates commonly used in good industrial practice are limited to about 15 kW/m$^2$ of tube surface at a furnace temperature of 982°C (1800°F). At lower temperatures, the dissipation could be as high as 31 kW/m$^2$ of tube surface.

#### Ceramic Radiant Tubes

If the tube in the above example had been made of alumina with a $K$ value of 12, the $\Delta T$ would have been 40.5°C (105°F), and the inner wall temperature 1073°C (1965°F). This, in turn, would have resulted in the products of combustion leaving the tube at a higher temperature. As the heat available for useful work inside the furnace is lower with higher exit gas temperatures, the overall efficiency of the furnace would have been adversely affected.

A study of Eq. (1) also shows that the $\Delta T$ through the wall of the tube is less with thinner walls. Thus, for overall furnace efficiency, it is desirable to use thin-walled tubes of high thermal conductivity. Lower temperature gradients also have a beneficial effect on tube life, as they result in less severe thermal stresses.

The thermal conductivities of some common ceramic materials are listed below, in Btu per hour per square foot per °F per inch thickness at a mean temperature of 1093°C:

| | |
|---|---|
| Silica | 12 |
| Alumina | 18 |
| Magnesia | 28 |
| Silicon carbide | 100 |

It is clear that silicon carbide is by far the best material for ceramic radiant tubes. Its use is expected to increase in the coming years for applications at temperatures above 1037°C, at which point metallic radiant tubes lose their practical value, because of the very low dissipation rates that would have to be used. At 1148°C (2100°F) and above, metallic radiant tubes cannot be used at all, as they would have no structural integrity.

### Applications for Ceramic Radiant Tubes

Possible applications for ceramic radiant tubes can be arbitrarily classified into two categories.

#### Large Diameter, High-Input Applications

Applications such as heating of steel for forging or melting of nonferrous alloys would fall in this category. For ceramic radiant tubes to be cost-effective in such cases, they should be available in large diameters of 6 to 10 in. and in effective heating lengths of 6 to 8 ft. Heat dissipation rates of 93 to 124 kW/m$^2$ of tube surface are desirable. Depending on the furnace temperature, these rates would translate into gross heat inputs of 145 000 to 435 000 W (500 000 to 1 500 000 Btu/h) per tube.

### Small Diameter, Low-Input Applications

Applications such as high-temperature carburizing and sintering would fall in this category. Low to moderate heat inputs are required in these cases, and the required tube sizes would range from 4 to 6 in. in diameter and 4 to 6 ft. in effective heating length. Gross heat inputs would range from 58 000 to 145 000 W (200 000 to 500 000 Btu/h) per tube.

### Cost of Ceramic Radiant Tubes

The present cost of an 8 in. OD by 7½ in. ID by 6 ft. long metallic radiant tube, with a dissipation capability of 15 kW/m$^2$ and an anticipated 3-year life, is around $400. As stated previously, the maximum practical use temperature of such a tube would be about 1037°C. A ceramic radiant tube would be required for operation at higher temperatures and/or higher dissipation rates. It is felt that a ceramic radiant tube of the same size, with a dissipation capability of 93 kW/m$^2$ and a 3-year life span, should be available in the price range $1000 to $1500, to be economically attractive to users. While this price might appear to be low, it should be recognized that a $1500 ceramic tube could turn out to be a total loss due to accidental mishandling.

At the present time, the prices for ceramic tubes of this size range from about $500 for an oxide-bonded tube (which is available in a minimum wall thickness of 1 in.) to about $1000 for a nitride-bonded tube (which is available in a minimum wall thickness of 3/4 in.). Tubes with thin walls get progressively more expensive. Recrystallized silicon carbide tubes, made by the slip casting technique, with a wall thickness of ¼ in., run about $4000. Extruded tubes, when available, are not expected to be any less expensive. Irrespective of the price and wall thickness, all currently available tubes seem to have a useful life of three to nine months.

The most promising development in the area of silicon carbide radiant tubes, at this time, appears to be the chemical vapor deposition (CVD) process of making tubes. More than one organization in the United States is currently engaged in making tubes by this process. It is expected that tubes made by the CVD process will be available in the not-too-distant future in wall thicknesses as low as 1/16 in. to 1/8 in. The CVD material is also expected to possess superior thermal conductivity and thermal shock resistance, and could open up the possibility of immersion heating of metal baths.

In the CVD process, silicon carbide is deposited in the vapor form on a substrate such as graphite or ceramic fiber. The substrate could be either stripped after deposition, as in the case of graphite, leaving a so-called free-standing tube, or it could form part of the final tube, as in the case of ceramic fiber. While work in this area is in a preliminary stage and much testing remains to be done, indications are that tubes with the substrate material left in are stronger and last longer in use than free-standing tubes. They also seem to be more resistant to thermal shock. Adequate data have not been collected so far on the thermal conductivity of these tubes, and it remains to be seen if the substrate material would detract from the otherwise excellent thermal conductivity that CVD silicon carbide is expected to possess.

### Existing Installations

Alumina radiant tubes have been used for a number of years by a U.S. heat-treating furnace manufacturer. The use of silicon carbide tubes in high-temperature industrial applications appears to be limited to the work done under the sponsorship of Columbia Gas System Service Corporation, Consolidated Natural Gas Service Company, and Southern California Gas Company.

A 3000 lb/h forging furnace, fitted with 10 oxide- and oxynitride-bonded silicon carbide tubes, 9½ in. OD by 7½ in. ID by 6 ft long, with a total gross input of 1 160 000 W (4 000 000 Btu/h), was built under the sponsorship of these companies. The thick walls of the tubes proved to be an insurmountable bottleneck in transferring the heat to the stock, which could be heated to only about 1024°C. The forging operation required 1287°C (2350°F) in the stock. The test was, therefore, terminated. Further work is currently in hand to retrofit this furnace with better quality, thin-walled tubes, and go through another test.

A 1500 lb/h aluminum melter, also built and tested under the sponsorship of these companies, proved to be a success. This melter was equipped with four nitride-bonded silicon carbide tubes, 9½ in. OD by 8 in. ID by 6 ft long, with a total gross input of 870 000 W (3 000 000 Btu/h). The unit has been in production at an aluminum foundry for over two years. The benefits of indirect firing have been amply demonstrated by the superior quality of the metal obtained from this unit. Figure 1 shows the aluminum melter.

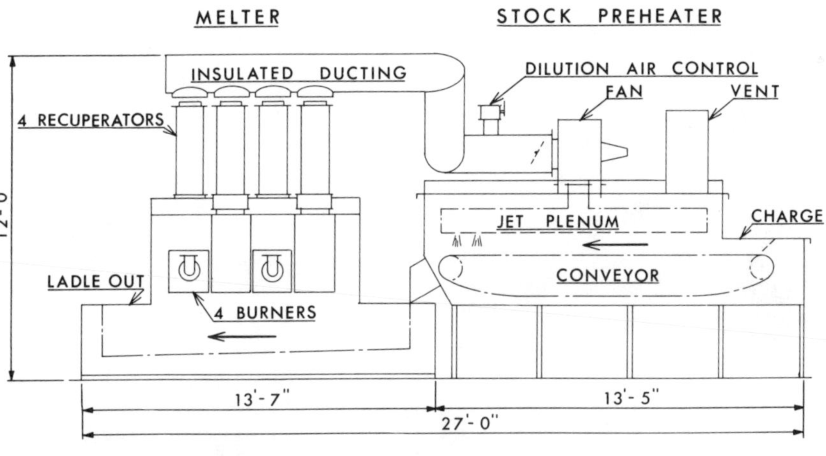


Fig. 1. Ceramic radiant tube-heated aluminum melter.

### Properties Required in Ceramic Radiant Tubes

For ceramic radiant tube heating to become successful in industrial practice, the following properties are essential in the tubes:

(1) Structural integrity at temperatures up to 1648°C (3000°F) and in horizontal, unsupported spans up to 8 ft.

(2) Thermal conductivity in the region of 150 to 200 Btu per hour per square foot per °F per inch thickness, at a mean temperature of 1371°C.

(3) Adequate thermal shock resistance to withstand the daily high-fire startup and total shutdown common in the industry.

(4) Impermeability under gas pressures up to 2 psig.

(5) Compatibility with commonly encountered gases, such as products of

combustion and prepared atmospheres containing CO, $H_2$, and hydrocarbons, with dew points down to −17°C (0°F).

(6) Wall thicknesses of ¼ in. or less.

(7) Reasonable strength to withstand normal handling.

(8) Reasonable cost.

(9) Acceptable life.

Also important is the simultaneous development of suitable burners and modular recuperators. The burners must be capable of releasing the heat uniformly along the effective lengths of the tubes, and the recuperators are an important prerequisite for obtaining an acceptable thermal efficiency from the inherently inefficient, indirect-fired, high-temperature heating processes.

# Ceramic Radiant Tube Systems In Use and Development in Europe

M. K. STEFFENS

Norton Co.
Wesseling, Federal Republic of Germany

Ceramic radiant tubes are in widespread use in Europe and are undergoing significant development in both design and materials. A review of the current status is presented in relation to systems design and materials selection for high-temperature heat-treating processes for metals.

Ceramic radiant tubes have been in service in Europe for many years. Their popularity is due to the fact that they allow the use of natural-gas fuel to provide heat for controlled-atmosphere processes. The alternative has typically been electricity at significantly higher cost.

Because of the long-standing high cost of energy in Europe, it is there that much heat-exchanger technology—including radiant tubes—has its roots and most established usage.

Some of the technology is not yet attractive in the United States based on lower energy costs, but a good base of experience has been established and newer materials are being incorporated for higher efficiencies. Therefore, an expanded role for radiant tubes is foreseen worldwide.

This paper refers to the development of furnaces with radiant tubes for high-temperature indirect heating in the metallurgical industry. Typically, radiant tubes for heat-treating, carburizing, or annealing have been metal, of closed-end design, and used in a furnace such as that shown in Fig. 1.

These high-temperature heat-treatment processes are under constant development for metallic materials requiring indirect heating for precise control of the atmosphere.

The general trend of energy savings has greatly influenced the development of heat-treatment furnaces with radiant tubes designed for traditional heat-treatment processes such as carburization, recarburization, carbonitriding, nonoxidizing annealing, normal nonoxidizing annealing, and bright annealing.

However, new technical fields such as space travel technology, semiconductor technology, and the development of new magnetic materials have promoted new heat-treatment processes.

### Development of Ceramic Radiant Tube Systems

Two main reasons are forcing a move away from metal radiant tubes: (1) the need for more uniform heat for increased production and (2) recuperation of waste heat which results in a higher flame temperature.

According to Fritsch,[1] there is—at a distance of 100 to 150 mm from the entrance point of the flame into the inner tube—a substantial excess temper-

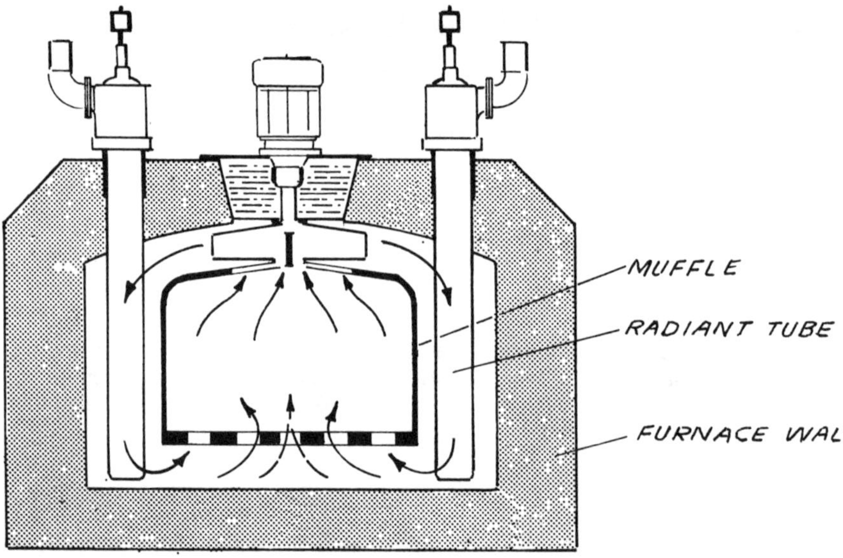


Fig. 1. Typical radiant-tube-heated furnace cross section.

ature which, under high operating furnace temperatures, can destroy the inner tube of the radiant tube system.

Development of the radiant tube systems must solve the following objectives:

(1) Testing and determining new materials which meet the functional requirements of the radiant tube systems for temperatures which exceed the application ranges/limits of the heat-resistant chrome nickel steel in consideration of the benefit/cost factor.

(2) Changes in construction and design improving the functioning of the radiant tube system and applicable for metallic as well as ceramic materials.

The two important functioning parts of the radiant tube systems can be classified based on the temperatures of application (furnace temperature)[1]:

(1) Up to 1100°C: Outer and inner tube of cast iron or heat-resistant chrome nickel steel.

(2) From 1100° to 1170°C: Outer tube of heat-resistant steel, flame (inner) tube of ceramic material.

(3) From 1170° to 1250°C (and higher): Outer tube and flame tube of ceramic material.

See Table I for comparative thermal properties of selected materials.

**Conventional Design**

The conventional radiant tube system, as illustrated in Fig. 2, is manufactured of cast iron or heat-resistant steel. The inner tube is more or less tightly connected to the burner. With well-adjusted combustion, two areas of higher temperature will be generated along the tube, the first 100–150 mm behind the burner and the second on the end of the inside tube—the region in which the exhaust gases turn back. Depending on the geometry of the tube

Table I. Comparison of the Thermal Properties of Selected Materials Utilized for the Construction of Inner and Outer Tubes of the Radiant Tube Systems

| | Heat-Resistant Steels e.g., G-X40CrNi24-5 | Pythagoras 1800 Z (78% Alumina) | Recrystallized SiC |
|---|---|---|---|
| Density g/cm$^3$ | 7.5 | 3.8 | 2.7 |
| Thermal expansion ($K^{-1}$) 400°–1000°C | 13–16.5×$10^{-6}$ | 4.8–6×$10^{-6}$ | 4.8×$10^{-6}$ |
| Specific heat (J/kg·K) | 0.50 | | 0.35 |
| Thermal conductivity 20°C (W/m·K) | 16.7 | 2.5 | 27 |
| Total emissivity | | ≈0.2 | 0.8 |
| Max. operating temp. (°C) | 1100°C | 1800°C | 1600°C |

system, the fuel type, and the useful output, there will be two over-heated regions in which the inner tube will be destroyed.[1]

#### Recirculating Design

The restoring radiant tube system (see Fig. 3) is provided with an impulse burner and the injective recirculation of some of the exhaust gases leads to a more uniform temperature. The outer tube is made of heat-resistant steel casting. The inner (flame) tube is made of recrystallized silicon carbide. This

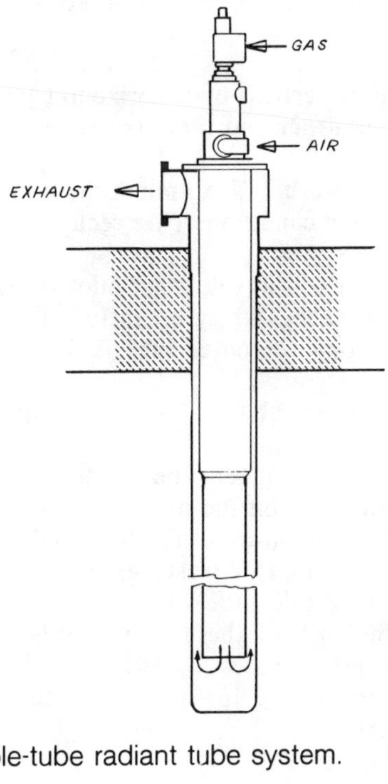


Fig. 2. Basic double-tube radiant tube system.

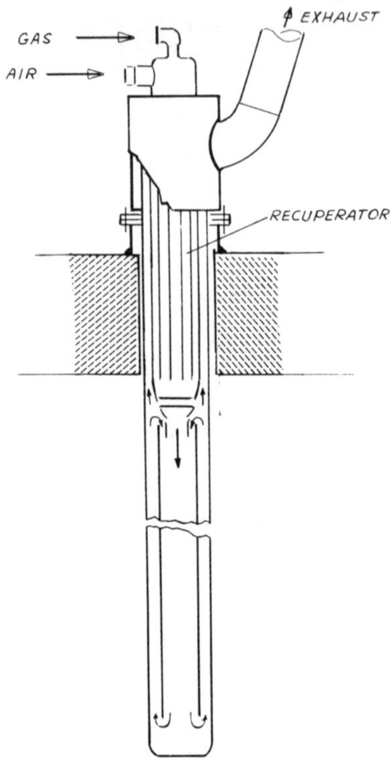


Fig. 3. Recirculating radiant tube system with integral recuperator.

system may be installed in a vertical or a horizontal position by ensuring a constant space between the burner and the inner tube. This figure also shows an integral recuperator.

Initially, the inner tubes were 78% alumina; these tubes did not meet the requirements of the application conditions after recuperative burners had been installed in the radiant tube system.

The preliminary one-year test in the Ipsen pilot station was confirmed by the application results of furnaces during operation. The basic results were[2]

(1) The temperature along the outer tube is determined exclusively by the inner tube.

(2) The inner tube transfers about 97% of the total heat transferred to the outer tube by radiation.

(3) The temperature along the inner tube is determined by the combustion process which—in turn—depends on the mixture of gas and air, the velocity of streams in the inner tube, the nature of the fuel, and the preheating degree of the combustion air. Restoration of waste gases leads to a more uniform temperature curve along the inside tube.

(4) The degree of efficiency of these radiant tube systems, considering also muffles in the heat transfer process, is 74%.

As an example of an operating furnace with radiant tube systems constructed with an inner recrystallized SiC tube:

Application: Annealing

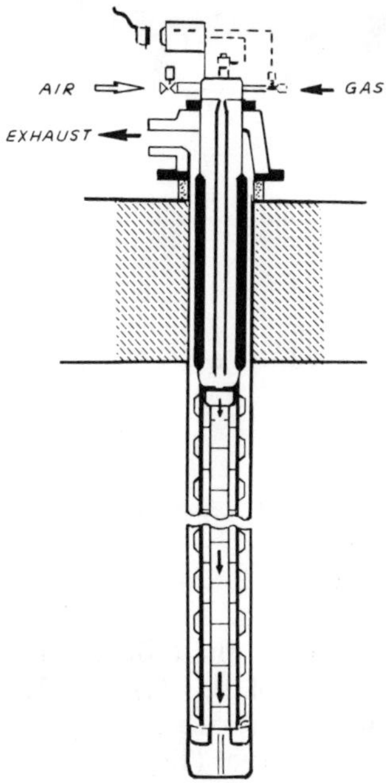


Fig. 4. Recirculating radiant tube system with "flexible" inner tube.

Operating temperature: Every cycle for 2 h at 950°C
Operating treatment cycle: 10 cycles/24 h
Operating period: Continuously over two years
Temperature gradient in the flame tube: 1450°–1100°C
Atmosphere: Oxidizing

This application demonstrates the capability and suitability of recrystallized SiC tubes as inner tubes for radiant tube systems.

### Flexible Inner Tube Design

Another type of radiant tube system uses interlocking segments for the inner tube, as illustrated in Fig. 4.[3,4] The "flexible" inner tube was designed to eliminate the mechanical stress caused by the plastic deformation of the steel outer tube during the service life of the tube.

The relatively thick wall design and the cost factor initially determined the choice of chamotte as the ceramic material. However, because its mass, poor conductivity, and premature breakage make it undesirable between the inner and the outer tube, the chamotte segments are being replaced with thin-wall silicon carbide segments.

### Improved Recuperative System Design

Newer radiant tube systems are currently being used and are optimized for the complete range of heat-treatment processes of ferrous and nonferrous materials, i.e., furnace operating temperatures above 1200°C are included: for processes up to 1100°C, the outer and inner tubes are made of cast iron; (1) for processes between 1100° and 1170°C, the inner tube is made of recrys-

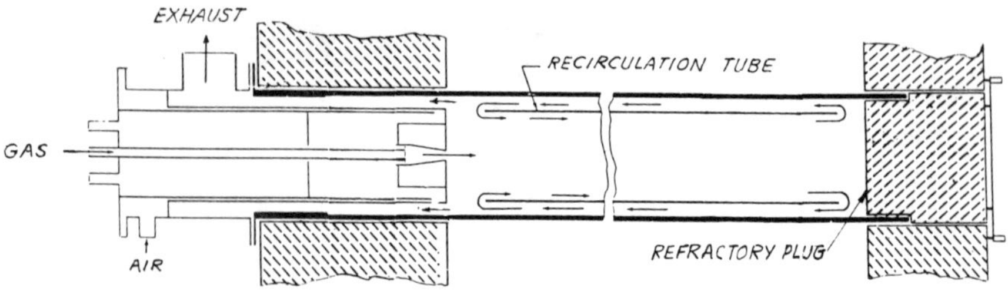


Fig. 5. Recirculating radiant tube system using an open-on-both-ends tube with a refractory plug.

tallized silicon carbide; (2) for processes above 1170°C, both outer and inner tubes are to be made of SiC.

This system presents the following innovations: (1) special construction of the burner for generating a very long, slim flame and (2) a supplementary recuperative source as a result of the tubular construction.

British Gas has a version of this type of design which uses a straight, open-on-both-ends tube with a refractory plug in one end, held in place by spring tension to ensure a good seal (Fig. 5).

These designs, as well as the recent designs of Ruhr-Gas, have demonstrated that an accurate geometry of the tubes leads to an integrated construction which can be manufactured in either steel or ceramic material (silicon carbide) as the process requires.

## Summary

Ceramic radiant tube systems are in active use and development in Europe and thin-walled SiC tubes and special shapes are helping improve the performance and energy savings of modern high-temperature systems.

## References

[1]W. H. Fritsch, "Design and Dimensioning of Industrial Burners," *Wärme-Gas Int.*, **32** [5] [6] (1983).

[2]H. Pfau, "Operating Principle of Innovative Recuperative Radiant Tubes under Actual Operating Conditions," *Wärme-Gas Int.*, **32** [10] (1983).

[3]J Wünning, "Radiant Tube Heating Optimized," *Wärme-Gas Int.*, **25** [11/12] (1976).

[4] h L. Wärne Firing of Industrial Furnaces with Rekumat Burners." *Wärme-Gas Int.*, **29** (1980).

# Natural Gas-Fired Immersed Ceramic Tube Aluminum Melter

BERNARD WRONA AND STEVEN SCHENCK

Babcock & Wilcox Co.
Lynchburg, VA 24505

DAVID KNOWLES

Gas Research Institute
Chicago, IL 60631

RICHARD SCHREIBER

Alzeta Corp.
Mountain View, CA 95050

CHARLES HART

Reynolds Metals Co.
Richmond, VA 23261

A program has been initiated to evaluate advanced ceramic materials and select a ceramic material that will survive (*a*) the hostile environment provided by molten aluminum and corrosive fluxes, and (*b*) natural gas-fired, high-temperature oxidizing conditions. Design and development efforts on a unique gas-fired fiber burner have been initiated to supply heat to the immersed ceramic tube aluminum melter. Initial results of the program are discussed.

Low thermal efficiencies of existing aluminum melting furnaces coupled with escalating natural gas prices are serious problems facing the aluminum industry. Industry estimates of reverberatory furnace efficiencies range as low as 18–25%. The combination of low furnace efficiency and high fuel costs produces significant economic incentives to develop a melting furnace to replace current, obsolete reverberatory aluminum melting furnaces.

Several approaches can be used to improve operating efficiency and production capability of gas-fired aluminum melting process equipment. Operating efficiency can be markedly increased by recovery of heat from flue gases for (*a*) preheating combustion air and (*b*) preheating metal charge. In addition, the inefficient radiant-heat transfer mode used in conventional reverberatory furnaces can be improved through use of immersed ceramic tube heaters. Specifically, an examination of the energy flow within a typical reverberatory melter furnace results in estimates that heat losses up the exhaust flue can range between 50 and 75% of heat input, 12 to 15% loss through walls of the furnace, and 8 to 10% in other losses. By using an immersed ceramic tube melter that provides heat by conduction into the melt, as opposed to radiation on the surface of the melt, substantial gas savings are possible due to decreased exhaust gas temperature. In theory, furnace efficiency can be further improved to 51% using waste-heat recuperation.

The first generation of advanced aluminum holding furnaces (using the immersed ceramic tube concept) is undergoing industrial development at the present time. Natural gas-fired, immersed ceramic tube burner-type holding furnaces appear to be a significant advance in the state of the art of holding-furnace technology. However, from a melting furnace point of view, major limitations to the emerging aluminum holding furnace exist, e.g., (*a*) brief lifetime of the ceramic tubes, typically 1–3 months and (*b*) insufficient heat flux to melt aluminum efficiently.

The Gas Research Institute (GRI) has therefore funded a multiparticipant program to develop a natural gas-fired immersed ceramic tube aluminum melter. Major thrusts of this program are to evaluate current and recently developed ceramic materials, and to identify materials that will survive in (*a*) the high-temperature oxidizing conditions on the inner surface and (*b*) the corrosive molten aluminum on the outer surface. In addition, development and testing of an improved burner is being pursued so that furnace heat flux may be increased significantly in order to melt aluminum at viable production rates. Other specific objectives of the program are to study the performance envelope of (*a*) current gas-fired aluminum holding furnaces and (*b*) commercially available ceramic tubes in a holding furnace environment. The overall, long-term objective of this effort is to demonstrate the technical and economic viability of a furnace concept using preheated air with natural gas-fired immersed ceramic tubes to melt aluminum, and to promote commercialization of such a furnace system.

Potential aluminum industry process improvements to be derived from this work include (1) reduced melt loss, (2) reduced air pollution treatment requirements, (3) decreased need for frequent fluxing, and (4) lower cost of furnace maintenance.

An added benefit of this program is that a materials properties data base will be developed for numerous ceramic materials for the application described above. The purpose of this report is to present an overview of the program and preliminary results of the work sponsored by the Gas Research Institute to advance aluminum melting technology.

**Project Organization**

To accomplish these objectives, the GRI project "Natural Gas-Fired Immersed Ceramic Tube Aluminum Melter" was divided into three phases:

Phase I: Advanced Ceramic Materials Evaluation and Selection
Phase II: Prototype Furnace Evaluation
Phase III: Large Scale Furnace Construction and Evaluation

Also, a project team was formed consisting of Babcock & Wilcox Company, Reynolds Metals Company, and Alzeta Corporation. In Phase I, Reynolds Metals Company will study the performance envelope of a currently available aluminum holding furnace. Alzeta Corporation will provide fiber burners as substitutes for conventional burners in the aluminum holding furnace. Babcock & Wilcox has the responsibility of program management and tubing material evaluation and selection. Specific tasks to be performed in Phase I are summarized below:

(1) Materials screening (Babcock & Wilcox).
(2) Combustion system burner retrofit (Alzeta).

(3) Advanced melting test-furnace combustion systems (Alzeta).
(4) Holding furnace performance evaluation (Reynolds).
(5) Melting furnace evaluation (Reynolds).
(6) System requirements.
(7) Project management.

With the project approximately four months into Phase I, the following sections highlight the results accomplished.

**Ceramic Tube Design Criteria**

Before candidate tube materials were selected for screening tests, design criteria were established. These requirements must be met if a ceramic material is to be used as a component for a natural gas-fired immersed heater system to melt aluminum. The design criteria established were as follows:

(1) Good combination of thermal, physical, and mechanical properties for the intended application.

(2) Compatibility with molten aluminum.

(3) Acceptable retention of properties after long exposure to molten aluminum.

(4) Minimal attack in dynamic, clean molten aluminum.

(5) Acceptable retention of properties after exposure to dynamic, clean molten aluminum.

(6) Minimal attack during exposure to dynamic molten aluminum with flux present.

(7) Acceptable retention of properties after exposure in dynamic, molten aluminum with flux present.

(8) Good resistance to oxidation and attack in a natural gas-fired high-temperature environment.

(9) Acceptable retention of properties after exposure to natural gas-fired high-temperature conditions.

(10) Adequate resistance to thermal shock.

(11) Adequate heat-transfer properties to achieve required melting rates.

This preliminary list of material acceptance criteria will be finalized during the program. Ease or cost of fabrication is not listed in the acceptance criteria because it is assumed that economical fabrication of advanced ceramics in large geometries is not an issue in this phase of the program, i.e., fabrication aspects will be considered in a later phase.

**Description of Candidate Materials**

According to the literature,[1–4] the primary wear mechanism of refractories in contact with molten aluminum is chemical reduction of the silicate components of the bond phase and/or aggregate to produce aluminum oxide and free silicon metal. This mechanism suggests that silicate (i.e., clay)-bonded materials are not suitable for this application, and attention must be directed to more chemically stable oxides, nitrides, and carbides. In addition, the expected thermal conditions indicate that nitrides and carbides would also be advantageous in terms of their resistance to thermal shock and thermal stress.

Selection of the list of candidate test materials was based on recommendations by commercial vendors for the immersed ceramic tube burner application. Because of silicon carbide's unique combination of strength, thermal properties, and resistance to aluminum corrosion, most of the materials were silicon carbide based. A wide variety of types of silicon carbides are undergoing fabrication development; therefore a brief description of silicon carbide ceramics is presented below.

Sintered alpha and sintered beta silicon carbide ceramics are prepared by sintering high-purity SiC with small additions (<1wt%) of sintering aids. The powders may be formed by cold-pressing, extrusion, slip casting, etc., into the required shape, followed by sintering to high density (i.e., greater than 98% of theoretical density). Because of relatively low porosity and the limited presence of second phases, the materials are typically stronger than other SiC ceramics and impermeable to gases. In sintered alpha SiC, the major crystalline phase present is hexagonal. Transformation of beta-phase SiC to alpha SiC occurs during extended firing at high temperatures (i.e., greater than 2000°C).

Recrystallized SiC, also referred to as pressureless sintered SiC, is considered to be high-purity material, i.e., greater than 99% SiC. Powders are formed into desired shapes, minus any sintering aids, then densified by pressureless sintering. However, the sintered part may contain as much as 20 vol% retained porosity, so the part may be permeable to gases. In addition, the high porosity results in substantially lower strength when compared to sintered alpha or sintered beta SiC.

Reaction-sintered or siliconized SiC is fabricated by sintering primary SiC with silicon and carbon. Sintering causes the silicon and carbon to form secondary SiC grains (i.e., reaction-sintered). As an alternative, a presintered SiC and carbon body may be infiltrated with silicon, followed by a subsequent sintering that causes the silicon to react with carbon to form secondary SiC grains (i.e., siliconized). The microstructures are characterized by primary SiC grains in a matrix of smaller, secondary SiC grains, free silicon, and porosity. The quantity of free silicon and porosity can be controlled somewhat with appropriate fabrication methods.

Nitride-bonded SiC is fabricated by consolidating a mixture of primary SiC grains and silicon, followed by heating in a nitrogen atmosphere to convert the silicon to silicon nitride. The microstructure is characterized by primary SiC grains in a matrix of silicon nitride that contains as much as 15 vol% pores. Strength of the material is due to self-bonding among silicon nitride grains and also between silicon nitride and SiC grains. However, the material is typically considerably weaker than the SiC ceramics described previously and may be permeable to gases.

Oxide-bonded SiC is fabricated by sintering previously mixed primary SiC grains with a silica-rich oxide phase. Bonding between silica grains and between silica and primary SiC grains occurs during the sintering operation. The microstructure is characterized by primary SiC grains in a highly porous (ranging from 14–20% pores) oxide matrix.

Chemically vapor-deposited (CVD) SiC composites are usually fabricated by depositing a SiC coating on an appropriate substrate, such as an alumina-silica fiber matrix or graphite. Two common gas precursors utilized for deposition are $CH_3SiCl_3H_2$ and $SiCl_4H_2CH$. The reactions are generally performed at $T > 1400°C$. Thermal decomposition of methyltrichlorosilane ($CH_3SiCl_3$), which contains both Si and C atoms, is achieved in a hydrogen atmosphere. Reduction of silicon tetrachloride ($SiCl_4$) by hydrogen is achieved through use of a hydrocarbon gas such as methane. Columnar grain structures are typical of CVD-processed coatings. When performed under optimum CVD process conditions, the SiC coatings are high in density and low in impurities.

Table I. Commercial Ceramic Materials for Use in Test Program at Babcock & Wilcox

| Tube ID | Vendor | Type | Fabrication | Density ($g/cm^3$) | Porosity (%) |
|---|---|---|---|---|---|
| A1 | Amercom | CVD | SiC-coated Nextel Fiber | 2.71 | 10.5 |
| B1 | B&W | Reaction-bonded | Rx-bonded $Si_3N_4$ | 2.35 | |
| B2 | B&W | Reaction-bonded | Rx-bonded $Si_3N_4/Al_2O_3$ | 2.59 | |
| B3 | B&W | Reaction-bonded | Rx-bonded $Si_3N_4$/SiC | | |
| C1 | Carborundum | Alpha SiC (Hexolloy) | Slip-cast sintered w/ fine grain | 2.99 | 1.1 |
| C2 | Carborundum | N-bonded SiC | Nitrided SiC+Si | 2.61 | 18.0 |
| D1 | Coors | N-bonded SiC(TN–15) | Nitrided SiC+Si (coarse grain) | 2.51 | 19.6 |
| D2 | Coors | Reaction-sintered (SCRB–205) | Siliconized recrystallized SiC (Si,C,SiC) | 3.05 | .4 |
| I1 | Ferro | $Al_2O_3$-coated SiC | | | |
| I2 | Ferro | Borosilicate glass-coated SiC | | | |
| E1 | I2R | N-bonded SiC | Nitrided SiC+Si | 2.26 | 26.6 |
| E2 | I2R | Alpha SiC | Sintered $\alpha$-SiC w/ coarse grains | 2.46 | 23.8 |
| E3 | I2R | Beta SiC | Sintered carbon+SiC | 2.71 | 0.8 |
| F1 | Mitsubishi (Asahi Glass Ceramics) | Alpha SiC | Sintered $\alpha$-SiC | 3.08 | 0.2 |
| G1 | Norton | N-bonded SiC (CN 983) | Nitrided SiC+Si | 2.44 | 21.0 |
| H1 | Thermo Electron | CVD | SiC-coated graphite tube | | |

In addition to the silicon carbide-based materials, several other reaction-bonded materials will be evaluated in the test program. These materials include reaction-bonded $Si_3N_4$, $Si_3N_4/20\%\ Al_2O_3$, and $Si_3N_4/20\%\ SiC$. These materials were prepared by nitriding slip-cast bodies of fine-grained silicon with 20% $Al_2O_3$ or SiC added, depending on the composition.

**Test Materials**

Table I is a list summarizing the types of materials purchased and being evaluated in the program. Comments on fabrication and some measured property information are provided where information is available from commercial suppliers. Figure 1 is a photograph showing the geometry and the approximate total number of tubes procured from various commercial vendors for the test program.

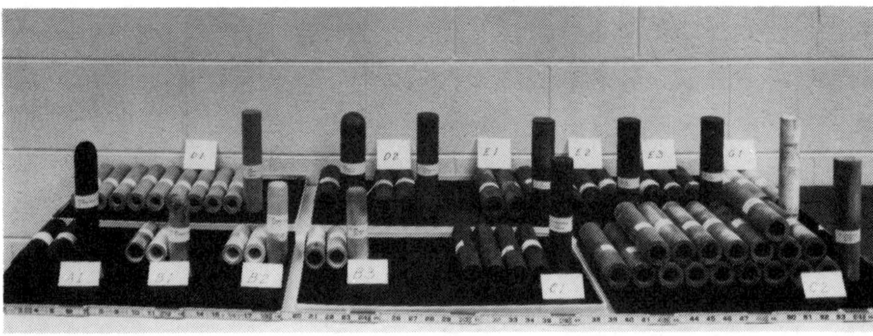

Fig. 1. Ceramic test specimens.

Fig. 2. Holding furnace for testing materials (manufactured by Burns Engineering).

## Description of Test Furnace

An aluminum holding furnace is basically a reservoir (with a limited heat source) for receiving molten metal from a melting furnace. A holding furnace will maintain the molten bath of metal in the liquid state until needed for pouring into molds for castings, ingots, etc. However, typical holding furnaces are not capable of supplying sufficient thermal energy to melt aluminum at acceptable throughput rates starting from room temperature.

Figure 2 is a photograph showing a recently developed, commercially available aluminum holding furnace that uses immersed ceramic tube, natural gas-fired burners as a source of heat. The furnace has a holding capacity of approximately 4000 pounds and utilizes two immersed ceramic tube burners as a heat source. The furnace was purchased primarily to serve as an apparatus for simultaneously exposing numerous, small, commercially available candidate ceramic tubes in screening tests in molten aluminum under controlled conditions (e.g., time, temperature, flux, etc.). Figure 3 is a schematic drawing showing more detail of the furnace interior. Furnace performance envelope information obtained with the furnace will also be used in the development of an optimized aluminum melter.

## Experimental Plan

Two types of tests for exposing ceramic specimens to molten aluminum are planned for Task 1, i.e., a static test and a dynamic test. The static test

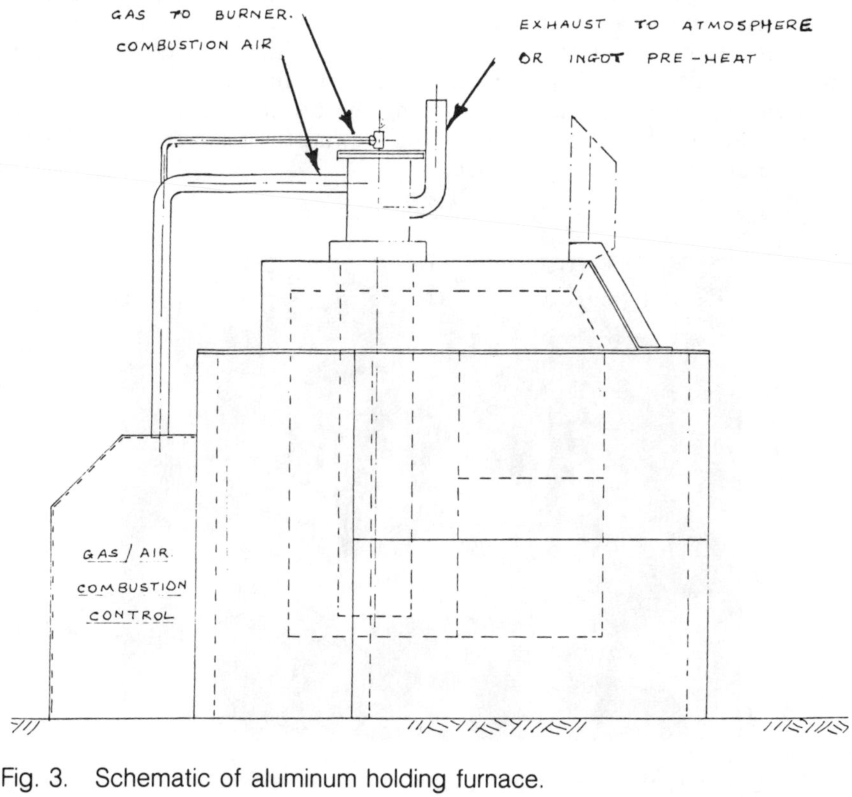


Fig. 3. Schematic of aluminum holding furnace.

Fig. 4. Static test specimen holder.

specimen holder is shown in Fig. 4 with a few test specimens in place. The specimen holder is basically a metal grate used to suspend ceramic tubes into the molten metal.

Dynamic testing of ceramic tubes is planned because industrial experience has shown that motion of molten aluminum past exposed ceramic material surfaces is an important factor contributing to premature failure of ceramic components in current aluminum reverberatory melting furnaces. Because of experimental difficulties (e.g., limited space in the furnace), this erosion/corrosion effect will be simulated by rotating the ceramic tube samples in the molten aluminum. Figure 5 is a photograph showing the dynamic sample holder with several tubes suspended in place.

Fig. 5. Dynamic test specimen holder.

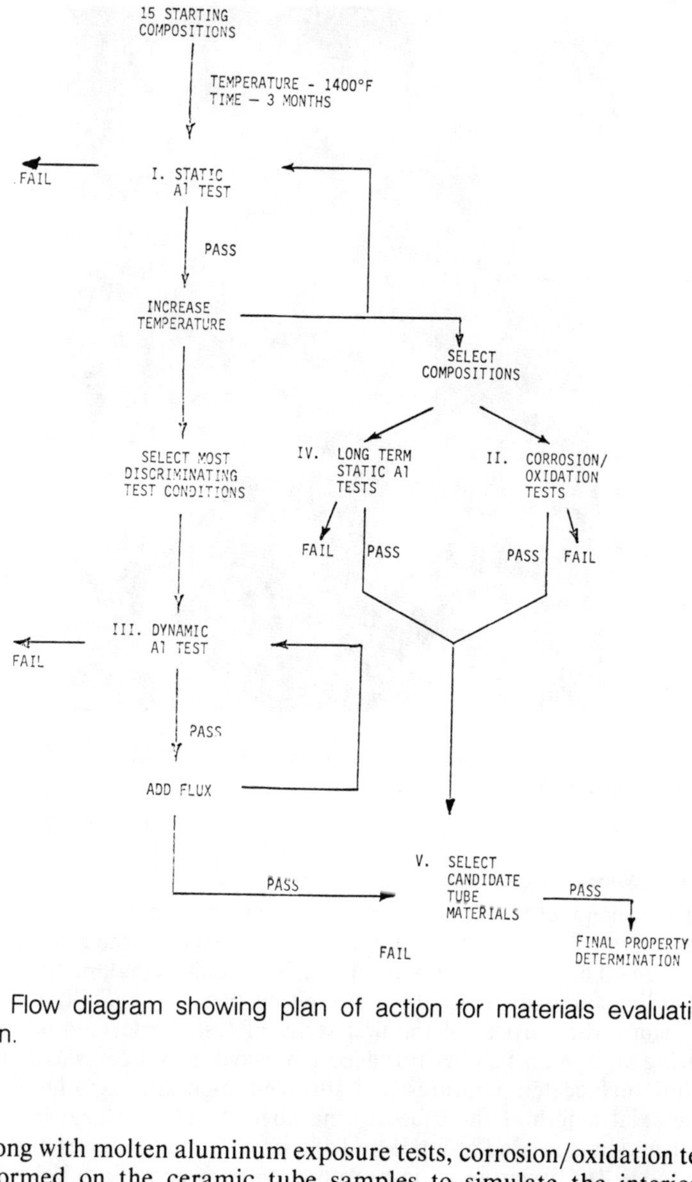


Fig. 6. Flow diagram showing plan of action for materials evaluation and selection.

Along with molten aluminum exposure tests, corrosion/oxidation tests will be performed on the ceramic tube samples to simulate the interior of an immersed burner tube. The methodology to be followed in conducting all of the screening tests is shown in Fig. 6. The end product from these tests will be one or two candidate tube materials.

**Results and Discussion**

In this section, results of initial tests performed at Babcock & Wilcox and Alzeta Corporation are described. Since this program is in an early stage, the results focus on characterizing the current state of the art of immersed tube burners and furnace operation.

Fig. 7. SiC burner tube instrumented with thermocouples on surface of tube (at Babcock & Wilcox).

**Babcock and Wilcox Baseline Tests**

Thermal conductivity and thermal shock are important considerations in evaluating performance of the ceramic tube component of the holding furnace burner system. They are also important considerations in making the transition from a holding furnace capability to a melting furnace capability.

Therefore, the purpose of the first series of tests performed in the Burns Engineering aluminum holding furnace at Babcock & Wilcox was to measure and record surface temperatures and surface-temperature gradients (in air) along the axial length of the tube at a number of different heat input levels. This information is needed as input for (*a*) thermal stress analyses of candidate ceramic tubes; (*b*) design of an improved, advanced ceramic tube component of the burner system; and (*c*) a reference data base for comparison of performance with alternative or future advanced burner system designs.

Original equipment ceramic burner tubes (supplied by Burns Engineering) were nitride-bonded silicon carbide manufactured by Carborundum Company. Figure 7 shows the burner (before installation in the furnace) inserted in an SiC tube instrumented with thermocouples cemented in place to measure tube surface temperatures during testing. Table II provides details of thermocouple location on the immersion tube and in the furnace for initial testing.

Several heat-up tests were performed in air with the furnace instrumented as in Table II. Results of one test are summarized in Fig. 8. This test began

Table II. Thermocouple (TC) Location in Babcock & Wilcox Tests

| TC No. | Location |
|---|---|
| 1 | Control thermocouple |
| 8 | Bottom of tube |
| 7 | 9 in. from bottom of tube |
| 6 | 18 in. from bottom of tube |
| 5 | 27 in. from bottom of tube (top) |
| 9 | Refractories, exit end |
| 10 | Refractories, load end |
| 11 | Ambient temperature |
| 12 | Exhaust temperature |

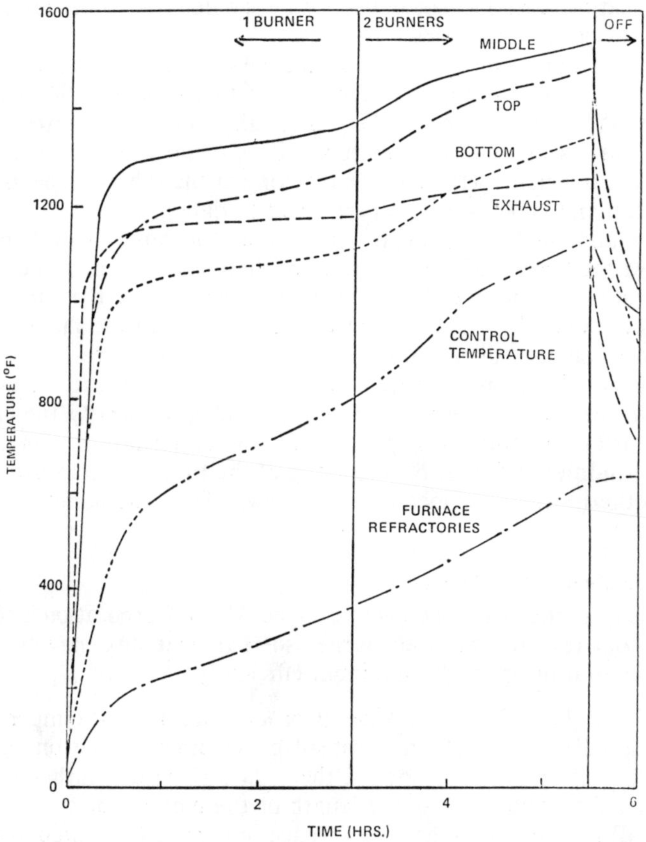


Fig. 8. Temperature data collected during heatup testing of instrumented furnace at Babcock & Wilcox.

by firing only the instrumented burner. Three hours into the test the second burner was started. Both burners were turned off after approximately 6 h. Because no aluminum was present in the furnace, the burners were operated on low fire (approximately 40% gas flow) during the entire test.

Using just one burner, the instrumented immersion tube approached peak operating temperatures in about 1 h. The temperature of the tube gradually increased by 26.6°C (80°F) from 1 to 3 h after startup. The temperature distribution over the length of the tube ranged from 593°C at the bottom of the tube to approximately 737° in the middle to approximately 687°C at the top. This spread represents a 126°C temperature gradient with one burner operating. This spread decreased to 93°C with both burners operating.

The furnace refractories attained an average temperature of 332°C (630°F) during the test. To ensure complete bakeout of the refractories prior to melting aluminum, both burners will be maintained on low fire for 24 h. Based on the measured heating rate, this bakeout schedule should remove any residual moisture in the furnace refractories.

Cooling of the immersion tube proceeded quickly after shutdown of the burners. In less than 30 min, the tube temperature dropped 204°–260°C. As in the case with rapid heating of the tubes, sufficient thermal stresses may be generated which could cause premature failure of the immersion tube during these thermal transients. However, no cracking of the tubes has been observed in the heatup and cooldown tests performed to date.

Examination of Fig. 9 provides insight into the cause of the temperature gradient. At the bottom of the tube interior there is a bottom spacer (part No. 3). This spacer deflects the flame away from the base of the tube and back up along the walls. Therefore, the hottest portion of the tube should be several inches above the base of the tube. The highest temperatures were measured with thermocouples 5 and 6, which were in the middle of the tube (9 and 18 in. up from the base). These thermocouples indicated nearly the same temperature during the entire test period and this uniformity is represented by one curve (middle) in Fig. 8. Near the top of the tube, temperature was lower as the top thermocouple is located farther away from the flame impingement area.

#### Alzeta Corporation Burner Development

In this task, the primary objective of the Alzeta Corporation is to develop, fabricate, and test an improved burner so that heat flux may be increased significantly in order to melt aluminum efficiently.

*Baseline Tests (Table III)*: At Alzeta Corporation, the first temperature data were measured on a metal tube instead of a ceramic tube on the Nikkan furnace. As will be described below, these data showed a rather nonuniform temperature distribution along the length of the metal tube.

The Nikkan aluminum holding furnace burner was operated at four loads (67, 84, 100, and 110% of fuel load) and 5% $O_2$. This $O_2$ level corresponds to an excess air level of about 27%. Although this level seemed rather high, it was decided to perform these first tests at the Nikkan specified conditions and compare the resulting temperature distribution to those taken later with the fiber burner at lower excess air levels. The Alzeta Pyrocore® fiber burner will operate initially at 10% excess air, as this is its recommended operating level. The Pyrocore fiber burner is a porous ceramic tube into which a natural gas/

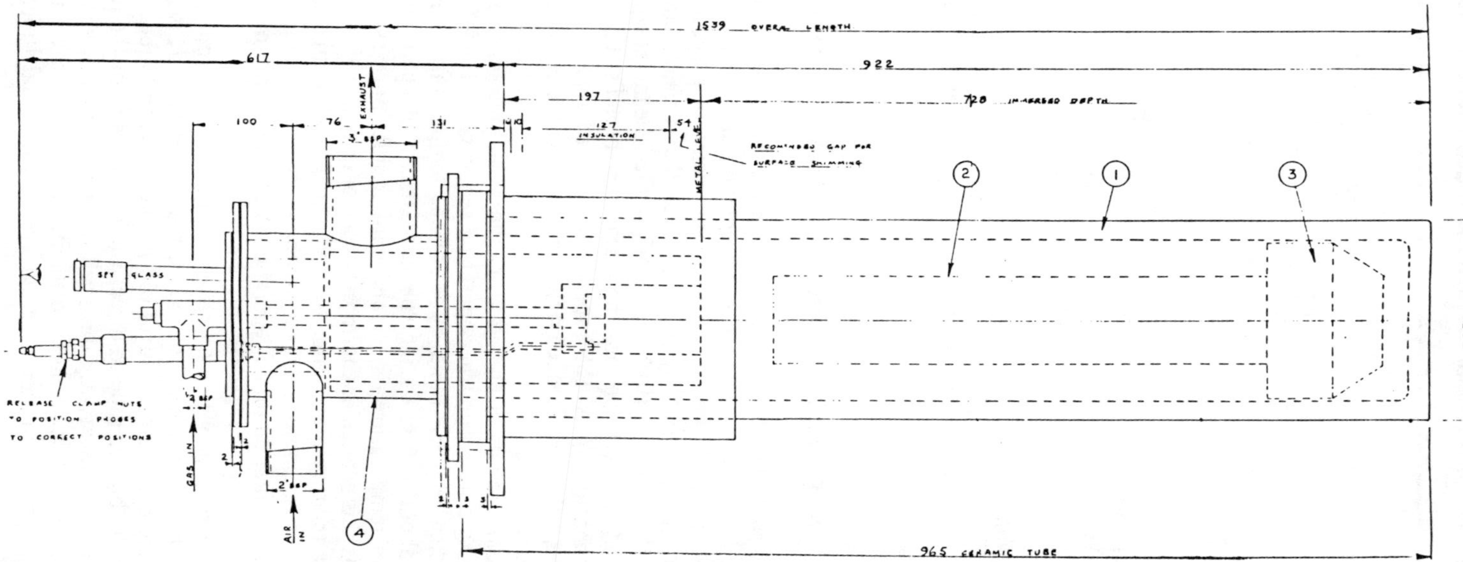


Fig. 9. Schematic of burner components and assembly.

Table III. Comparison of Aluminum Holding Furnace Burner Efficiency Data

| Measured Data | Nikkan | Burns† | |
|---|---|---|---|
| | | No. 1 | No. 2 |
| Load | 100 | 37* | 38* |
| Heat input (MW) | 14.5 | 36.7 | 37.6 |
| % Oxygen, dry air | 5 | 1.7 | 1.9 |
| Exhaust temperature, °C | 593 | 648 | 648 |
| Air temperature, °C | 32 | 37 | 37 |
| Calculated Data | | | |
| Excess air, % | 27.5 | 8.2 | 8.5 |
| Sensible heat loss (W) | 4103 | 8526 | 8787 |
| Latent heat loss (W) | 1450 | 3654 | 3770 |
| Total heat loss (W) | 5553 | 12180 | 12557 |
| Thermal efficiency, % | 62 | 67 | 67 |
| Useful heat flux (MW) | 8.9 | 24.5 | 25 |

†Burns Engineering reports that each burner provides 51.5 W (177.5 MBtu/h), releasing 31.7 MW (109.2 MBtu/h) while melting aluminum at 698°C, giving an efficiency of 62%.
*Low fire.

air mixture passes through the porous ceramic tube and burns on the surface of the tube.

The automatic control system was used to operate the burner during these baseline tests. The thermostat's control thermocouple was welded to the mid-point of the metal tube. The temperature at this point then controlled the burner's firing rate. It was found, by trial-and-error, that a thermostat control setting between 487° and 498°C (910° to 930°F) yielded the target equilibrium temperature of 648° to 704°C on most points on the shroud's surface.

Two rows of thermocouples, 180° apart, were welded along the length of the metal tube. The welds were spaced 4 in. apart and staggered by 2 in. from those on the opposite side of the tube.

Figure 10 shows the measurements from these thermocouples at various time intervals during a normal burner operating sequence at 100% load (14.5 MW). The data show that the metal tube surface temperature rises rapidly from ambient to about 704°C during the first 20 min of operation. Subsequently, the rate of temperature rise slowed until equilibrium was reached in about 1.5 to 2 h. At this particular thermostat control setting, which is believed to be realistic for melting aluminum, the measured peak temperature was 712°C. This figure also shows that the temperature distribution was quite nonuniform during the first 20 min of operation. Uniformity improved at equilibrium.

The peak temperatures, which contribute to this nonuniformity, occurred on the sides of the metal tube near the exit of the burner tube. It is somewhat surprising that these peak temperatures did not occur at the tube's tip, since one would expect the burner tube to direct the hot gases from the flame onto this area. The burner is apparently designed to prevent this by the presence of a series of holes (on 10-cm centers) drilled in the side of the burner tube and by radiation from the burner tube surface to the metal tube. The low

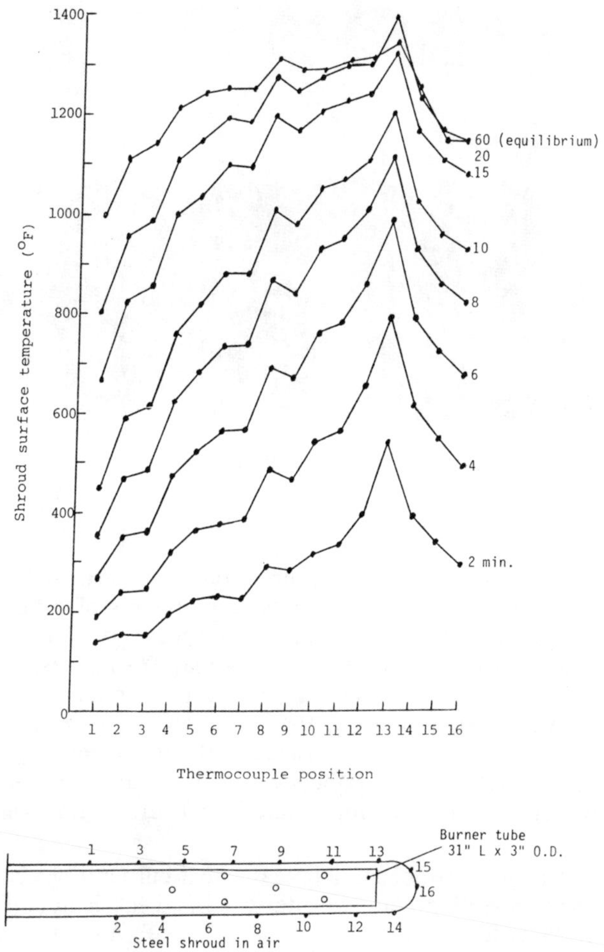


Fig. 10. Metal tube surface temperature distribution (100% load).

temperatures on the metal tube's tip may mean that this is a "dead" volume; i.e., one containing a stagnant pocket of gas, similar to the burner in the Burns Engineering furnace.

Temperature distribution data for the 67, 84, and 110% loads were also taken. Development of the temperature distributions over time for these firing rates was essentially the same as that for 100% load. Of course, the peak temperatures reached at equilibrium were higher or lower than those shown in Fig. 10, depending on the firing rate. For 67% load, this temperature was 579°, for 84% load it was 668°, and for 110% load it was 768°C.

The last exercise performed during baseline testing was the calculation of thermal efficiency. At the 100% load condition, the stack temperature was approximately 593°C, which corresponds to an efficiency of 62%. In other words, the useful heat flux to the load was 8.9 MW (31 MBtu/h).

Fig. 11. Ceramic immersion tubes used in Burns and Nikkan aluminum holding furnaces.

Figure 11 shows the difference in burner tubes that are used in the Nikkan furnace and the Burns furnace. The outside diameters of tubes used in the Nikkan and Burns furnace are 4½ and 8 in. respectively. Wall thicknesses of the Nikkan and Burns furnace tubes are 1/4 and 3/4 in., respectively.

Shakedown tests and fine tuning of the Burns aluminum holding furnace have been completed. Subsequently, approximately 1400 pounds of aluminum ingots have been charged into the furnace in the initial melting tests. At the present time, all operations at Babcock & Wilcox are going smoothly and screening tests of candidate ceramic materials should begin soon.

*Fiber Burner Retrofit:* To accommodate the fiber burner in the Nikkan burner hardware, a new mounting plate was designed and fabricated. This plate includes penetrations for the burner pipe (1 in. NPT), uv flame scanner, and peepsight. This plate will be attached to the end of the shroud's exhaust plenum. Figures 12 and 13 show the arrangements of the conventional Nikkan and Alzeta fiber burners, respectively. Fortunately, no modification of the Nikkan plenum (other than removing one cover plate, the burner, and inner burner tube) was necessary for incorporation of the fiber burner hardware.

The cylindrical wire screen forms for the fiber burners have been fabricated. These were produced by a machine shop specializing in forming custom shapes from wire cloth. These burner forms were designed to resist crushing, bending, and torsional forces that could lead to separation of the fiber layer from the screen.

Altogether, 10 burner forms were fabricated. All of the burners were sized to provide 14.5 MW, but at different specific heat release rates. That is, one-half of the burners were physically smaller than the others and so will operate at a higher specific heat release rate—100 MBtu/h per $ft^2$ vs 80 MBtu/h per $ft^2$.

The two burner sizes are shown in Fig. 14. When formed with the fiber layer, the dimensions of the longer burner are 15 in. long by 2 in. OD, and the dimensions of the shorter burner are 12 in. long by 2 in. OD.

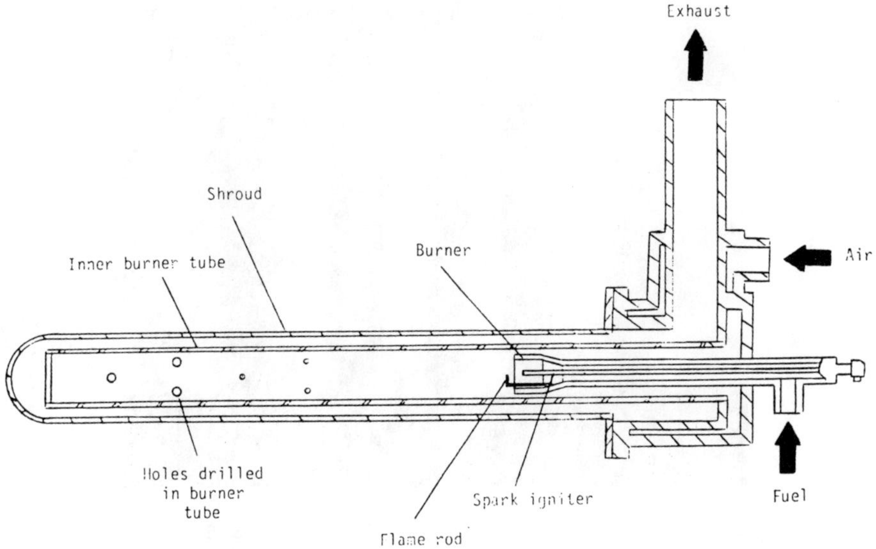


Fig. 12. Conventional Nikkan burner system.

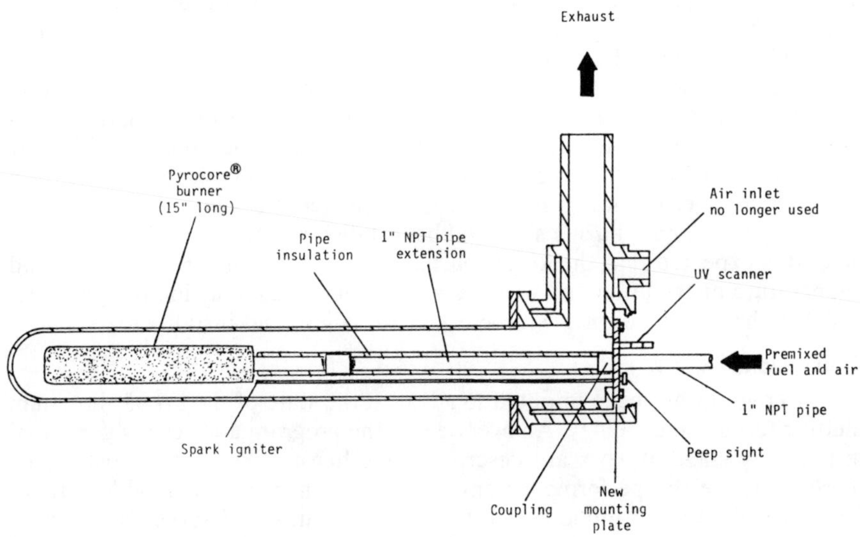


Fig. 13. Alzeta Pyrocore® burner retrofit in Nikkan immersion tube.

The fiber burner shown in Fig. 13 is the longer burner. This figure shows that the fiber burner is quite short compared to the length of the immersion tube. The reasons for this relatively short burner length are two-fold. First, the heat input rate for this holding furnace system is really quite small, leading to a surprisingly small burner surface area requirement. Second, the fiber burner's length/diameter ratio should ideally be less than 10 to produce uni-

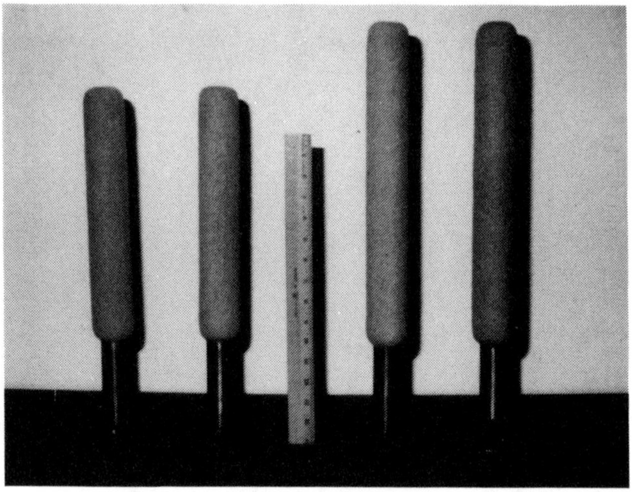

Fig. 14. Alzeta fiber burners for aluminum holding furnace.

form combustion on the burner's surface. The longer burner's $L/D$, 7.5, incorporates a safety factor to ensure uniform surface combustion. Burners with an $L/D>10$ can be produced, but the internal reactant flow must be altered to provide uniform radial flow.

Due to these constraints, obtaining a uniform temperature distribution along the immersion tube with this fiber burner may be problematical. Upcoming tests will provide the answer. However, the fiber burner will most certainly be of benefit in the future melting furnace application where the burners can meet the much higher heat flux demand easily.

In any case, both burner sizes will be installed and tested in the immersion tube. It is expected that the longer burner will provide a more uniform shroud temperature distribution and will operate at a lower combustion temperature, but that the shorter burner will be easier to handle and install.

## Conclusions

A program has been initiated to advance the state of the art of aluminum melting furnace technology. An overview of the program and results generated in the early stage of work are described. Baseline tests have been performed to characterize the performance envelope of commercially available furnace burners and ceramic immersion tubes. These results will serve as reference data for assessing future improvements to be achieved during performance of this program in (*a*) burner design, (*b*) furnace design, and (*c*) ceramic materials development and selection. Therefore the stage is set for making significant advances in aluminum melting furnace technology.


## Acknowledgments

The authors gratefully acknowledge the technical advice of W. Liang and E. Tabb of the Gas Research Institute; and R. Komoroski, J. Baroch, and E. Merriwether of the Babcock & Wilcox Company for the performance of numerous experimental tasks.

## References

[1]K. J. Brondyke, "Effect of Molten Aluminum on Alumina-Silica Refractories," *J. Am. Ceram. Soc.*, **36** [5] 171–74 (1953).

[2]J. G. Lindsay, W. T. Bakker, and E. W. Dewing, "Chemical Resistance of Refractories to Al and Al-Mg Alloys," *J. Am. Ceram. Soc.*, **47** [2] 90–94 (1964).

[3]A. E. Standage and M. S. Gani, "Reaction Between Vitreous Silica and Molten Aluminum," *J. Am. Ceram. Soc.*, **50** [2] 101–105 (1967).

[4]K. Prabriputaloong and M. Piggott, "Reduction of Silica by Molten Aluminum," *J. Am. Ceram. Soc.*, **56** [4] 184–85 (1973).

# Ceramic Heat Exchanger/Acid Vaporizer

MURRAY COOMBS AND PAUL VIDANO

AiResearch Manufacturing Co.
Torrance, CA 90509

This paper presents the design, development, and testing of a single-tube ceramic heat exchanger capable of vaporizing concentrated sulfuric acid in a solar flux. The vaporizer is a critical component in a solar-driven, hydrogen-producing process under investigation by the Department of Energy. The development work included the laboratory demonstration of ceramic-to-metal seals capable of operating at 10 atm and 537°C (1000°F). Compatibility tests with sulfuric acid for NC-430, SASC, and SC-2 have demonstrated acceptable levels of corrosion. The test rig developed for evaluating the vaporizer performance is discussed.

The design, development, and testing of a 10 kW ceramic heat exchanger used in thermochemical hydrogen cycles is described. This heat exchanger vaporizes concentrated sulfuric acid (98 wt%) using a solar heat source. Based on previous studies, ceramics are the best candidate materials for this highly corrosive environment.

The vaporizer is a critical component in a solar-powered thermochemical process that produces hydrogen by splitting water. A process schematic, including the vaporizer, is shown in Fig. 1. The process requires water and solar energy to produce hydrogen and oxygen using a sulfur-iodine cycle.

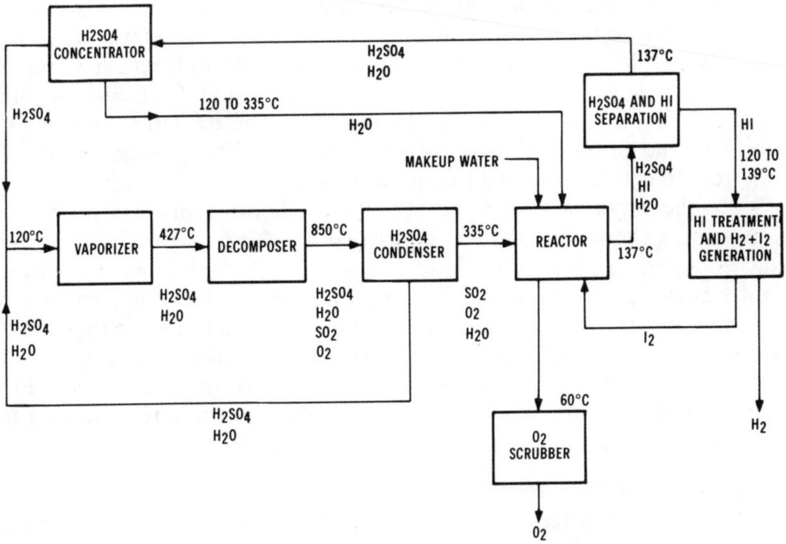


Fig. 1. Thermochemical water-splitting cycle for hydrogen production.

The vaporizer performs a critical step in the cycle, the vaporization of sulfuric acid which partially decomposes to sulfur trioxide and water. These products are further decomposed in a catalytic reactor downstream of the vaporizer. This sulfur-iodine process was developed by General Atomic Technologies and is currently under investigation by the Department of Energy.

### Thermal Design of the Vaporizer

The ceramic vaporizer consists of a single 1.0-in. tube (0.125-in. wall) constructed of sintered $\alpha$-SiC from The Carborundum Company. A 0.625-in. diameter alumina rod is located in the center of the tube. This rod creates an annular passage that improves the heat-transfer characteristics. The vaporizer operating conditions are summarized in Table I.

Table I. Vaporizer Operating Conditions

| | |
|---|---|
| Power input, kW | 10 |
| Flow, lb/h | 43.7 |
| Inlet temperature, °C | 27 |
| Outlet temperature, °C | 427 |
| Saturation temperature, °C | 400 |
| Operating pressure, atm | 5 |
| Acid concentration, % | 98 |
| Outlet composition, mol% | |
| $H_2SO_4$ | 31.3 |
| $H_2O$ | 37.8 |
| $SO_3$ | 30.9 |

The vaporizer was initially sized using a direct solar flux of 170 $kW/m^2$ impinging on the tube and enclosure. The tube and enclosure were concurrently reradiating to an ambient environment. A finite element model was constructed which divided the tube into four regions representing different flow regimes. These regions, shown in Fig. 2, were (1) subcooled liquid, (2) slug-flow boiling, (3) dry-wall flow, and (4) superheated vapor flow. The transition between slug-flow boiling and dry-wall flow occurred at 70% quality. A tube length of 48 in. was required to achieve 10 kW performance.

Since the vaporizer will be tested in the laboratory instead of in a solar environment, a thermal model was generated which represented the tube completely enclosed in insulation. Power was supplied to the vaporizer by four radiant heating elements located around the single tube. The finite element model included axial and radial tube conduction, radiation exchange, and the heat-transfer effects of the four flow regimes. The predicted average tube-wall temperatures for the solar and laboratory conditions are compared in Fig. 2. The temperature levels expected in the laboratory testing are predicted to be similar to the levels achieved in a solar environment.

### Material Compatibility Testing

Three types of silicon carbide were evaluated to determine the compatibility with boiling sulfuric acid (98 wt%). Samples of NC-430 (Norton), sintered $\alpha$-SiC (Carborundum), and SC-2 (Coors) were placed in boiling sulfuric

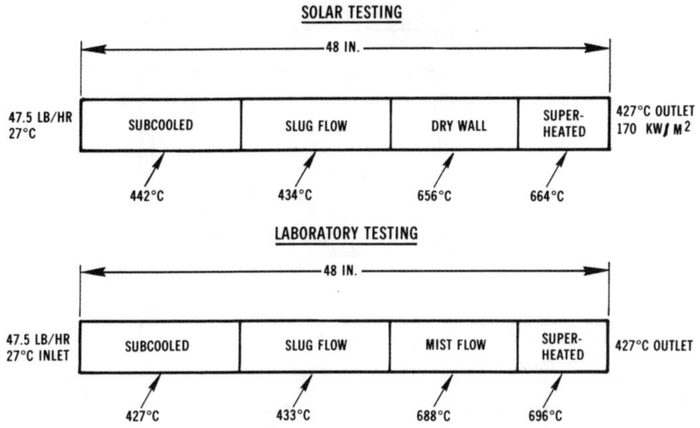


Fig. 2. Tube wall temperatures for solar and laboratory operation.

acid at atmospheric pressure. The results of five months of testing are shown in Table II. All three types of silicon carbide exhibit excellent compatibility with sulfuric acid in both the liquid and vapor phases.

**Seal Design and Testing**

Ceramic-to-metal seals for both the hot and cold ends of the vaporizer tube were developed and demonstrated in the laboratory. Both the cold and hot-end seal designs were experimentally evaluated in the seal test rig shown in Fig. 3. Pressurized nitrogen filled the ceramic tube, while heating elements and a cooling water jacket maintained the cold and hot seal temperatures to design conditions. The force required to set the hot-end seal was transmitted by a loading piston using pressurized nitrogen. A movable carriage simulated

Table II. Material Percent Weight Gain

| Time (months) | Material Type | | |
|---|---|---|---|
| | SC-2 | SASC | NC-430 |
| | Vapor specimens | | |
| 1 | −0.0191 | −0.0706 | −0.0234 |
| 2 | −0.0340 | −0.0776 | −0.0258 |
| 3 | −0.0392 | −0.0989 | −0.0281 |
| 4 | −0.0570 | −0.1811 | −0.0535 |
| 5 | −0.1253 | −0.2709 | −0.1154 |
| | Liquid specimens | | |
| 1 | 0.0099 | 0.7730 | 0.0908 |
| 2 | 0.1692 | 0.7153 | 0.2396 |
| 3 | 0.4978 | 0.9782 | 0.6093 |
| 4 | 0.8040 | 1.833 | 1.2807 |
| 5 | 2.6969 | 4.6939 | 3.061 |

NOTES
1. Numbers represent average percent weight gain of three samples.
2. Negative sign represents weight loss.
3. Numbers represent percent of original weight of specimen.

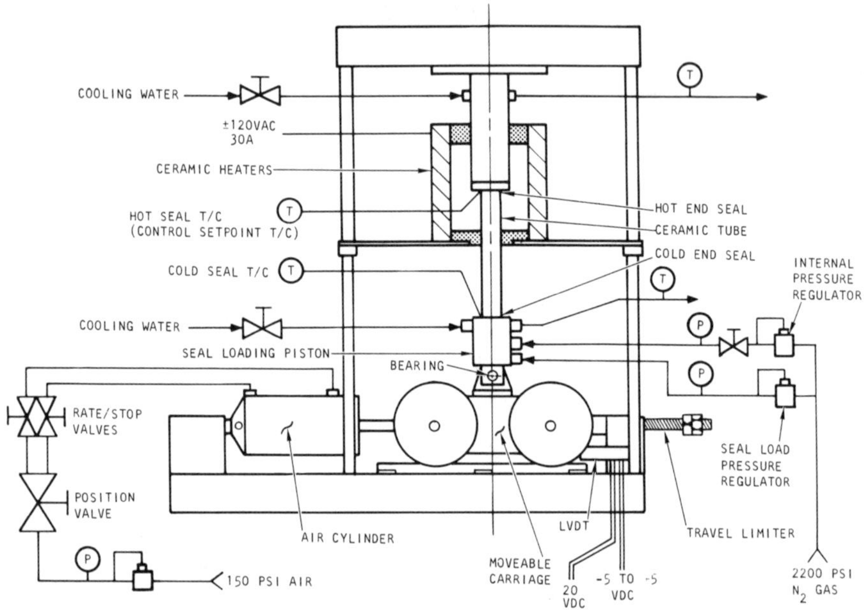


Fig. 3. Seal test rig piping and wiring schematic.

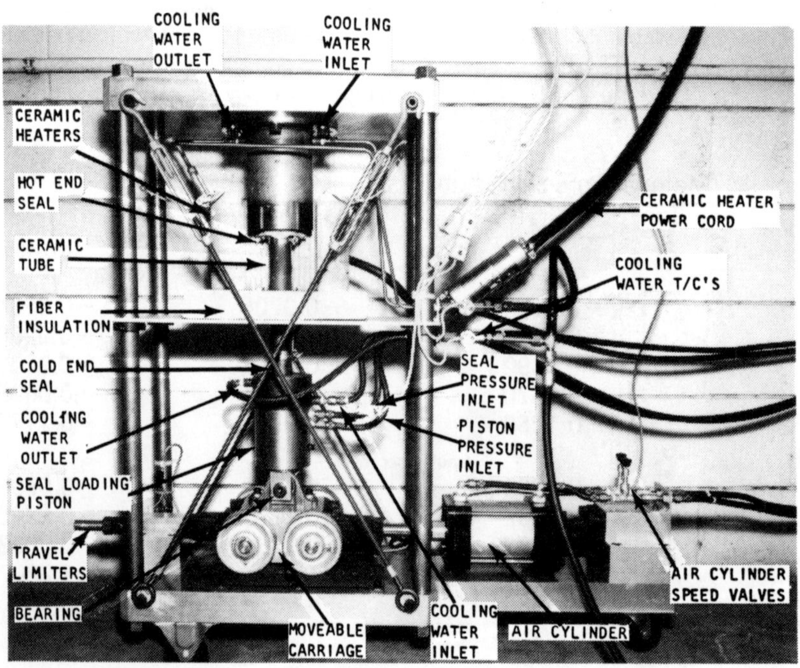


Fig. 4. Laboratory seal test rig.

tube deflections expected in a multitube heat exchanger. A photograph of the assembled seal test rig is shown in Fig. 4.

The cold-end seal was designed to operate at 66°C and up to 10 atm by supporting the ceramic tube in a 316 stainless steel cup. A Kalrez (Fluoro Elastomer) O-ring, located between the ceramic tube and metallic cup, effected the seal. A Teflon* pad at the base of the tube prevented point-contact stresses. This entire assembly was mounted in a bearing which allowed rotational movement.

The hot-end seal was designed to operate at 538°C and allowed rotational movement. This seal uses a "ball-and-cone" design, in which a ceramic tube is seated in a metallic socket located in the outlet manifold. Compliant gold foil was placed between the ceramic tube and metallic socket. The loading force required to set the seal was applied by a piston/cylinder assembly supporting the cold-end seal. This assembly also accommodated any axial thermal expansion.

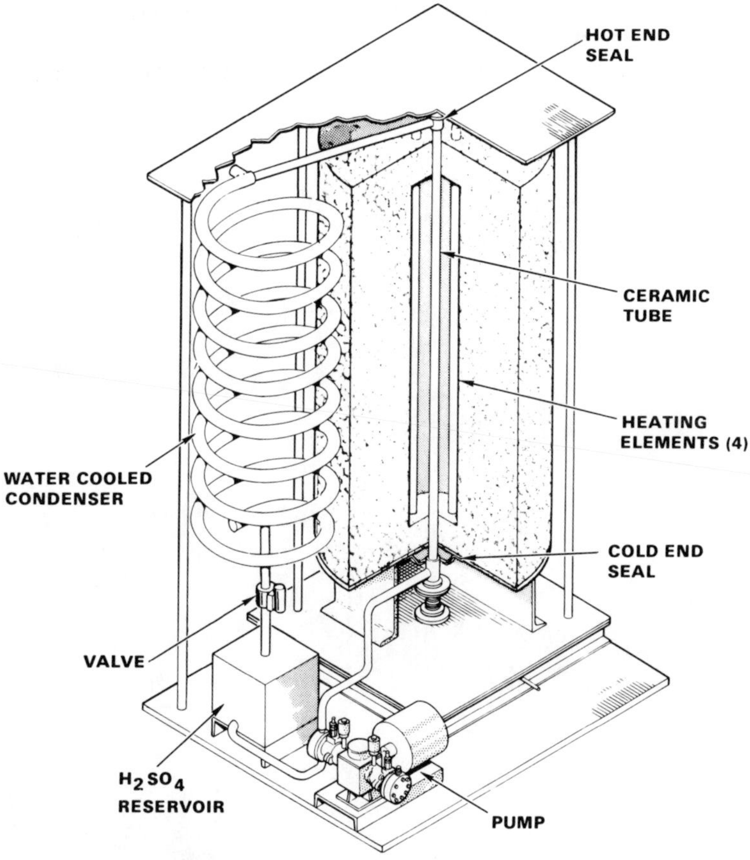


Fig. 5. 10 kW laboratory experiment ($H_2SO_4$ Vaporizer).

*E. I. du Pont de Nemours & Co., Inc., Wilmington, DE.

Leakage rates for both the hot- and cold-end seals were measured by monitoring the pressure decay at design temperatures and pressure. For both seals, the pressure decay was within the uncertainty of the instrumentation during a 10-min test (±1 psi at 150 psig).

### Single-Tube Test Rig

The performance of the single-tube vaporizer will be experimentally evaluated in the test rig shown in Fig. 5. The test rig is a closed-loop system which circulates acid through the vaporizer and support equipment. Sulfuric acid (98 wt%) is collected in a reservoir at atmospheric pressure and slightly elevated temperatures (nominally 32°C). The $H_2SO_4$ is pumped by a Teflon diaphragm metering pump to 5 atm internal pressure and then vaporized in the tube. Power to the tube is supplied by radiant heating rods. The operating pressure level is maintained by the control valve downstream of the vaporizer. The outlet temperature of the vaporizer is held at 427°C by varying the flow rate. The hot $H_2SO_4/SO_3/H_2O$ vapor exiting the ceramic tube is condensed in a water-cooled heat exchanger and collected in the sulfuric acid reservoir.

System testing will initially use water to establish proper operation. Sulfuric acid will then be used to evaluate the vaporizer performance during steady state and transient operation.

# Development of a High-Pressure Ceramic Heat Exchanger

D. M. KOTCHICK AND M. G. COOMBS

AiResearch Manufacturing Co.
Torrance, CA 90509

W. T. Bakker

Electric Power Research Institute
Palo Alto, CA 94303

The results of a program to develop the technology to design, fabricate, and operate high-temperature, high-pressure ceramic heat exchangers are described. An extensive materials data base on siliconized silicon carbide was established. The data indicate that this material will have adequate strength up to 1260°C (2300°F) in clean combustion environments and does not exhibit slow crack growth. A high-temperature, high-pressure ceramic-ceramic seal system was developed based on a glass-encapsulated graphite-based material. A complete heat-exchanger module was fabricated and tested under simulated engine conditions at wall temperatures up to 1260°C.

In many advanced systems associated with the utility industry, the performance of the system is highly limited by temperatures achievable with current metallic components. The use of ceramic elements for some of the critical components can often increase operating temperature up to 537°C and result in highly improved performance. Typical applications include direct- and indirect-fired gas turbine engine systems, ultrahigh-temperature solar receivers, and high-temperature heat-recovery/cogeneration installations. Ceramics may be beneficial wherever higher temperatures provide efficiency or process advantages.

The Electric Power Research Institute (EPRI) recognized the potential of ceramic materials applied to advanced utility systems and initiated several projects directed to the development of such systems. One project was the coal-fired, prototype, high-temperature, continuous-flow heat exchanger project. Under this program, a closed-cycle gas turbine power conversion system coupled to a coal-fired cumbustor via a ceramic heat exchanger was studied. The heat exchanger operated at a wall temperature in excess of 1093°C (2000°F) and produced a working fluid turbine inlet temperature of 982°C (1800°F). In addition to the design study, an extensive effort was undertaken in the areas of ceramic materials characterization, fabrication techniques, and testing to develop the technology base for such a heat exchanger. A small ceramic heat-exchanger module (Fig. 1), incorporating many of the design features of the full-scale heat exchanger, was developed, fabricated, and tested at material temperatures up to 1260°C. The study was completed in 1977; the results are given in Ref. 1.

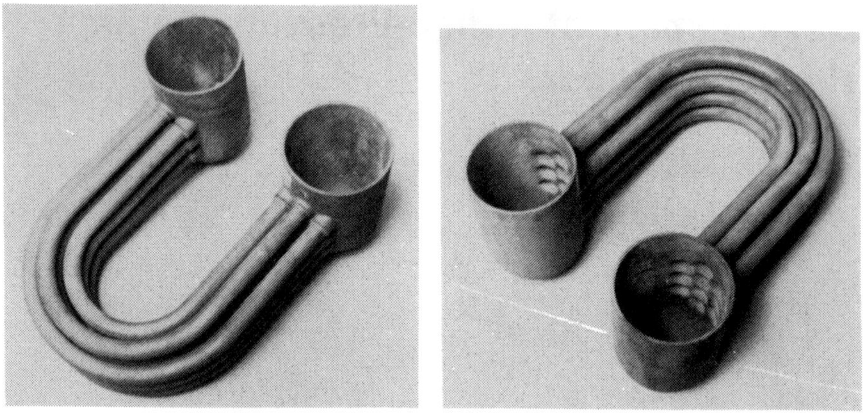

Fig. 1. Ten-tube SiC heat-exchanger module.

In a follow-up project conducted by AiResearch, effort was concentrated on the establishment of the design philosophy, material data base, and fabrication technology required for the successful demonstration of a ceramic heat exchanger. The design requirements for the CCPS-40-1 engine were selected to establish design, reliability, and performance goals for the ceramic heat exchanger. The CCPS-40-1 is a 200-hp, closed-Brayton-cycle engine and represents an uprated version of a currently existing engine. The ceramic heat exchanger designed for this engine is shown in Fig. 2. It consists of an assembly of six heat-exchanger modules coupled with a series of V-band clamps, also known as Marmon clamps. Each heat-exchanger module comprises 30 ceramic finned U tubes bonded to inlet and exit ceramic manifold sections. The ceramic material is siliconized silicon carbide.* The hot gas flow is confined to the finned portion of the heat exchanger by a series of ceramic flow diverter plates; firebricks block the gap created by the manifold joints. Insulation is placed between the top diverter plate and manifolds, and between the lower diverter plate and the U-shaped portion of the tubes. Bellows are used on the inlet and exit manifolds to provide for thermal growth. Based on this design, a second-generation 10-tube heat exchanger was designed, fabricated, and tested. Some of the ceramic hardware produced in this program is shown in Fig. 3. The results of this program are presented in Ref. 2 and summarized in Ref. 3

At this point the program was directed to the development of the high-temperature, high-pressure ceramic seals required to join the six ceramic modules into a complete heat exchanger. In addition, the data base of the NC–430 material was significantly expanded. This seal development effort and the further material characterization are summarized in this paper. A complete description of the program is presented in Ref. 4.

**Materials**

The purpose of the basic materials investigation was to generate engineering design data for the siliconized silicon carbide under test conditions that would simulate the operating heat exchanger. The purpose was to complete

*NC–430, Norton Co., Worcester, MA.

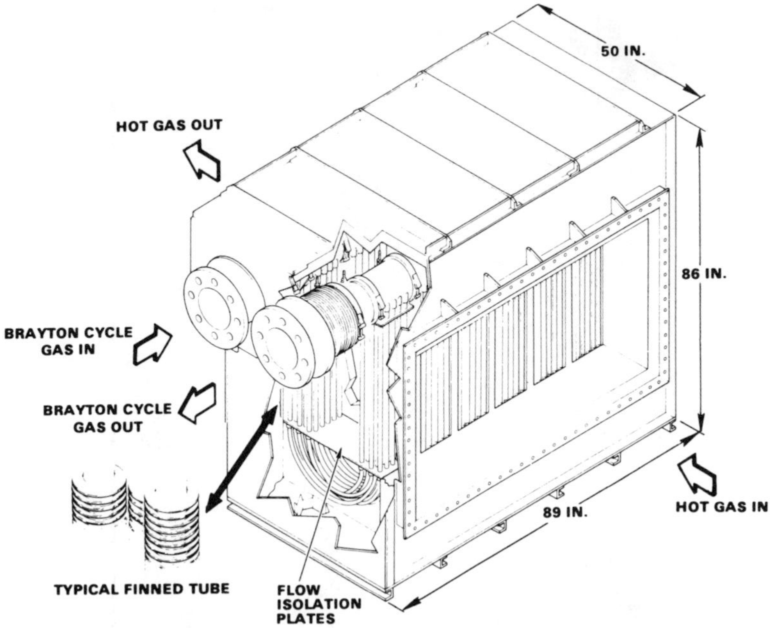


Fig. 2. Brayton-cycle ceramic heat-exchanger design.

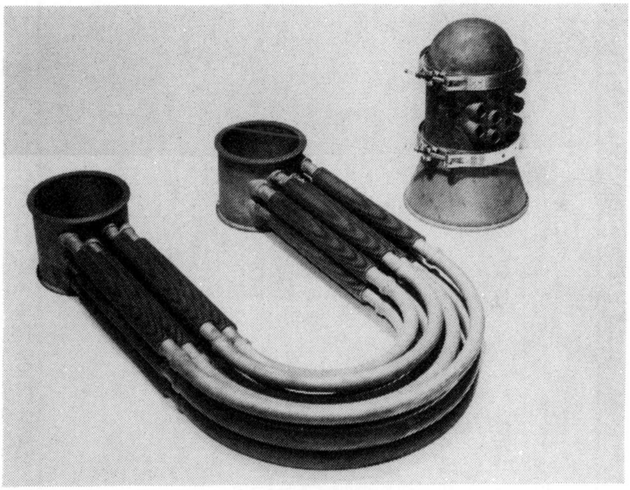

Fig. 3. Heat-exchanger module and assembled transition section, open hole manifold, and end cap.

Table I. Test Plan for Materials Properties (Norton NC-430)

| Test | Specimen | Specimen Condition | Test Furnace | Test Temp. °C (°F) | Displacement Rate (in./min) | Min. Qty. | Group No. | Tested |
|---|---|---|---|---|---|---|---|---|
| Strength | Standard bar (⅛×¼×2 in.) | As-processed | Keith | 1260 (2300) | 0.02 | 30 | 1 | 30 |
| | | | | | 0.0002 | 30 | 2 | 30 |
| | | As-processed, uncoated* | | | 0.02 | 20 | 114 | 20 |
| | | As-processed, slag-coated* | | | | 9 | 115 | 9 |
| Stress rupture | | As-processed | GA | 1260 (2300) | Static loading | 59 | | 59 |
| | | As-processed, slag-coated | | | | 16 | | 16 |
| Cyclic fatigue | | As-processed | Keith | 1260 (2300) | Cyclic (triangular loading) | 13 | 111 | 10 |

*Specimens survived 500-h/2300°F stress rupture loading prior to strength tests.

design data collected on NC–430 during the previous phase of the program (Ref. 2). This approach required the formulation of a detailed test plan and the design and fabrication of specialized high-temperature test equipment and fixtures.

Siliconized silicon carbide was chosen for the high-temperature ceramic heat exchanger due to its good thermal and mechanical properties and its relative ease of fabrication into the required complex shapes. Because the melting point of the silicon phase is lower than that of SiC, Si-SiC materials cannot be used above 1343°C (2450°F) for a long time. Standard flexural test bars ($\frac{1}{8}$ by $\frac{1}{4}$ by 2 in.) were obtained with one as-processed surface on one of the larger faces. The as-processed surface was used as the test surface to ensure that data collected during the investigation would be indicative of the flaw distribution and severity on as-processed heat-exchanger components. Detailed specifications for fabrication and inspection were used to ensure specimen quality. These specifications indicate the dimensional and standard nondestructive tests that each sample must pass before it is accepted for testing. Nondestructive tests included visual inspection, fluorescent dye penetrant (per MIL-I-6866 Type 1, method B) and radiographic (per ASTM-E-94-68 level 2T-2) inspection. Samples with flaws detected with these techniques were rejected.

The basic material test plans are shown in Table I. Tests were classified into three categories: (1) strength, (2) stress rupture, and (3) cyclic fatigue. They were conducted using the ceramic quarter-point flexure fixture shown in Fig. 4. All tests were conducted at 1260°C.

The strength tests were conducted in a mechanical test machine[†] at two loading rates and using an SiC resistance-heated furnace. Specimens tested

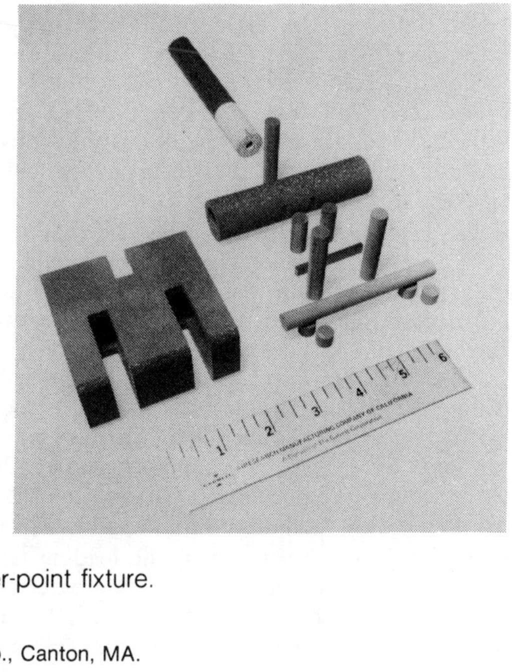

Fig. 4. Quarter-point fixture.

[†]Instron Corp., Canton, MA.

at 0.02 in./min were preoxidized for 1 h at 1260°C before testing. These results were compared with those for specimens that were unoxidized initially and tested at 0.0002 in./min and 1260°C, since testing time allowed about 1 h for oxidation before failure occurred. Even under these conditions, the flexural strength increased by about 10% from 28 to 31.4 ksi. This is approximately the same amount as in tests conducted at constant temperature during the previous phase of this program when the displacement rate decreased from 0.02 to 0.0002 in./min. The formation of an oxide layer on this material had previously been assumed to increase strength, at least in short-term tests. These results suggest that another, less understood strengthening mechanism may be acting in addition to the formation of the oxide layer.

For the stress rupture tests, an eight-position furnace with a 5- by 5- by 36-in. hot zone was designed and constructed. The furnace uses overhead, deadweight loading arms that contact microswitches to record specimen failures. All stress rupture tests were conducted at 1260°C for 500 h at stress levels from 70 to 98% of the average fracture strength of similar samples tested in the previous program. Some samples with $Na_2SO_4$-coated surfaces were tested. For both coated and uncoated sample tests, no apparent slow crack growth was detected; that is, no natural delayed failures were observed over the 500-h test with the exception of the first few hours. For both coated and uncoated samples that survived the 500-h test, permanent mid-span deflections of 0.0010 to 0.0017 in. were measured. Micrographic analysis of sectioned specimens showed extensive microcracking in the tensile zone of the test bar. Samples that survived the stress rupture test were tested at 1260°C in quarter-point tests (0.02 in./min). Modified Weibull plots were used to compare the results to baseline data. In general, the average strength of uncoated samples was greater than the baseline, whereas the average strength of the salt-covered samples was less than baseline (Fig. 5).

Microscopic examination of the sodium sulfate-coated samples showed intergranular penetration of 5 mils on the coated surface and 3 mils on the uncoated surfaces. Figure 6 compares the as-processed surface of a test bar, with the same surface after the 500-h stress rupture test with and without sulfate coating. Without the sulfate, a glossy oxide coating forms on the surface. The sodium sulfate preferably attacks the silicon phase, but the large SiC grains are also attacked.

For cyclic fatigue tests, samples were loaded into the flexure fixture so that the maximum outer fiber tensile stresses were placed on the as-processed surface of the bar sample. The fixture was placed in a high-temperature SiC resistance-heated furnace. Samples were cyclically loaded between 0.1 and 1.0 times approximately 70% (19.0 ksi) of the average fracture strength (28.0 ksi) of samples tested at 0.02 in./min displacement. A triangular loading pattern (with the same loading and unloading rate) was used at a frequency of 1 Hz. The loading and unloading rate (effectively 8000 lb/min) was determined by the 1-Hz frequency. Tests were run for 1 h (3600 cycles) at 1260°C. Samples that did not fail were loaded to failure at 300 lb/min so that failure stresses could be compared with data collected at 0.01 in./min displacement rate.

None of the samples broke during the cyclic loading test. The average fracture strength of the cycled samples was slightly greater than the average fracture strength of the baseline data, indicating no effect of cycling at these conditions. The results were compared to cyclic fatigue tests conducted in the previous phase of the program and to the baseline strength data, as shown in Fig. 7. Cyclic fatigue damage is probably not a major design consideration at these conditions until the maximum use limit of the material (1343°C) is approached.

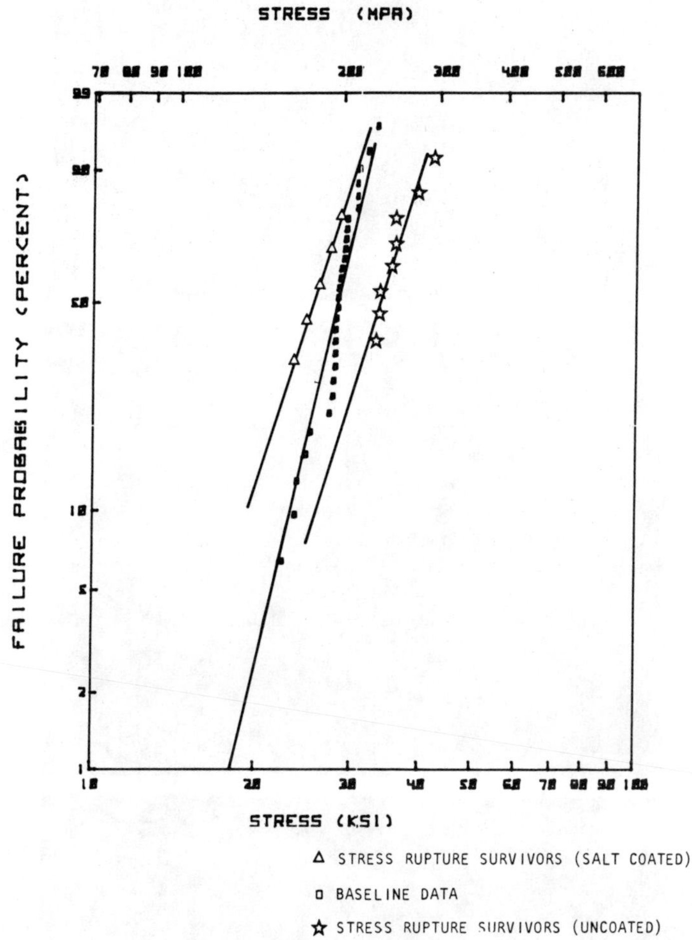


Fig. 5. Weibull plots for NC-430 SiC before and after 500-h stress rupture test.

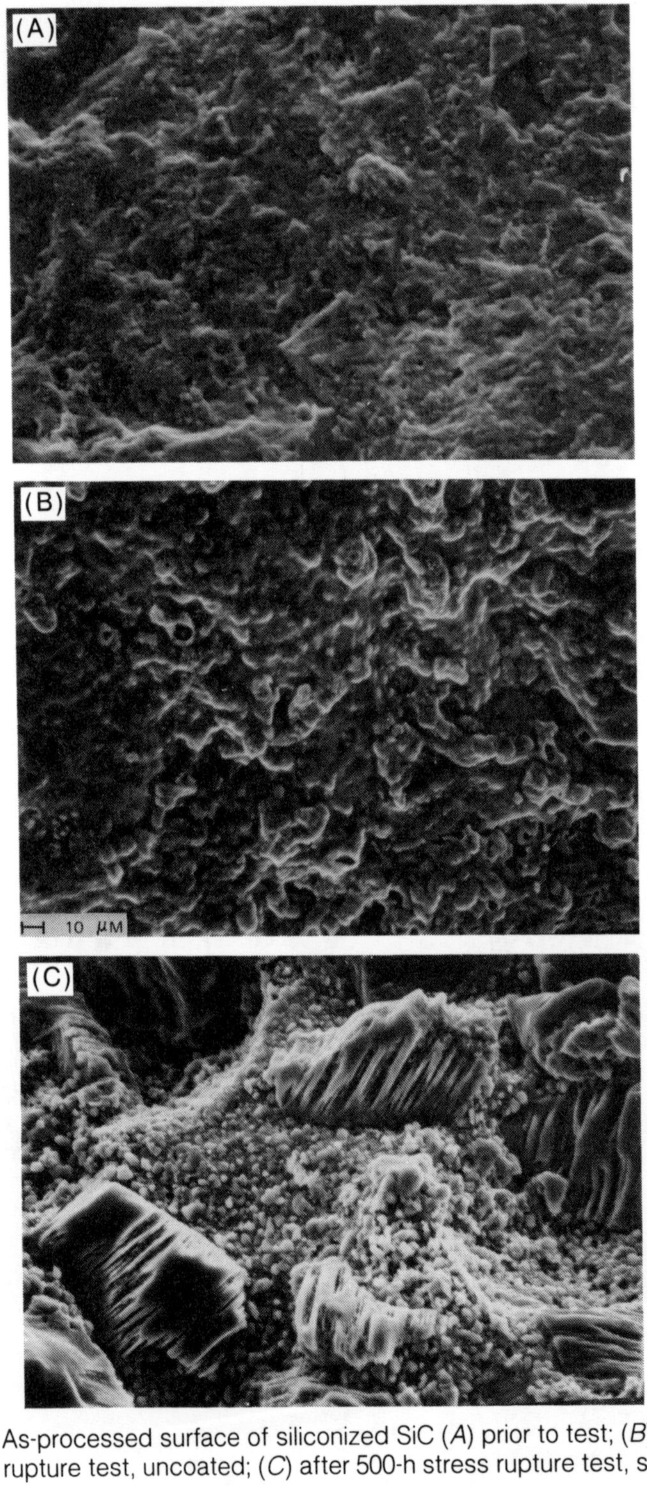


Fig. 6. As-processed surface of siliconized SiC (*A*) prior to test; (*B*) after 500-h stress rupture test, uncoated; (*C*) after 500-h stress rupture test, salt-coated.

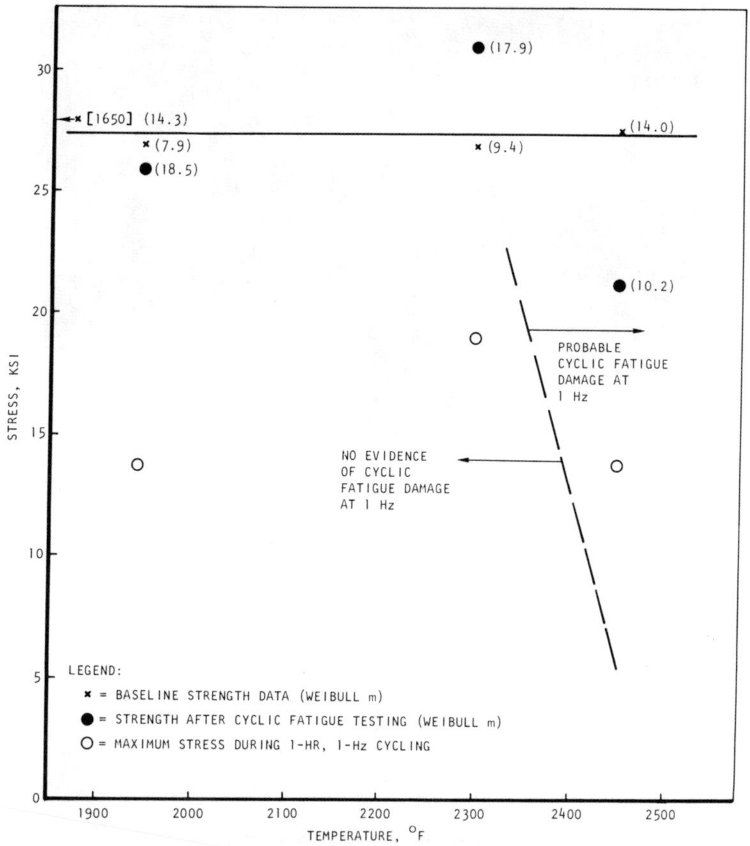


Fig. 7. Summary of cyclic fatigue results for NC–430.

**Seal Development**

A series of tests was performed on subscale heat-exchanger components (Fig. 8) to develop the seals required at ceramic-to-ceramic and ceramic-to-metal joints, as identified in the CCPS-40-1 reference design. The design goals for the seals are continuous operation at 280 psig internal pressure and 982° and 332°C, respectively, for the ceramic-to-ceramic seal and ceramic-to-metal seal. Figure 9 shows the apparatus designed to test both seals. It consists of a ceramic end cap and manifold segment mounted against a metal piston. Pressurization of the metal piston provides a sealing load along the component axis. Electrical and internal pressure feedthroughs are located on the metal piston. A single bayonet-type SiC resistance heating element was used for heating.

The types of seals that were tested unsuccessfully included various metal alloys in the shapes of O rings, C rings, spiral windings, and foils. These seals were largely unsuccessful because the sealing force required placed very high line or point contact stresses on the ceramic components, leading to failure. A successful seal was formed using a carbon gasket.‡ To use the carbon seal

‡Grafoil III (2D), High-Temperature Materials, Inc., Lowell, MA.

Fig. 8. Typical ceramic heat-exchanger components.

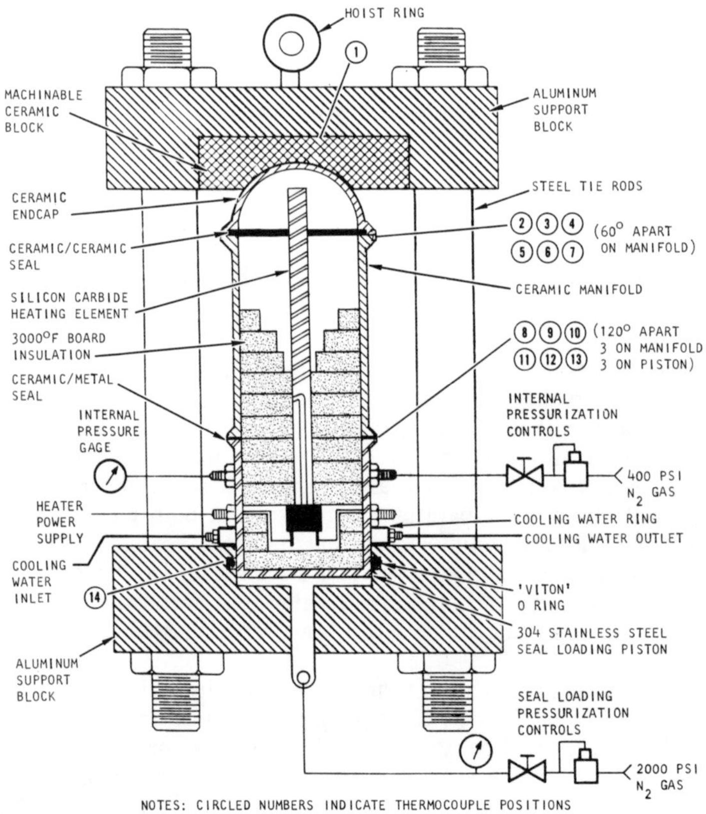


Fig. 9. Cross-sectional schematic of seal-test fixture and support systems.

at high temperatures, however, required oxidation protection, which was provided by a glass coating.

Many glass compositions from the leaded frit and aluminoborosilicates were investigated. The best results were obtained with an aluminoborosilicate bonding glass[§] and tetraethyl orthosilicate (TEOS). A 0.03-in.-thick Grafoil seal was used between the ceramic components. It was sealed with a slurry mixture of the aluminoborosilicate glass and TEOS. This system was then wrapped in fiber insulation and heated to 1800°F in air. The system was brought to an internal pressure of 150 psig and then sealed off. A pressure decay of less than 0.7 MPa was measured over 6 h.

**Subscale Ceramic Heat-Exchanger Module**

To prove the glass-protected graphite seal concept, a subscale heat-exchanger module was tested. The intention was to operate a heat-exchanger module with the end cap seals developed in the previous effort at design pressure (1.9 MPa) and temperature (982°C). The module built for this test is shown in Fig. 10. It is a U-tube configuration with the tubes bonded to the ceramic manifolds. The unit was fabricated by the Norton Company from their NC-430 siliconized silicon carbide.

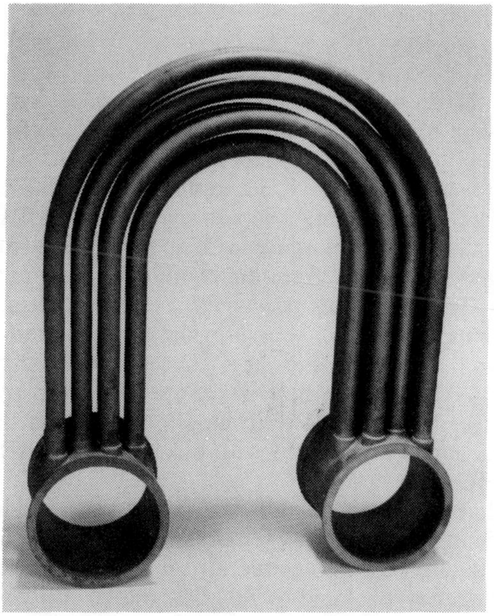

Fig. 10. Ceramic heat-exchanger test module.

[§]3GF–272B, O. Hommel Co., Pittsburgh, PA.

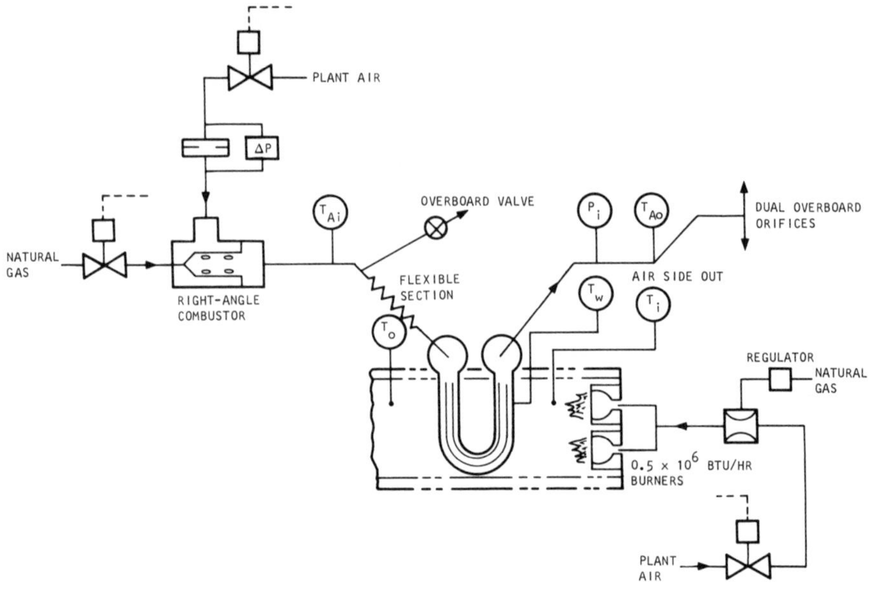


Fig. 11. Test schematic.

A schematic of the test setup is shown in Fig. 11. The heat-exchanger test furnace consists of a rectangular steel box open at the top and on one end.

The furnace was lined with K-30 insulating firebrick to provide an L-shaped tunnel. Two commercial, $0.06\times10^6$ W ($0.5\times10^6$ Btu/h) natural gas burners were installed at the closed end of the brick-lined tunnel. The heat-exchanger module was inserted from above into the tunnel around the corner from the burners. This technique was used to prevent the module tubes from seeing the burner flame directly, thus providing a more uniform temperature distribution across the leading edge tubes. The stoichiometric burners provide approximately 1648°C (3000°F) flue gas at the heat-exchanger hot face. The heat released by the burners was controlled by regulating the air flow to the burners. Fuel was added and mixed with air automatically to achieve a prescribed constant fuel/air ratio.

A right-angle natural gas combustor was used to provide a source of hot air to the cold-side inlet of the heat-exchanger test module. The airflow through the combustor is fixed, and the desired air inlet temperature to the cold side of the heat exchanger is regulated by fuel flow that is automatically controlled by temperature selection. Flow rates of up to 40 lb/min at 662°C (1225°F) were achieved by manually operating an overboard valve located between the right-angle combustor and the heat-exchanger module.

Two NC-430 hemispherical end caps were sealed, one cap per manifold, to the heat-exchanger test module, using 0.030-in.-thick Grafoil seals (Grafoil III) and 3GF-272-B bonding glass mixed with tetraethyl orthosilicate (TEOS). The wetting characteristic of the sealing glass was improved by preoxidizing the heat-exchanger manifolds and end caps at 982°C for 4 h. The bonding glass was applied to both the interior and exterior peripheries of the ceramic-

Fig. 12. Module glass seal joint.

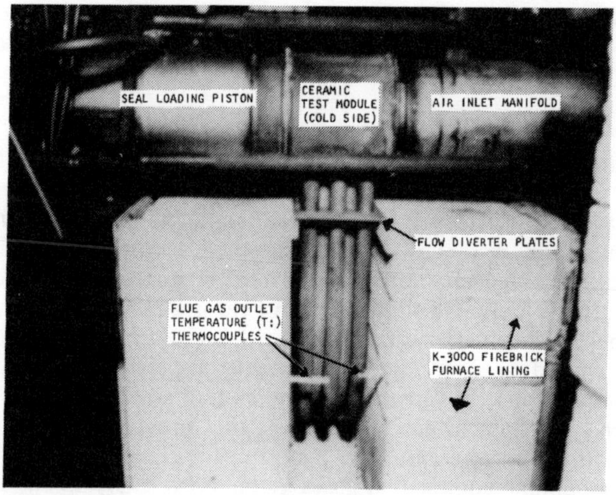


Fig. 13. Heat-exchanger module installed in test rig.

to-ceramic joint to prevent oxidation of the encapsulated Grafoil seal. The glass was fired onto the heat-exchanger manifolds, one at a time, at 1800°F. A close-up view of the exterior periphery of the ceramic-to-ceramic seal joint is shown in Fig. 12. The remaining open manifold ends were sealed to the heat-exchanger module test furnace stainless steel duct work with Inconel 601-jacketed Grafoil seals. This type of seal configuration was used at the ceramic-to-metal interface since the bonding glass would not wet the stainless steel.

The heat-exchanger test module is shown installed in the test fixture in Fig. 13. The heat-exchanger manifolds were supported between a flanged stainless steel duct system and a nitrogen-gas-pressurized, ceramic-faced piston assembly. The ceramic¶ piston faces were machined with impressions to match the hemispherical ends of the heat-exchanger manifold end caps. By pressurizing the piston assemblies, a compressive force was generated between stainless ductwork and the piston that provided the ceramic-to-ceramic and ceramic-to-metal seal loads. The stainless steel inlet duct was fixed to the top edge of the test furnace framework. Both the seal-loading piston assemblies and the stainless steel outlet duct were free-floating to allow for unrestricted thermal expansion of the ceramic test module. Thus, the hot manifold is free to move with respect to the cold manifold.

The ceramic test module was pressure-proof-tested to 140 psia at ambient temperature for 5 min in the test furnace without incident. The first high-temperature test run was conducted immediately following the proof test. Tunnel burner pilots were ignited and allowed to idle for 30 min at a minimum fire temperature of 704°C (1300°F). At this point, the right-angle natural gas combustor was ignited and the manifold internal pressure was brought to 30 psia to initiate gas flow through the heat-exchanger tubes. Tunnel temperature and manifold internal pressure and flow were gradually increased to a tunnel inlet temperature of 1204°C and an internal pressure of 55 psia. The tunnel inlet temperature was increased to 1315°C, the manifold pressure to 60 psia, and the cold-air inlet temperature to approximately 537°C when the ceramic test module ruptured.

Postmortem examination of the heat-exchanger module assessed the probable cause of failure to a section of one of the tubes which was thinner (i.e., 0.055 in. vs the design thickness of 0.125 in.) and was incompletely siliconized.

## Discussion and Conclusions

The materials studies reported here reinforce the favorable data obtained previously on siliconized silicon carbide. At 1260°C this material does not suffer from slow crack growth and can tolerate a moderate amount of deformation without failure. Investigations by others[5] indicate that the fracture toughness increases considerably at this temperature. In clean combustion gases the strength increases slightly when held under stress for long times at 1260°C. These data compare favorably with those of most other structural ceramics. It should be noted, however, that the properties of siliconized silicon carbide decrease drastically at the melting point of silicon 1426°C. This limits the use of this material to heat exchangers that provide clean working fluids with temperatures up to about 1093°C.

Contaminants in combustion gases (such as $Na_2SO_4$) do not appear to be highly detrimental because they did not adversely affect the stress rupture life; however, a slight decrease in short-term strength was noted, and a reacted zone was present in the specimens. These results need to be quantified, and their effect on the long-range durability of structural ceramics should be assessed.

Providing effective high-temperature, high-pressure seals for large-diameter components proved to be more challenging than anticipated. The usual metallic devices, such as O rings, C rings, flexitallic gaskets, etc., proved to be ineffective. Graphite gaskets were found to be effective for short periods

---

¶Marinite, Manville Corp., Denver, CO.

but suffered from excessive oxidation. Encapsulation of the graphite gaskets in a specially formulated glass frit was found to prevent oxidation and was successfully used for tests up to 10 h at 1.030 kPa (150 psi). It is believed that this system can be further refined to provide a commercially acceptable solution to the sealing problem. Long-term tests are required to prove the stability of this sealing system under simulated operating conditions.

Testing of the subscale module was disappointing because it failed before design operating conditions were reached. Postmortem analysis revealed undetected flaws, thinning, and incomplete siliconizing as the most likely causes of the failure. This result indicates that further advances in ceramic manufacturing technology and nondestructive examination are needed before heat exchangers with relatively thin walls (1/8 in.) can be used with confidence. Pressurized heat-exchanger segments, tested successfully for longer times in other programs, generally operated at lower pressures (690 kPa, 100 psi) and used tubes with 1/4-in. walls.

These and other studies indicate that additional development is needed before the design, construction, and operation of large pressurized ceramic heat exchangers can be recommended for field evaluations.[6,7] Most of the missing technology is under development. The Gas Research Institute is sponsoring an extensive program on ceramic recuperators which includes significant funds for manufacturing development and which will include long-term field exposures. EPRI's basic materials research program is sponsoring a study on the corrosion of structural ceramics by impure combustion gases. The DOE is sponsoring further development work on high-pressure heat exchangers. Thus, it is expected that the ceramic technology data base needed to demonstrate pressurized heat exchangers will be available in the near future.

## References


[1]A. Pietsch, et al., "Coal-Fired Prototype High Temperature Continuous Flow Heat Exchanger," AiResearch Mfg. Co. of Arizona, EPRI Report AF-684.

[2]M. G. Coombs, et al., "High-Temperature Ceramic Heat Exchanger," EPRI Report FP-1127, Research Project 545–2, July 1979.

[3]M. G. Coombs, "Development of a Ceramic Heat Exchanger for a Closed Cycle Gas Turbine," ASME 79–GT–59.

[4]M. G. Coombs, et al., "High-Temperature Ceramic Heat Exchanger Development," EPRI Report AP–3019, Research Project RP–545–2, May 1983.

[5]G. Popp and R. F. Pabst, "Fracture Behavior of Ceramic Materials at High Temperatures," *Met. Sci.* 130–32 (1981).

[6]M. E. Ward and A. J. Kubasco, "Ceramic Heat Exchanger Technology," EPRI Report AP–2883, March 1983.

[7]M. Coombs, D. Kotchick and H. Strumpf, "High Temperature Ceramic Recuperation and Combustion Air Burner," Final Report, February 1984, GRI Report 83/0039.

# Ceramic Heat Exchangers for Closed-Cycle Gas Turbine Systems

H. W. CARPENTER, J. CAMPBELL, JR., L. H. RUSSELL, AND D. E. WRIGHT

Rockwell International/Rocketdyne Div.
Canoga Park, CA 91304

High-temperature, coal-fired combustors with ceramic heat exchangers were designed for CCGT systems. The objective in evaluating CCGT systems is to convert U.S. coal to electricity with higher efficiency. Higher temperatures are required to accomplish this goal and ceramic heat-exchanger surfaces allow the use of working fluid temperatures to 1371 °C (2500 °F) and higher. The results of a comprehensive government study are described in which an atmospheric fluidized bed and cyclone fired combustor/heat exchanger were designed for operation at 954° and 1232 °C (1750° and 2250 °F).

Studies conducted at Rocketdyne, sponsored by the Department of Energy, examined a variety of coal-fired, closed-cycle gas turbine power generation systems. The primary emphasis was on the preliminary design of the fired heat exchangers, which had been identified as a major cost item. The study ground rules specified that several coal-fired heater preliminary designs be prepared for CCGT working fluid temperatures of 954 °C or higher for both a 700 MWe power station containing two 350 MWe units, and for a 25 to 50 MWe cogeneration station. The 954 °C or higher working fluid temperature required the use of ceramic heat-exchange surface in the fired heaters.

This paper is concerned principally with the fired heater ceramic design concepts developed during the study. Two heater concepts were explored in depth: (1) a high-temperature, atmospheric fluidized bed (AFB) operating with a 954 °C working fluid, and (2) a slagging cyclone combustor (SCC) operating with a 1232 °C working fluid temperature.

### Designing Ceramic Fired Heat Exchangers

A general review of "The Role of the Ceramic Heat Exchanger in Energy and Resource Conservation" was presented in Ref. 1. That paper describes the many present and potential applications for ceramic heat-exchange surface and provides an extensive bibliography.

The current studies deal specifically with the design of internally pressurized ceramic heat exchangers for closed-cycle gas turbine heat input service when utilizing coal firing. The CCGT cycle working fluid temperatures in the ceramic heat-exchange surface will be higher than those for which metal surface may be utilized, i.e., above 815 °C and typically in the range of 954° to 1232 °C maximum temperature. The internal pressure will be about 3.4 MPa (500 psi) or higher. The fired heater designer must face many constraints in designing heaters for this service.

As expected in developing new designs and in using advanced technology, there are risks. Designing combustor/heat-exchanger systems that are made of ceramics is more complex because ceramics are completely brittle, i.e., unforgiving of any strain beyond the elastic limit. The designer must use new design tools and approaches, and he must work closely with materials and stress engineers.

Ceramic heat-exchanger hardware is restricted in shape and size, due to limitations of the fabrication process. The resultant properties are directly dependent on the fabrication history. Flaws are always present in ceramics and a probability of failure in service must be accepted in the design. Allowable design stresses must provide a wide margin of safety to achieve the required high reliability, and structures must be designed to avoid the many small stress conditions that in metal designs are simply relieved by a small amount of yielding. Since the ceramics are completely brittle, much design and analysis must be done to eliminate stress risers, to keep thermal stresses low by prudent thermal management, and to minimize vibration-induced stresses. *All* of the stresses, combined, must be controlled.

An analysis of the constraints inherent in pressurized ceramic heat-exchanger design led to setting up the following guidelines for the fired heater designer.

**Ceramic Constraints**

(1) The size of ceramic tubing, i.e., length and diameter, should be kept relatively small, about 1 inch in diameter and a few feet in length. The smaller hardware keeps thermal stresses lower due to a lesser thickness requirement and it is much more practical to fabricate.

(2) It is desirable for the ceramic heat-exchanger surface to be designed in the form of relatively small complete modules. This allows for the complete module to be made in a furnace, so that on-site, ceramic-to-ceramic joints are not required. Additionally, it is advantageous to be able to replace the small complete modules when one of the inevitable, but hopefully infrequent, failures occurs.

(3) Pressure-containing, ceramic-to-metal joints should be located outside of the heat-exchanger enclosure, where the thermal strain may be readily computed on the basis of the temperature of the working fluid in the vessel and where analysis is not complicated by external heat transfer and/or a corrosive/erosive environment. Metal-to-ceramic joints should be minimized.

(4) Tubular ceramic modules should be designed to minimize thermal strains that result from bringing the equipment up to temperature. Manifold and tube supports should permit strain-free expansion where at all possible.

(5) Excessive vibrational strains should be avoided. Relatively short, stiff heat-exchanger loops will have a high natural frequency and thus tend to be out of synchronism with the relatively low frequencies, i.e., usually below 10 cps, encountered in large fired heaters.

(6) Because of the brittle nature of ceramics, and the probability that a failure under pressure will be accompanied by the flying about of pressure-propelled pieces of ceramic, all ceramic heat-exchanger surface should always be contained within sturdy furnace or heat-exchanger walls that are fabricated of metal. Additionally, any exterior ceramic manifolding should be contained within a metal shell that will prevent fragmentation. Ceramic failures should not result in structural collapse.

(7) The fired heat exchanger should be capable of operation without mechanical abuse of the ceramic tubular heating surface. Fired heat-exchanger

designs which require rodding to dislodge obstructive deposits of semifused coal ash are almost certain to sustain extensive ceramic failures. Conventional soot blowing with large powerful retractable blowers discharging relatively cold steam on air at several hundred psi pressure differential and impinging directly upon the hot ceramic tubes will be unacceptable. Water lancing, or the use of water in soot blowers, is also unacceptable.

**Coal-Firing/Ceramic Constraints and Interactions**

The material above dealt principally with constraints and guidelines for the designer which result from the properties of ceramic heat-exchanger materials in an internally pressurized heat exchanger. This section discusses the interactions between the hot ceramic surfaces and the environment of the coal-fired combustor.

Three coal combustion concepts are suitable for fired heaters in the size range of interest; the dry-bottom pulverized coal (or radiant furnace) concept, the slagging combustor, and the fluidized bed combustor.

An American coal's ash content usually ranges between 5 and 15% of its weight as-fired. The ash content exercises a strong influence on many aspects of fired heater configuration and operation. The design responses to the challenges of direct coal firing of steam boilers have been developed over many years, and are still in the process of active development. Even today, the design procedures for matching a large conventional boiler to the characteristics of a particular coal are inexact. It is not unusual to find that the operators of such equipment encounter much difficulty with unmanageable ash deposits in various sections of the fired heater, which prevent continuous operation at full capacity, and in extreme cases, cause unscheduled shutdowns. Changes in the coal supply may also be troublesome.

The large fired heaters required for heat input to CCGT powerplants and cogeneration installations will tend to be similar in many respects to steam boilers. They will rely upon the combustion technology, heat exchange surface cleaning technology, and combustion control technology developed for steam boilers. A significant difference will be the temperature of the heat-exchange surface, which rarely goes above 593°C in steam boilers, but will be 871°C to perhaps as high as 1260°C in coal-fired ceramic heaters.

The dry-bottom pulverized coal-fired heater is the standard of the industry in steam boiler design. Dry-bottom pulverized coal furnaces are, in principle, designed to cool the suspended ash particles by radiation to the containment surface to a temperature below the ash particles softening point before the ash particles can impinge on the walls. Thus, the ash that does reach the wall (which is a small fraction of the total ash), is in principle, dry and does not strongly adhere to the walls. Additionally, the gas temperature entering the convection surface is also arranged so that the ash particles are dry enough so that they do not strongly adhere to the tube banks. The accumulation of unmanageable deposits on both furnace walls and convection banks is prevented by frequent operation of powerful soot blowers, which dislodge the deposits before they can grow thick enough to insulate the walls, unmanageably raise the furnace exit temperature, and/or plug up the tube banks.

At the lower end of the temperature region of interest with fired ceramic heaters, i.e., about 982°C wall temperature, it seems likely that the differential between wall temperature and the ash softening temperature of many coals will be so small that operation of the dry-bottom concept will not be practical.

At the upper end of the region of interest, the wall temperature is so high that operation on the dry-bottom principle will be impossible with most American coals.

On the basis of the coal/heating surface temperature interaction, it appears that in the low-temperature range, the available combustion concepts for high-temperature ceramic heat exchangers will include fluidized bed combustion at a bed temperature sufficiently low that the ash particles do not soften and agglomerate to an unacceptable degree. In the higher temperature range, the only available combustion concept is the slagging combustor, either pulverized coal-fired or cyclone-fired, in which the ceramic portion of the heat exchanger is operated at a temperature so high that the ash is in a molten state and runs off the high-temperature ceramic heating surface of its own weight.

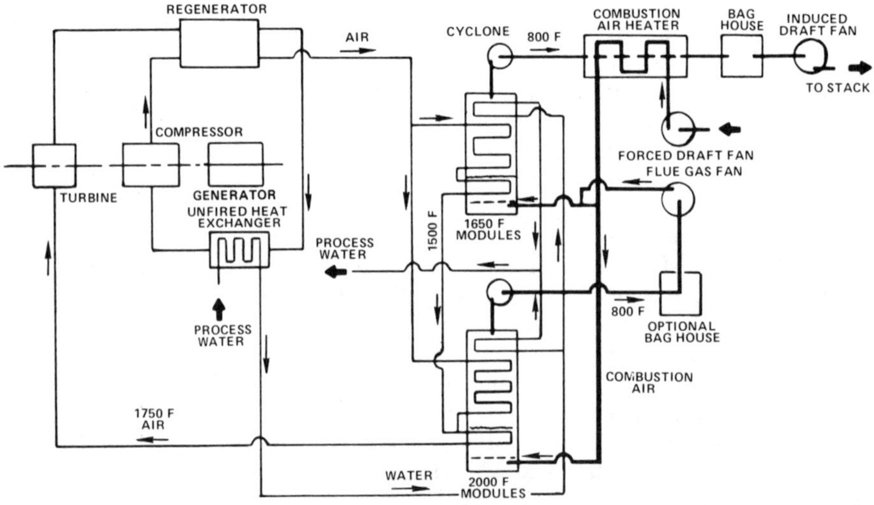


Fig. 1. Flue gas schematic.

## Descriptions of Fired Heater Designs

### Atmospheric Fluidized Bed Fired Heater for 1750°F CCGT Working Fluid Temperature

This design concept utilizes atmospheric fluidized bed combustors operating in series to provide a CCGT working fluid temperature on the order of 954°C. The flow paths for both the flue gas and the working fluid are per the typical schematic of Fig. 1. The design concept includes the following elements:

(1) The CCGT working fluid is heated to about 815°C, i.e., the dividing line between metal and ceramic heat-exchange surface, in fluidized beds that operate between 871° and 926°C and contain only metal heat-exchange surface.

(2) The working fluid is heated to the final temperature of about 954°C in fluidized beds that operate in the region of 1010° to 1093°C and contain only ceramic heat-exchange surface.

(3) All beds operate at atmospheric pressure.

(4) The combustion gases from the high-temperature beds are routed through the low-temperature beds to permit absorption of the sulfur oxides.

(5) The flue gas temperature from the high-temperature beds is reduced to about 426°C by suitable convection sections prior to introduction of the gases into the bottoms of the low-temperature beds.

(6) A fan is provided to boost the flue gas from the high-temperature beds up to the windbox pressure of the low-temperature beds, permitting balanced draft operation of the high-temperature beds.

(7) Dust collection equipment is positioned between the high-temperature beds and the boost fan to minimize fan erosion and gas distributor problems in the low-temperature beds.

The low-temperature, all-metal heat-exchanger bed design concepts are essentially identical to those described in a companion paper (Ref. 2). The description will not be repeated here.

The high-temperature, ceramic heat-exchange surface, bed design concept, is illustrated in Figs. 2 and 3. All of the ceramic heat-exchange surface is submerged in the high-temperature fluidized bed. The convection surface of the high-temperature bed modules is cooled either by lower temperature CCGT working fluid or by process fluids.

The design concept employs a modular approach to the ceramic in-bed heat exchanger. The relatively small modules of Fig. 3 would be factory assembled, including the ceramic inlet and outlet headers. Each module is 4 feet 9 inches high×12 inches wide×12 inches deep. The tubes are 1.5-inch OD by approximately 1/8-inch thick. They are projected to be of one of the several varieties of silicon carbide which are presently available, i.e., siliconized silicon carbide, sintered silicon carbide, or chemically vapor deposited silicon

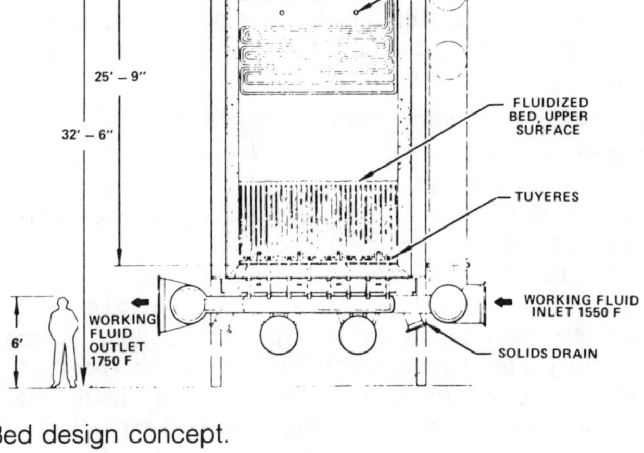


Fig. 2. Bed design concept.

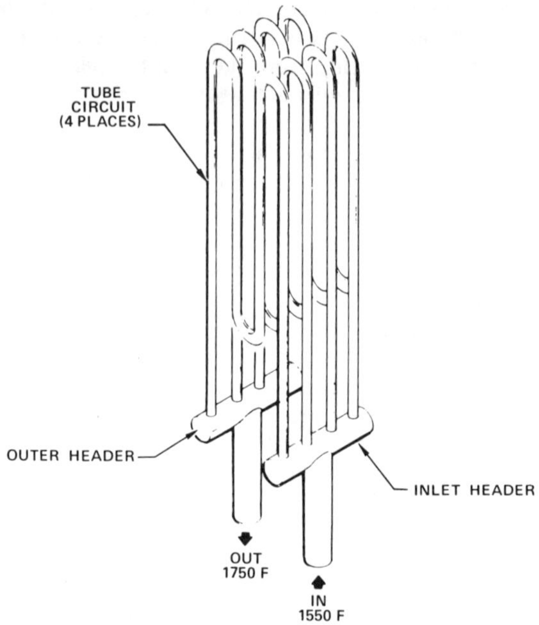


Fig. 3. Ceramic heater module.

carbide. The inlet and outlet headers of Fig. 3 are buried in inactive bed material below the level of the air distributor inlets of Fig. 2. The stub inlets and outlets of the headers project through the baseplate of the fired heater, where they are sealed relative to the baseplate to avoid leakage of combustion gases.

The stub inlets and outlets are then connected to the large circular manifolds below the combustor by means of clamp-type, ceramic-to-metal joints. The large manifolds, and their connections to the stub inlet and outlet headers of the ceramic heater modules, are of internally insulated metal construction. By arranging the support elevations of the inlet and outlet headers of the ceramic modules at different elevations, it is possible to minimize thermal strains in the ceramic tubing due to the temperature differential between the inlet and outlet.

**Discussion of Fluidized Bed Ceramic Heater Design Concepts**

The design concepts embodied in Figs. 2 and 3 are in general responsive to the design constraints previously discussed. It is recognized that there are very many unknowns, and much development has to be accomplished prior to the achievement of commercially successful designs.

The entire concept of series fluidized bed operations, with sulfur oxides originating in the high-temperature beds being captured in the low-temperature beds, needs extensive development to determine its technical, operational, and economic feasibility. The technical questions relate to whether the sulfur oxides will be absorbed. The operational questions relate to whether the gas distribution system of the low-temperature beds can be kept reliably functioning while accepting dusty flue gas from the high-temperature beds. The economic questions relate to the justification of the flue gas transfer fan, the dust removal equipment, and the greater low-temperature bed area required by the series bed concept. With the series bed concept, all of the flue gases are passed through the low-temperature beds, so it is likely that the low-temperature bed plan area will be the same regardless of the amount of coal burned in the high-temperature beds. This factor tends to make it economically advisable to limit the amount of coal burned in the high-temperature beds to the minimum required to heat the working fluid from the maximum temperature at which metal heat-exchanger surface can be used to the target final temperature.

Several alternatives to the series bed concept can be visualized in the event that one or more of the factors discussed above makes their use impractical. The high- and low-temperature beds could be run in parallel, and the flue gases leaving the high-temperature beds could be cleaned of sulfur oxides either by wet scrubbing or by the developing technology of dry scrubbing via spray-drying techniques. It seems likely that the bed drain material from both the high- and low-temperature beds, but especially the high-temperature bed, will be sufficiently high in calcium oxides to provide the reagent for either wet or dry scrubbing. Proposing wet or dry scrubbing for fluidized bed flue gases may seem like heresy, but it should be recognized that with high-sulfur coals, each of these techniques requires very much less limestone consumption than does fluidized bed SOX removal, and this offsets the extra capital involved.

The high-temperature bed operations depend on staying sufficiently below the bed temperature at which an unmanageable agglomeration of large ash particles occurs. That critical bed temperature is probably a function of the coal ash characteristics and, to a lesser degree, of the limestone characteristics. Development will be necessary to identify proper design and operating techniques, and to categorize coals and limestones for suitability for this type of operation.

The ceramic aspects of the design concept will also require extensive development. The resistance of the silicon carbide family of ceramics to oxidation is very good, but the resistance is dependent on the presence of a protective layer of silicon oxide on the surface. If the continual peppering of the surface with particles of bed material rapidly erodes the protective $SiO_2$ layer, then tube life may be unacceptably short. Development in this area will identify interrelationships between the ceramic surface temperatures, and bed velocity. If an economically acceptable compromise between bed velocity and silicon carbide corrosion/erosion cannot be achieved under oxidizing bed conditions, it may be necessary to operate the high-temperature bed under slightly reducing conditions.

The modular approach to the ceramic in-bed heat-exchange surface resolves many of the identified problems in ceramic design. Ceramic-to-ceramic field joints are unnecessary. Maintenance, i.e., replacement of failed tubes, can be achieved by replacement of the entire module. The placement of the ceramic module inlet and outlet headers in inactive bed material permits their relatively unrestricted expansion and yet shields them from temperature gradients that would be experienced if they were in the active bed. The serpentine design of the module, and the support arrangements for the headers, will permit the minimization of thermal strains due to heatup, cooldown, and working fluid temperature rise through the ceramic circuit. The relatively small size module and small tube sizes do not greatly exceed present-day manufacturing capabilities. The modular approach permits factory quality control testing of each module prior to shipment, including hydrostatic testing. And finally the design permits compliance with the safety criteria that all pressurized ceramic circuits be completely enclosed and that no significant structural collapse accompany ceramic circuit failures.

Analysis, development, and design refinement of the ceramic heat-exchange surface configuration is needed in several areas. Metal-to-ceramic joint configurations need development to minimize cost, leakage, and unreliability. The support system needs thorough and sophisticated analysis, and design refinement, to minimize thermal strains. The support system and the fluidized bed combustion system need joint development to avoid excessive vibrational strains. The module concept of Fig. 3 shows no tube spacers, but they may be required. The designer has many options with respect to module configuration to control vibrational behavior, including tube sizing, tube lengths, and various spacer concepts. However, one adds spacers at the price of complicating the thermal strain picture. Considering the vibratory nature of many fluidized bed operations, the development of satisfactory design configurations may well be a difficult development task.

#### Cyclone Combustor Fired, Slagging, Ceramic, Fired Heater Design

This preliminary design was intended to explore the area of the presently foreseeable highest possible CCGT working fluid temperature, i.e., on the order of 1232°C. The combustion concept selected to meet this objective is the cyclone combustor, slagging (i.e., wet-bottom) furnace.

An artist's isometric concept of a 350 MWe embodiment of the design is presented in Fig. 4. The single 350 MWe unit is approximately 175 feet high, 60 feet wide, and 125 feet deep. Crushed coal and hot air are admitted to the cyclones where combustion is intensive enough to raise the gas temperature to over 3000°F. These hot gases pass over ceramic tube elements in which the CCGT working fluid is heated from 843° to 1232°C. Molten slag accumulates on the boundary walls and the ceramic heating surface and drains off (by gravity) to the slag tank. Approximately 50% of the ash content of the coal is collected this way. The hot combustion gases, after having been cooled to approximately 1426°C in the ceramic heater zone, are discharged into the tall radiant furnace, and travel up the furnace to the exit at the top, by which time their temperature has been reduced to 1148°C. Conventional metal CCGT and steam working fluid cooled convection surfaces are provided in the pendant and horizontal convection banks to cool the gaseous products of combustion from the 1148°C to approximately 426°C prior to their discharge to the air heater.

This design is dominated by the requirement for the 1232°C working fluid temperature. The ceramic materials will operate at temperatures up to about 1260°C and will be continuously exposed to a running slag condition at that temperature. Soot blowing of the ceramic tube sections is not planned and is contraindicated. The demands on the ceramic tubing materials will be more severe than those of the fluidized bed concept discussed above. The design choice is sintered or chemically vapor deposited silicon carbide.

Two alternative concepts for the arrangement of the ceramic surface in the slagging heat exchange zone were defined. In one concept, ceramic tubular modules, similar in many respects to those described for the high-temperature

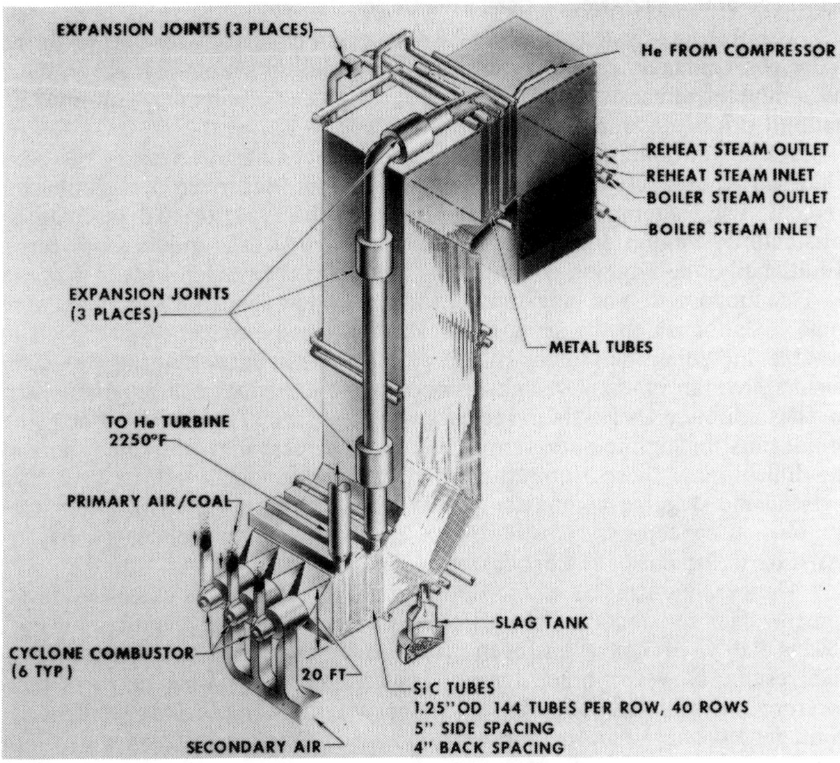


Fig. 4. 1232°C cyclone fired combustor heat exchanger.

fluidized bed design, were arranged to protrude horizontally into tunnels connecting the cyclone combustors with the vertical radiant furnace. This system provides many of the advantages of fabrication, installation, maintenance, and limitation of thermal strains that have been described for the fluidized bed system. However, the containment is complex, the ratio of containment surface to ceramic surface tends to be relatively high, and there are many containment wall penetrations and much manifolding to accommodate. Also, there is no inert bed material equivalent with which to insulate the ceramic module inlet and outlet headers from the hot combustion gases. The alternative concept is shown in Fig. 4. It incorporates much longer ceramic tubing, which is hung in pendant loops from the roof of the containment.

In each instance, the containment of the cyclone combustors themselves, and of the slagging zone which encloses the ceramic surface, is constructed of water-cooled tubing similar to that utilized in steam boiler construction. The steam or hot water produced in these containment structures is utilized in a bottoming Rankine cycle. The design concepts provide that all ceramic pressure parts are contained within the water-cooled containment walls and that failure of any ceramic pressure part will not result in either flying ceramic pressure part fragments in the working spaces of the plant nor the structural collapse of any of the containment structure.

### Discussion of Slagging Cyclone Combustor Design Concepts

The slagging cyclone ceramic heater design concept is responsive to the major constraints on very high-temperature coal fired heaters. However, there are a number of serious questions and development problems which must be resolved prior to commercial application.

For steam boiler applications, the slagging combustor concept has been just about completely displaced by the dry-bottom pulverized coal combustor concept. The slagging combustor is limited in the types of coal that can be satisfactorily combusted (many coals have too high an ash liquid temperature). Additionally, the slagging combustor is limited in the range of loads that can be accommodated. The slag tends to freeze at low loads. While there were some coals for which the slagging cyclone combustor concept was especially suitable, including the famous Illinois coals, it also appears that slagging combustors have an inherently high nitrogen oxide emissions characteristic, and this has inhibited their sale in recent years. The special attributes of slagging combustors for high-temperature CCGT fired heaters may overcome some of the difficulties with coal limitations and load range, but it will be necessary to bring the slagging combustor's NOX emissions on a par with alternative combustion concepts. A possible NOX emissions control technology may be based on treatment of the exhaust gases with ammonia.

The ceramic aspects of the design concept will require extensive development. The resistance of the silicon carbide ceramic to degradation in a flowing slag environment has been preliminarily explored (Ref. 3) with favorable results. However, much longer term testing over a wide range of temperature and coal ash conditions is obviously required. The long pendant ceramic loop concept requires the development of fabrication technology. Both concepts require extensive design refinement in the areas of ceramic surface support, avoidance of thermal and vibrational strains, and achievement of viable wall penetration details.

In the combustion technology and heat-transfer area, development is needed on the arrangement of the ceramic surface in the slagging zone to ensure a free-draining slagging environment that will not plug with solidified ash deposits which would limit capacity and/or require mechanical cleaning.

### General Ceramic Design and Fabrication Needs

A number of general design and fabrication needs became apparent during this study.

*Fabrication*: Technology for manufacturing longer tubes, U tubes, and serpentine tubes needs to be developed. And these tubes need to be fabricated with a higher assurance of adequate quality control.

*Joining*: Techniques for joining SiC-to-SiC and SiC-to-metal on-site are required for assembling combustor/heat-exchanger (C/HX) modules.

*Nondestructive Evaluation*: Techniques for identifying critical defects from 10 μm on up in size are required to ensure the necessary high reliability for commercial plants. These techniques must be economical and efficient because the surface area in large C/HX facilities will be extensive.

*Design Property Data*: Design data on thermal and mechanical properties are inadequate. Strength data in particular are required for determining a safe design stress level. The available data have been obtained on small test bars broken in bending, and with a diamond-ground surface that is totally different from that of the as-fabricated surface. Tensile data on specimens with as-fabricated surface is needed for developing reliable design strength levels. It is also necessary to determine the extent of variation from batch to batch of different materials.

*In-Service Strains*: Some knowledge of the actual strains generated on heat-exchanger tubes in service must be obtained. Examples include 20-foot long U tubes suspended in an SCC and serpentine tubes protruding into a fluidized bed.

*Durability*: Much more information in regard to the durability of SiC tubes under C/HX service conditions needs to be known. Little, if anything, is known, for example, on whether there is any effect of static or cyclic fatigue on the strength of the different types of SiC. Little or no slow crack growth data have been obtained in any C/HX environment. Resistance to thermal stress of different types of SiC, of different tubing shapes and sizes, and of joints, is difficult to predict. Other phenomena that have not been evaluated are the oxidation-erosion mechanisms on the surface of SiC in the abrasive AFB environment, and the long (in terms of years) corrosion resistance of SiC to molten coal slag.

### Conclusions

(1) Coal-fired high-temperature ceramic heat exchangers for CCGT service will require quite different designs than those for the lower temperature, all metal HEX:

(*a*) The heat-exchanger surface temperature will approach, or exceed, the coal ash fusion temperature, limiting the available combustion concepts.

(*b*) The negligible ductility of ceramics, plus limitations in their fabrication, will require novel configurations.

(2) Coal-fired CCGT heater design concepts are available for the 871° to 1232°C range.

(3) Sufficient study background is now available to support a CCGT fired heater ceramic component development program.

### Acknowledgment

This study was funded by the U.S. Department of Energy, Heat Engines and Heat Recovery under Contract Number EF–77–C–01–2612. The Program Manager was Mr. Carey Kinney.

Reprinted by permission of The American Society of Mechanical Engineers.

### References

[1]C. F. McDonald, "The Role of the Ceramic Heat Exchanger in Energy and Resource Conservation," ASME Paper #79-GT-106, 24th Annual International Gas Turbine Conference and First ASME Solar Energy Conference, March 11–15, 1979, San Diego, CA.

[2]D. E. Wright, J. Campbell, and J. C. Lee, "Coal-Fired, Closed-Cycle Gas Turbine Cogeneration Systems"; presented at the 25th Annual International Gas Turbine Conference, March 9–13, 1980, New Orleans, Louisiana.

[3]R. A. Perkins and H. W. Lavendel, "Reaction of Silicon Carbide with Fused Coal Ash," EPRI–294, November 1976.

# Ceramics in Heat Exchangers—An Overview

Bryan D. Foster

Norton Co.
Worcester, MA 01606

This paper presents a broad review of current applications of ceramics in heat exchangers in the United States. The various programs and design approaches are described and discussed and are compared where appropriate.

There is a growing acceptance of ceramics in heat-exchanger applications where, previously, only metals were used. Further, there are many new opportunities emerging for heat exchange that were not possible with nonceramic materials. This has led to a proliferation of applications and design approaches which suggests a need for a score sheet to put the situation in perspective. For example, we have identified some 15 general types of heat exchangers described and nearly 40 combinations of vendors with competitive approaches. In addition to the reference aspect of this review, it is hoped that broader awareness of current successful use of ceramics in this field will encourage more experimentation and application.

The applications have been divided into three broad categories: (1) waste heat recovery, (2) process heat muffling, and (3) energy conversion.

### Waste Heat Recovery

Probably the best-known use of ceramics in heat exchangers is that of regenerators for blast furnaces and glass tanks. These regenerators incorporate checker brick and cycle back and forth, soaking up heat from exhaust gases and later giving it up to preheat combustion air for more efficient burning of fossil fuels. Later regenerators have used rotating disks of ceramic honeycomb or moving beds of ceramic grain to increase efficiency of automotive gas turbines or of gas-fired furnaces.

Recuperators represent an alternative way to recover waste heat from exhaust gases and give it up to preheat combustion air. They typically operate at a constant temperature—conducting the heat through the wall of a tube or finned plate which separates the clean air from the dirty flue gases—or conducting the heat along the length of a heat pipe which passes through a barrier separating the air from the exhaust.

#### Tube-Type Recuperators

Tube-type recuperators are experiencing the greatest amount of activity at this time—and the most competition—with five well-known vendors currently having products available, or in advanced stages of development. The five design approaches are illustrated in Fig. 1. All are capable of providing

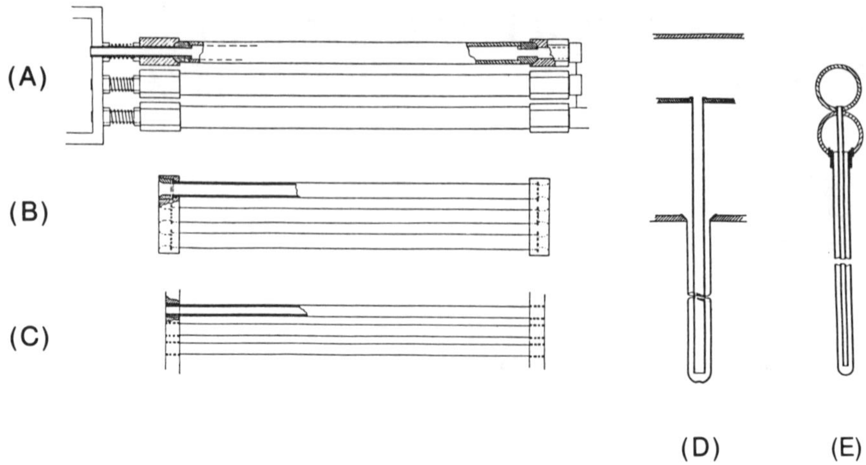


Fig. 1. Schematic illustrations of five design approaches of ceramic tube-type recuperators.

preheated combustion air at 537°C or higher from dirty flue gases at temperatures as high as 1537°C, and without excessive leakage. All use some form of internal surface extension, such as ceramic or metal inserts, "core busters," or integral molded fins or cruciforms. All are described in detail elsewhere in this book.

**Tube-in-Shell**

Design A has been promoted and sold by Hague International for the past several years. The side walls are constructed of hexagonal blocks which nest to form a containment shell for the exhaust gases. Tubes are inserted between the two walls and are held in place by end pieces with spherical surfaces seated in hemispherical recessions in the hexagonal blocks. Each assembly is kept in compression by a heavy metal spring. Tubes are silicon carbide with various bonds, depending on the application and position in the recuperator. They are typically 3 to 4 in. OD by 48 in. with 1 to 2 in. spacing between tubes. Some tubes may have external fins.

Design B is more recent but is now commercially established by C & H Combustion Co. It has no springs and the side walls are constructed of tongue-and-grooved tile with threaded plugs holding the tubes in place and providing a means for tube replacement or hole plugging without shutting down the recuperator. Fiber ceramic gaskets enhance butt joint seals. Tubes are typically 2 in. OD by 48 in. long or larger with minimum 2 in. spacing between tubes. A preheat air temperature capability of 2400°F is said to be feasible.

Design C has been introduced by Garrett AiResearch and will soon undergo its first field scale trial in a forge furnace flue gas. Side walls are large, 2-in. thick plates with holes molded in. It has neither springs nor screw threads, but relies on a narrow annulus ring of specially formulated glass to provide the seal and to hold the tubes in place. Tubes are replaced at room temperature by removing the seal glass with a hole saw. The initial tube set-up uses sintered $\alpha$-SiC with integral cruciform internal surface extension. Tubes are 1.5 in. OD by 52 in. long with 1 in. spacing between tubes.

Another tube and shell recuperator—made by Encomech—has been successfully used in the United Kingdom for some time. It is said to use large molded blocks and plugs for end walls and, typically, uses 6 in. OD by 72 in. long tubes made of silicon carbide, or other ceramics as the application and economics dictate. The distance between the tubes is typically 20 to 40% of tube diameter and they can be replaced without major shutdown.

#### Tube-in-Tube

Turning to the bayonet or tube-in-tube style of recuperator, design D by Babcock & Wilcox has undergone laboratory prototype testing and is scheduled for full-scale testing in a B & W steel soaking pit flue sometime in 1985. The tubes are suspended vertically in the flue with gravity pressure on gasketed flanges providing the seal. Outer tubes (3.75 in. OD by 76 in. long with a wall thickness of 0.25 in.) are made from slipcast and siliconized SiC. Inner tubes can be metal, but silicon carbide was chosen for the hottest part of the recuperator for added life and safety. Tubes can be replaced by pulling from overhead when the unit is at room temperature.

Design E was developed by Solar Turbine International and has undergone extensive prototype testing. The outer tubes are 2¾ in. OD by 60 in. and are made of sintered $\alpha$-SiC; the inner tubes are metal. The seal is provided by a proprietary bonding of the silicon carbide to a threaded metal sleeve which is then screwed into a metal header. Each header has a series of tubes which hang down into a horizontal flue; they can be replaced by removing individual header assemblies.

Other tube-type recuperator designs and other vendors certainly exist, but the preceding descriptions give a good idea of the range of current offerings.

Note that tube-type recuperators are typically designed to handle 5 to 20 or more million Btu's/h (designs have been developed for up to 100 million Btu's—the major difference being that on-site construction would be required). The openness of tube spacing reduces the possibility of clogging when used with dirty exhaust such as that from aluminum remelt furnaces. Open spacing also offers the opportunity for soot blowing to remove buildup of deposits.

#### Finned Plate Recuperators

Finned plate recuperators have relatively small air and flue gas passages, are typically sized for only 0.2 to 1.45 mm W (0.5 to 5 mm Btu/h), and are limited to clean flue gases such as from forge furnaces. Three current designs are illustrated in Fig. 2.

Design No. 1 has been promoted and sold by GTE and others for several years. It is made of bonded layers of finned cordierite plates in a typical 12-in. cube with a maximum flue-gas temperature of 2500° C.

Design No. 2 was recently introduced by Coors Porcelain Co. after extensive field testing. It is slightly larger and reportedly has better thermal shock resistance and higher temperature capability (to 2700°C) owing to a proprietary mixed-oxide formulation. Cubes are again constructed from bonded, stacked, finned plates to form a monolithic structure.

Design No. 3, developed by AiResearch, uses larger, individual Si/SiC plates with freestanding fins. The plates are held together mechanically with thin gaskets providing separation and seal on the closed edges, allowing movement of individual plates and permitting the use of large plates without excessive thermal gradients or stress. This also would allow disassembly for cleaning or repair. A 1.45 mm W unit is scheduled to be field-tested in 1985.

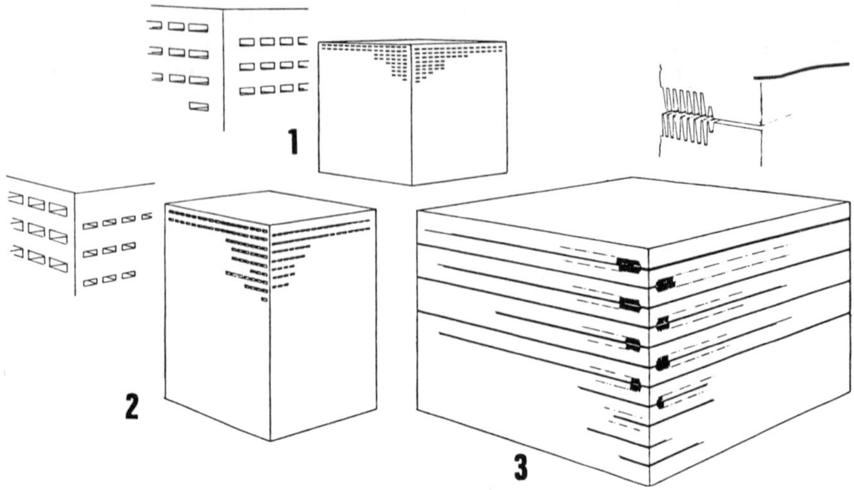


Fig. 2. Schematic illustrations of three designs of ceramic finned plate-type recuperators.

#### Heat-Wheel Recuperators

Heat-wheel recuperators based on honeycomb structures of cordierite and other ceramic materials have been under trial and development for several years, with Corning Glass Co. leading much of the effort. No known waste-heat recovery applications have been achieved due primarily to seal problems, but Midland Ross recently developed an improved system design and an active test and development program is underway.

#### Heat-Pipe Recuperators

The use of heat pipes as a potentially practical technology for recuperation is a relatively new development. Thermo Electron Co. reports elsewhere in this book on working models produced by chemical vapor deposition (CVD) techniques which have achieved some remarkable heat transfers. Much work remains to be done before commercialization, but an active program is underway.

#### Moving Bed Regenerators

Moving beds recover waste heat by passing flue gas through a material (typically alumina grain) or vice versa and then moving the material to the clean side of the process. Three such process approaches are currently in development or trial.

Aerojet General has a trial operating in an aluminum melting environment where the waste heat is used to heat water for process steam and the thermal transfer process is enhanced by fluidizing the bed of grain around the boiler pipes. See the detailed report elsewhere in this book.

Thermo Electron Co. has a field trial preheating combustion air for an aluminum melting furnace, also reported in this book.

A significant problem with both of the above units is the distributor plate—now metal—which feeds the hot and sometimes dirty flue gas into the bottom of the fluidized bed. These plates obviously limit the corrosive atmo-

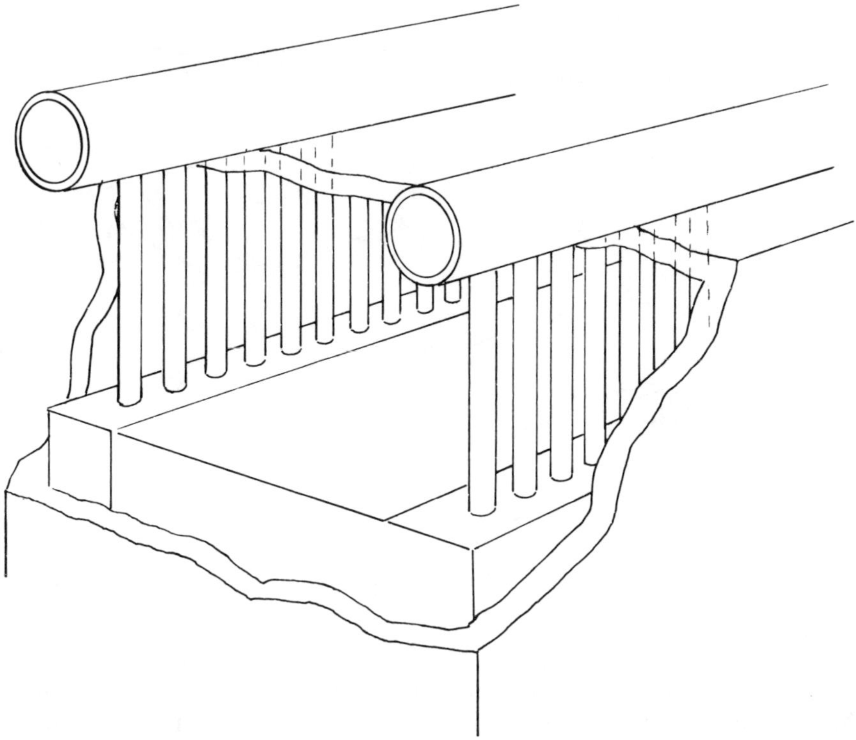

Fig. 3. Schematic illustration of vertical radiant tubes in a gas-fired heat-treating furnace.

spheres and maximum temperature of the flue gas and hence the effectiveness of the recuperator. The use of ceramics for distributor plates is under consideration.

Southwest Research Institute has proposed a raining bed approach which eliminates the distributor plate but introduces some other challenges.

Another problem with distributor plates is clogging. Wire brushes are being used with some success, but a better way is needed. Aerojet is pursuing a ceramic rod-type distributor plate arrangement with a series of rigid and moving rods—just getting started in 1985.

**Process Heat Muffling**

The second major category of heat exchangers is process heat muffling. This includes ceramics firing, metal heat-treating, metal melting and holding, calcining powders, and coal gasification and petrochemical reaction vessels. Here the concern is to get heat into a product or process while keeping the heat-source atmosphere separated from the process atmosphere for various reasons.

**Muffle Kilns**

Ceramic firing kilns often use muffles. These range from huge sanitary-ware tunnel kilns with large 2-in. thick muffle slabs to very small bench-type

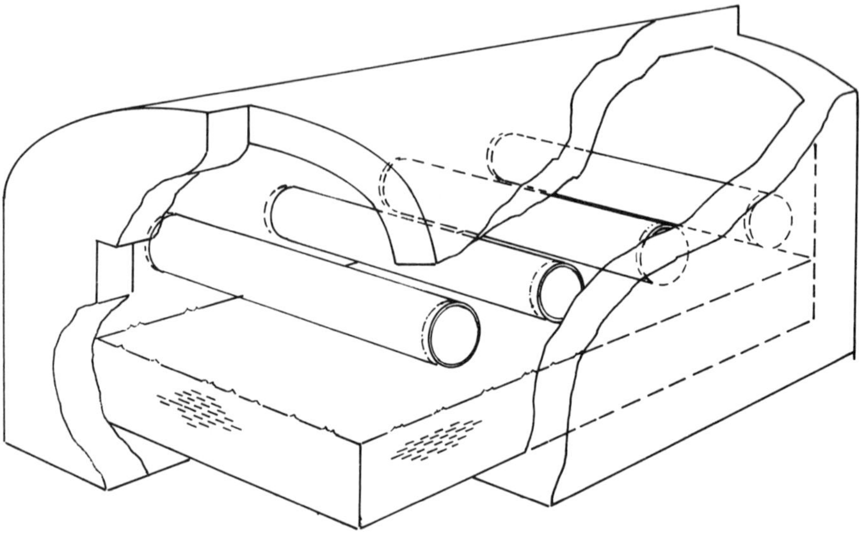

Fig. 4. Schematic illustration of horizontal radiant tubes in a gas-fired aluminum melting furnace.

lab kilns with dense, high-purity muffle tubes. They are another form of ceramic heat exchanger that has demonstrated dependable, economical performance over the years—often in competition with metals. Another example is the recent successful substitution of silicon carbide for alloy tubes for heat-treating continuous saw blades. In these situations the requirement is usually to contain the product being heated and keep the heat source on the outside.

**Radiant Tubes**

In many furnace situations the heat source must be contained and still provide heat to the workstuff located near the muffle. Radiant tubes are a good example of this approach.

Ipsen gas-fired heat-treating and carburizing furnaces have long used 3-in. diameter ceramic muffle tubes mounted vertically in an arrangement something like that shown in Fig. 3.

More recently, Holcroft has developed an aluminum melting furnace using 8-in. diameter SiC tubes mounted horizontally over the bath with gas burners firing through the tubes, as illustrated in Fig. 4. Combustion products are kept away from the metal for less dross and greater productivity.

Midland-Ross has been working on a closed-end tube design for higher temperatures and/or greater throughput in heat-treat furnaces. A smaller-diameter inner tube introduces burning gases and enhanced heat transfer is achieved by use of a proprietary arrangement in the annulus. Here again the competition is alloy radiant tubes, which have been used for many years. A typical arrangement would look something like Fig. 5.

North American is also working on an advanced radiant tube burner design and Aichelin, LOI, and Ipsen and others in Europe all have well-established designs in operation.

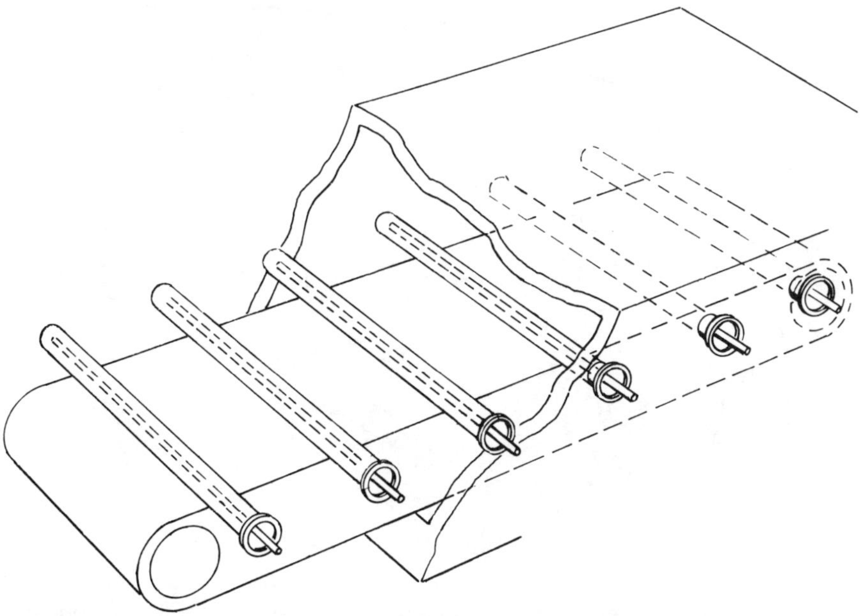

Fig. 5. Schematic illustration of closed-end gas-fired radiant tube system.

#### Roof Muffles

Another type of muffle is the roof or arch, by which burners are fired overhead and the material being heated needs either a special protective environment or protection from products of combustion. Examples of powder calcining and aluminum melting applications are illustrated in Figs. 6 and 7. Powder calcining is a proven application in which a silicon carbide system replaced an alloy metal system and saved about 12 in. of height and allowed a much wider span as well as good protection against damage in case of an inadvertent temperature overshoot. This is a seven-foot wide span operating at 1204°C.

#### Immersion Heaters

For many years, immersion heater tubes made of silicon carbide have been in service in furnaces made by Lindberg, Warwick, Holden, and others. Typically, they have electric resistance heaters inside. Babcock & Wilcox is now working on an innovative gas burner to be used in an immersion heater tube for aluminum melting and holding furnaces.

#### Reactors

Reactors for coal gasification or chemical products can benefit from the high thermal transfer and temperature and corrosion resistance of ceramics. Exxon has a pilot operation using alumina tubes (silicon carbide is also a candidate) in conjunction with their "flexicoking" process.

#### Energy Conversion

The third category of heat exchangers relates to energy conversion such as solar collection or burning of low-grade fuels and even more efficient power

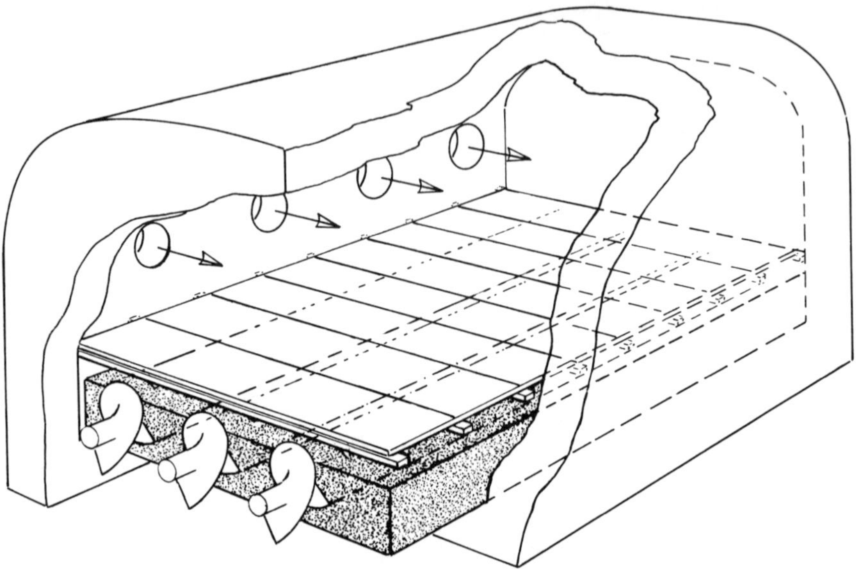

Fig. 6. Schematic illustration of a beam and plate ceramic muffle in a powder calcining furnace.

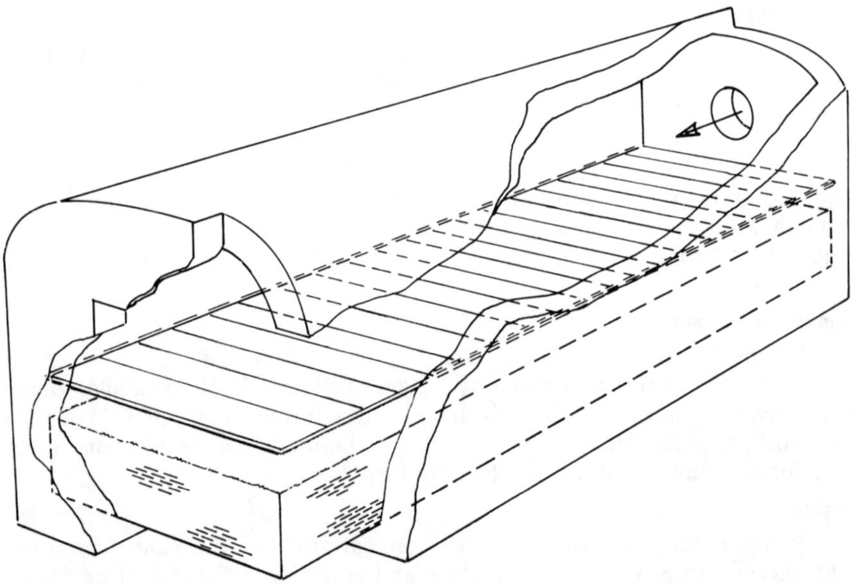

Fig. 7. Schematic illustration of a thin-plate ceramic muffle in an aluminum melting furnace.

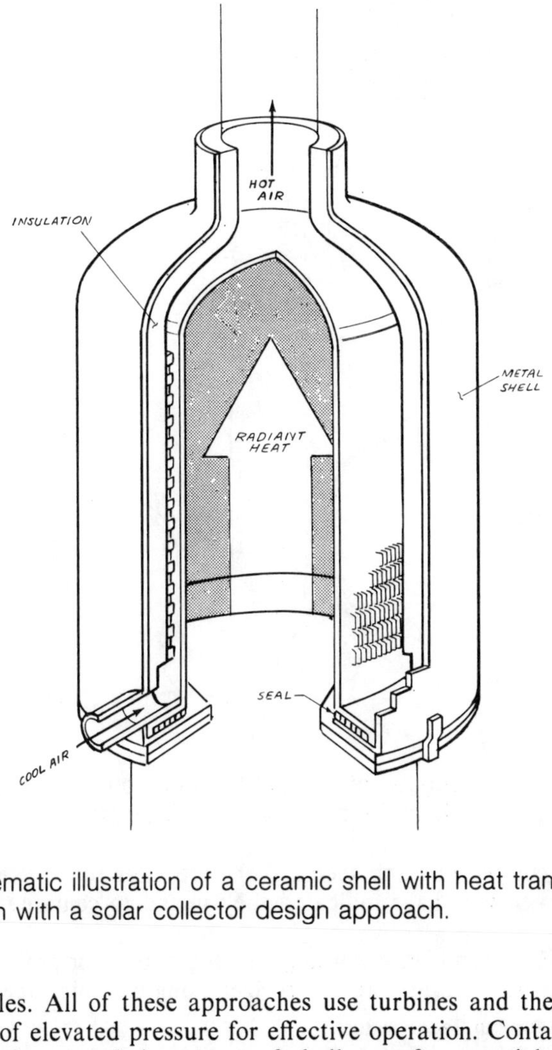


Fig. 8. Schematic illustration of a ceramic shell with heat transfer fins used in conjunction with a solar collector design approach.

for automobiles. All of these approaches use turbines and therefore require some degree of elevated pressure for effective operation. Containment of this pressure introduces a whole new set of challenges for material properties and joint and seal designs.

**Solar Collection**

There are three solar collector projects using ceramics of which we are aware. The first involved a transfer shell with fins, which was developed by AiResearch and Norton Co. in conjunction with the Jet Propulsion Laboratory in Pasadena (Fig. 8). The working fluid was to be air at 0.3 to 1.04 MPa (50 to 150 psi) in an open cycle to drive a turbine for electric power. Design and construction were completed a few years ago, but no funds have since been made available for testing and operation.

The second solar collector involved long SiC tubes in a 1 MW bench model solar receiver designed by Black & Veatch under contract to EPRI. The working fluid was helium at 1.04 MPa and temperatures up to 1537°C (also for turbine electric power). The project never achieved successful oper-

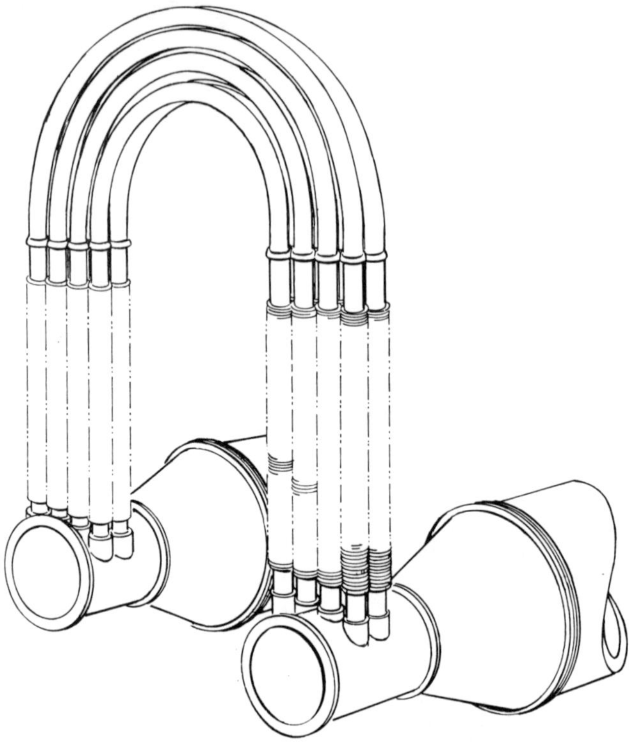

Fig. 9. Schematic illustration of integral finned ceramic tubes and manifold assembly designed for heat transfer from a fluidized bed of burning coal.

ation and the components and technology were later sold to the Weizmann Institute in Israel, who have indicated they intend to continue development efforts.

Third, AiResearch is now working on a solar collector project for making hydrogen fuel with a sulfuric acid vaporizer using ceramic tubes.

**Automotive Gas Turbine Regenerators**

Regenerators, as used in gas turbine engines for vehicular power, are usually heat-wheels using cordierite honeycomb with very small cells (0.020 in.) and very thin walls (0.003 in.). Heat is taken from exhaust gases and used to heat the compressed air before it goes into the combustor. Pressures range up to 1.04 MPa so leakage around this moving wheel is a major concern. Temperatures are relatively low at 815°C, but ceramics are less expensive than the metals that might otherwise be used. Ford, General Motors, and AiResearch all have programs with the heat-wheel technology pretty well in hand. What is needed is an adequate ceramic turbine—a subject beyond the scope of this paper.

**Combustion Of Low-Grade Fuel**

Heat extractors or heat-source heat exchangers typically involve burning low-grade or dirty fuel—often in a fluidized bed—to heat a pressurized working fluid to drive a turbine for power generation. High pressures are attractive from the standpoint of allowing smaller turbines, but pressure-tight joints with other components are a major technical problem for ceramics. However, ceramics are essential for temperature and corrosion (slag attack) resistance, so a strong development effort continues.

AiResearch conducted a program on a unit such as that shown in Fig. 9 which proved the heat transfer and joint seals concepts and Norton Co. proved that the shape could be made and be gas-tight. But at present no additional work is underway.

Rocketdyne is currently working with straight tubes about 1 in. in diameter by 48 in. long but they are not expected to be commercial for some time due to ceramic-to-metal seal problems.

Solar Turbine International has a program with Argonne using 10-ft. long tubes joined to SiC headers with a pressurized working fluid in a coal-gasification application. No immediate commercial application is anticipated, but seal experience is being gained.

Hague International is also involved with a pressurized heat-extractor application in their contract with the Bonneville Power Authority in Oregon. The program is on hold due to lack of need for the electric power, but principals believe it will be built and will perform. The fuel is to be wood chips. Temperature is to be around 871° to 1093°C with pressure around 1.04 MPa. The same basic tube and shell construction is used as was described earlier. Encomech in the United Kingdom is also working on a higher pressure capability, which could possibly make it viable for gas-turbine service.

### Summary

As can be seen from the extensive array of heat-exchanger applications involving and depending on ceramics, we can look forward to an increasingly important role in energy conservation and industrial development. New awareness, coupled with new materials and applications engineering, will help continue and expand this role.

Section III

# Materials Development and Evaluation

## PRODUCT MANUFACTURING



## PRODUCT TESTING AND EVALUATION PROGRAMS

# Development of Cost-Effective Ceramic Tube Process for High-Temperature Heat-Recovery Systems

D. W. Roy, K. E. Green, and M. Rivkin

Coors Porcelain Co.
Golden, CO 80401

A cost-effective material/process has been developed for fabrication of SiC tubes which satisfy many of the structural and operating environment requirements for high-temperature heat-recovery systems. Available property and corrosion data plus cost projections are presented.

In recent years much attention has focused on ceramics for incorporation in future high-temperature heat-recovery systems that can operate above the 926°C temperature limits of metallic recuperators without the use of dilution air.[1-3] Recognizing that cost-effective and reliable ceramic is an essential element of the overall program, the Gas Research Institute has contracted with several ceramic companies, including the Coors Porcelain Company, to address the challenges in supplying ceramic components. The basic goal is to develop a ceramic formulation and process which would yield a material with high reliability, that could be sold for about $15 per linear foot for an unfinned one-inch diameter tube, with an ⅛-inch wall.

### Procedure

A four-phase, cost-sharing contract, initiated at Coors in 1982, is outlined in Table I. Phase I has been completed and we are now in the final stage of process development. Concurrently, process demonstration and material testing (Phase III) have been initiated.

Table I. Program Outline for Development of Cost-Effective Ceramic Tube Process for High-Temperature Heat-Recovery Systems

| | |
|---|---|
| Phase I | |
| | • Establish material specification |
| | • Establish design specification |
| | • Preliminary material development |
| Phase II | |
| | • Material formulation and development |
| | • Cost-effective process development |
| Phase III | |
| | • Process demonstration |
| | • Testing |
| Phase IV | |
| | • Commercialization |

Phase I was dedicated to establishing material and design specifications through personal contacts with designers, system fabricators, and end-users. The material and design specifications are summarized in Tables II and III, respectively. Based on prior test programs conducted by other contractors, reaction-bonded silicon carbide (RBSC) was selected as the candidate material with the highest probability of success. During Phase I, a limited effort was applied to material development. Process development was initiated in Phase II.

Table II. Material Specification for Low-Cost RBSC Tube Process

| Property | Specification |
|---|---|
| Specific gravity ($g/cm^3$) | $3.00 \pm 0.10$ |
| Apparent porosity (%) | $\leq 3$ |
| Flexural strength (3 pt. loading) | $\geq 25\,000$ psi |
| Coef. of thermal expansion ($\times 10^{-6} \cdot {}^{\circ}C^{-1}$) | $4.5 \pm 10\%$ |
| Modulus of elasticity ($10^6$ kpsi) | $50 \pm 15\%$ |
| Thermal conductivity (W/(m·K) | $80 \pm 20\%$ |

Table III. Design Specification for Low-Cost RBSC Tube Process

| Dimension | Specification |
|---|---|
| Outside diameter | $\leq 2.00$ in. |
| Wall thickness | $0.125 - 0.250 \pm 0.010$ in. |
| Length | $\leq 50 \pm 20\%$ in. |
| Camber | 0.006 in./in. (or 1%) |
| Roundness | $\pm 0.010$ in. or 3% |

For Phase II process development, extrusion was selected as the forming method which would maximize reliability with respect to dimensional control and ceramic quality as well as providing the most cost-effective high-volume process.

There is a wide selection of SiC raw materials as regards cost, purity, particle-size distribution, source, and availability. Therefore, a major effort was expended during Phases I and II to identify cost-effective formulations which could be extruded and fired, and which subsequently yielded tubes that met or exceeded the material and design specifications noted previously. Concurrently with the formulation and processing studies, in-process quality assurance procedures were established to ensure quality, reproducibility, and reliability of the ceramic components.

As a result of the Phase I and II efforts, three formulations have been developed. Each formulation can be extruded and fired with high yields of tubes which meet the design specification. Typical microstructures of the fired ceramic for the three formulations are shown in Fig. 1. Note that there is a significant difference in the average crystal size of the silicon carbide phase as well as in the size and distribution of the silicon metal phase.

The microstructural differences result in some variation in the mechanical properties, as will be shown later. It is anticipated that the microstructure may

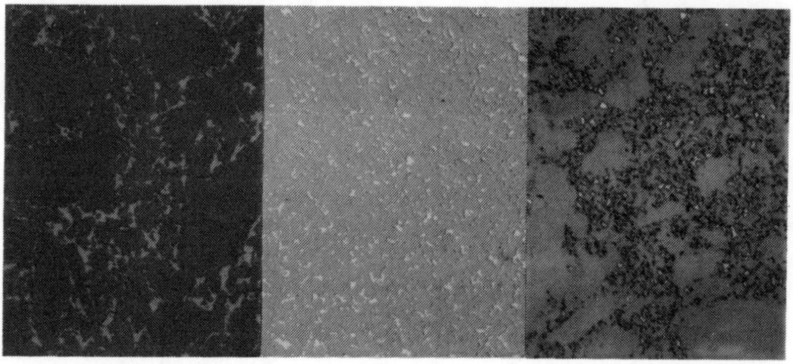

Fig. 1. Microstructure of three Coors/GRI formulations.

also have an effect on flue-gas compatibility. Whereas it is expected that all three formulations will perform satisfactorily in many high-temperature environments, differences in grain-boundary area as well as variations in the silicon carbide and silicon metal phases may affect the resistance to corrosion and/or erosion by particle impaction.

The program is now in Phase III (process demonstration and testing). The process demonstration consists of fabricating small quantities of extruded tubes (up to five feet long, diameters up to a two-inch maximum). (Larger diameter tubes can be fabricated using slip cast technology.)

The test program which has recently been initiated is outlined in Table IV. It is anticipated that, if the results of the initial tests are encouraging, the program will be broadened to determine the compatibility of the silicon carbide with environments other than aluminum remelt.

Table V is a summary of the physical properties of the three formulations which have been developed on the GRI contract. Also included, for comparison, are the material specification from Table II and the properties of Coors type SC-2, a cold-pressed, fine-grained, high-strength reaction-bonded silicon carbide which is widely used in mechanical shaft seal applications. Very few property tests have been made as a function of temperature; however, it is expected that the data will loosely parallel the temperature curves available for other similar materials.

Table IV. Material Characterization Test Program

| | |
|---|---|
| I | Coors Porcelain Company—In-House<br>• Characterization of physical properties |
| II | AiResearch Manufacturing Company (California)<br>• C-ring tests comparing Coors/GRI materials with control (NC-430) supplied by AiResearch.<br>• Flexural strength tests on as-fired tubes (10 in. long×1 in. diam.×1/8 in. wall) |
| III | Solar Turbines Incorporated<br>• Flue-gas compatibility tests in aluminum remelt environment |

Table V. Physical Properties of GRI Formulations Compared with Specification and SC-2

| Property | Units | Specification Target Properties | GRI-A RBSC-205 | GRI-B RBSC-105 | GRI-C RBSC-305 | SC-2 |
|---|---|---|---|---|---|---|
| Grain size distribution | | | Bimodal | Monomodal | Trimodal | Monomodal |
| Density | $g/cm^3$ | $3.00 \pm 0.10$ | $3.05 \pm 0.03$ | $3.05 \pm 0.03$ | $3.05 \pm 0.10$ | $3.10 \pm 0.02$ |
| Flexural strength (3 pt.) | $\times 10^3$ psi | $\geq 25$ | 41 | 47 | 47 | 75 |
| CTE (20°–1000°C) | $10^{-6} \cdot °C^{-1}$ | $4.5 \pm 10\%$ | 4.4 | 4.3 | 4.5 | 4.3 |
| MOE | $10^6$ psi | $50 \pm 15\%$ | 53.6 | 51.2 | 53.4 | 57.3 |
| Poisson's ratio | | | 0.16 | 0.20 | 0.15 | 0.19 |
| Thermal conductivity (200°C) | W/m·K | $80 - 20\%$ | | | | 93.4 |
| Shear modulus | $\times 10^6$ psi | | 21.3 | 21.4 | 23.2 | |
| Bulk modulus | $\times 10^6$ psi | | 26.3 | 28.4 | 25.4 | |

A C-ring diametric compression test was employed by AiResearch as one of two methods for characterization of as-fired tubes. "The main advantage of using the C-ring compressive test is that it results in fracture initiation due to tensile stress at the outside of the tube, which is the area most subject to strength reduction due to reaction with the system environment. The results of C-ring compressive tests of as-received and exposed tubes provide a good indication of the relative change in strength due to reactions at the outside of the tube specimens."[4] The configuration of the C-ring specimens is shown in Fig. 2. The analytical expression for the C-ring compressive test is[5]:

$$\mathrm{MOR} = \frac{P}{wD_o\left[\left(\frac{D_o - t}{2t}\right)\ln\left(\frac{D_o}{D_o - 2t}\right) - 1\right]} \tag{1}$$

where $P$ = breaking load (lb), $w$ = length of C ring (in.), $D_o$ = outer diameter of C ring (in.), and, $t$ = wall thickness (in.).

The C rings were set between an SiC pedestal and an $Al_2O_3$ push rod, and vertically loaded at constant crosshead speed using a mechanical loading machine.* AiResearch also is characterizing the Coors/GRI material by testing as-fired tubes at room temperature and 2300°F.

For a tube in four-point loading the equation for modulus of rupture (psi) is:

$$\mathrm{MOR} = \frac{16\ PbD_o}{(D_o^4 - D_i^4)} \tag{2}$$

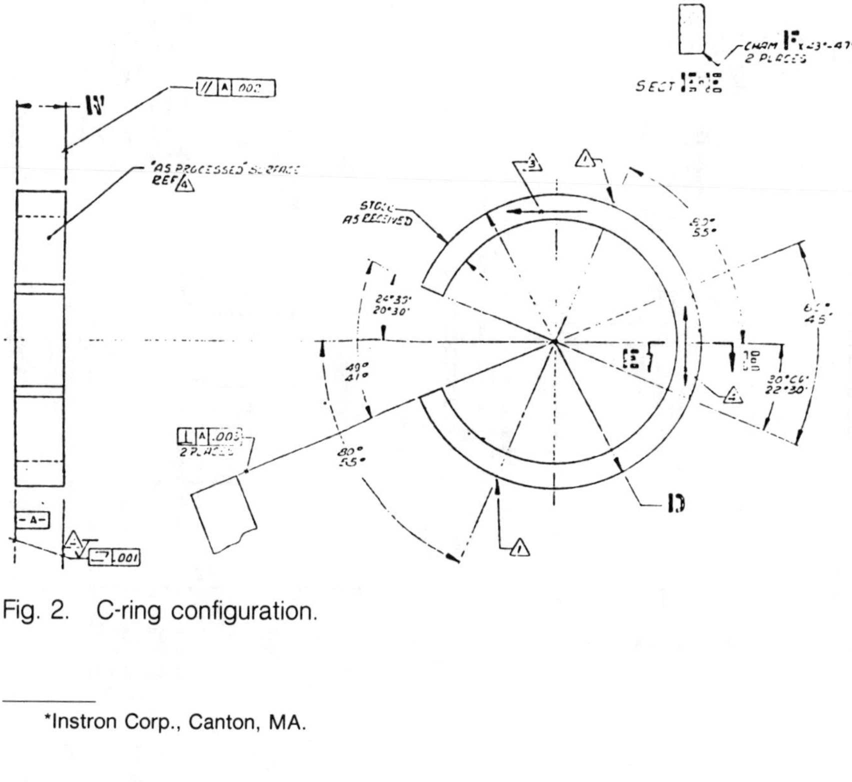


Fig. 2. C-ring configuration.

*Instron Corp., Canton, MA.

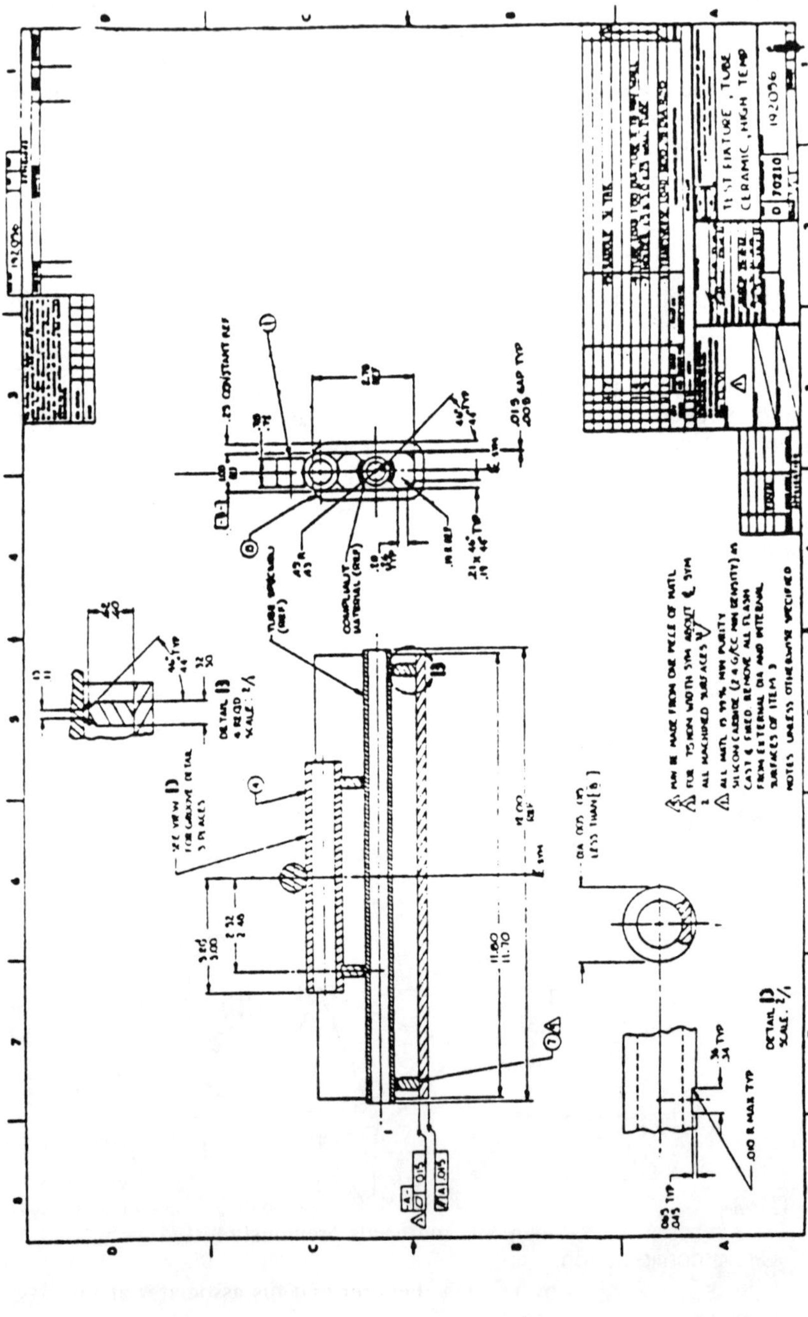


Fig. 3. Schematic diagram for tube flexure tests.

where $P$ = breaking load (lb), $b$ = ¼ loading span of the tube (in.), $D_o$ = outer diameter of tube (in.), and $D_I$ = inner diameter of tube (in.).

Tube samples were tested for flexural strength using a four-point loading method shown schematically in Fig. 3. Loading fixtures were designed to accommodate tubes 10.5 to 12.0 inches long with a 1.00-inch outer diameter. For high-temperature testing the loading fixture was made out of silicon carbide. The test set-up was designed to eliminate off-axis stresses so that full loading is vertical and perpendicular to the sample surface. The test fixture was mounted in a high-temperature furnace which in turn was located in the same Instron testing machine as used for the C-ring tests.

"The loading rates for both the C-rings and the tube tests were based on standard loading rates for 4-point loading of standard rectangular test bars (⅛×¼×2″). By setting the strain rates of the tube loading and C-ring loading equal to that of the bar loading, the strength data of the different shapes can be validly compared."[6]

All test specimens were inspected prior to testing by fluorescent penetrant techniques and by radiography.

A statistical analysis was applied to the MOR data for both the C rings and tubes. The average and median strength, standard deviation, and Weibull modulus were determined and are summarized in Table VI. A plot of the Weibull strengths is shown in Fig. 4.

The test results for formulation "A" SCRB 205 were summarized by AiResearch as follows:

"The strength data generated at room temperature for Coors material SCRB 205 indicates the C-rings have a higher average strength (27.6 ksi) than the tubes (24.6 ksi). This can be expected because the tubes consist of a greater testing surface area (and volume) which would necessitate having a higher number of total flaws—any one of which could cause failure. The closeness of the Weibull modulus for both C-ring (13.0) and tube (11.3) indicates a relative consistency between the material used for the C-rings and the tubes.

The room temperature strength of Coors SCRB 205 at 27.6 ksi compares very favorably to Norton NC–430 at 21.8 ksi for the C-ring test. Since the two materials were tested under identical conditions, this comparison is the most valid for all data obtained.

High temperature strength tests of Coors SCRB 205 tubes were hindered by fixture malfunction. For only three samples tested, the average strength was 28.5 ksi. This compares to 30.8 ksi of Norton NC–430 ¾″ tubes tested at 2450°F in an earlier AiResearch study. However, at room temperature, the strength of Coors SCRB 205 at 24.8 ksi compares favorably to Norton NC–430 at 22.3 ksi and 21.3 ksi in two different tests."[7]

Flue-gas compatibility tests in a simulated aluminum remelt environment will be performed by Solar Turbines Incorporated under contract to GRI. A total of six two-inch-diameter, four-foot-long tubes (two from each of the three formulations) will be tested. While no "in-stack" performance data on these materials are yet available, it is expected that they will behave in a manner similar to other reaction-bonded SiC materials previously tested by Solar Turbines Inc.[8] and AiResearch of California. The tubes which were tested at AiResearch were analyzed by Dr. J. I. Federer and his associates at ORNL.[9] Although there may not be total agreement on the long-term effectiveness, the tests which have been conducted provide solid evidence that reaction-bonded

Table VI. Modulus of Rupture Data for Tube and C-Ring Tests

| Material | Specimen Type | Test Temp. °C (°F) | Strain Rate in./min | Strength (MPa) Mean | Std. Dev. | Med. | Weibull Modulus | Test Quan. |
|---|---|---|---|---|---|---|---|---|
| SCRB-205-"A" | Tube | 21 (70) | 0.0067 | 171 | 15.1 | 172.5 | 11.8 | 28 |
| NC-430* | Tube | 21 (70) | 0.00009 | 153 | 10.4 | 154.6 | 14.3 | 9 |
| NC-430* | Tube | 21 (70) | 0.0009 | 147 | 16.6 | 148.4 | 9.2 | 16 |
| SCRB-205-"A" | Tube | 1260 (2300) | 0.0067 | 196‡ | | | | 3 |
| NC-430* | Tube | 1343 (2450) | 0.00009 | 212 | 18.6 | 214.6 | 11.3 | 10 |
| SCRB-205-"A" | C-ring | 21 (70) | 0.0067 | 160 | 13.8 | 162.2 | 12.9 | 39 |
| NC-430† | C-ring | 21 (70) | 0.0067 | 129 | 21.3 | 131.1 | 6.5 | 22 |

*High Temperature Ceramic Heat Exchanger Development, Final Report, July 1979, EPRI–FP1127.
†From AiResearch, 1983.
‡27.5, 27.7, 30.3.

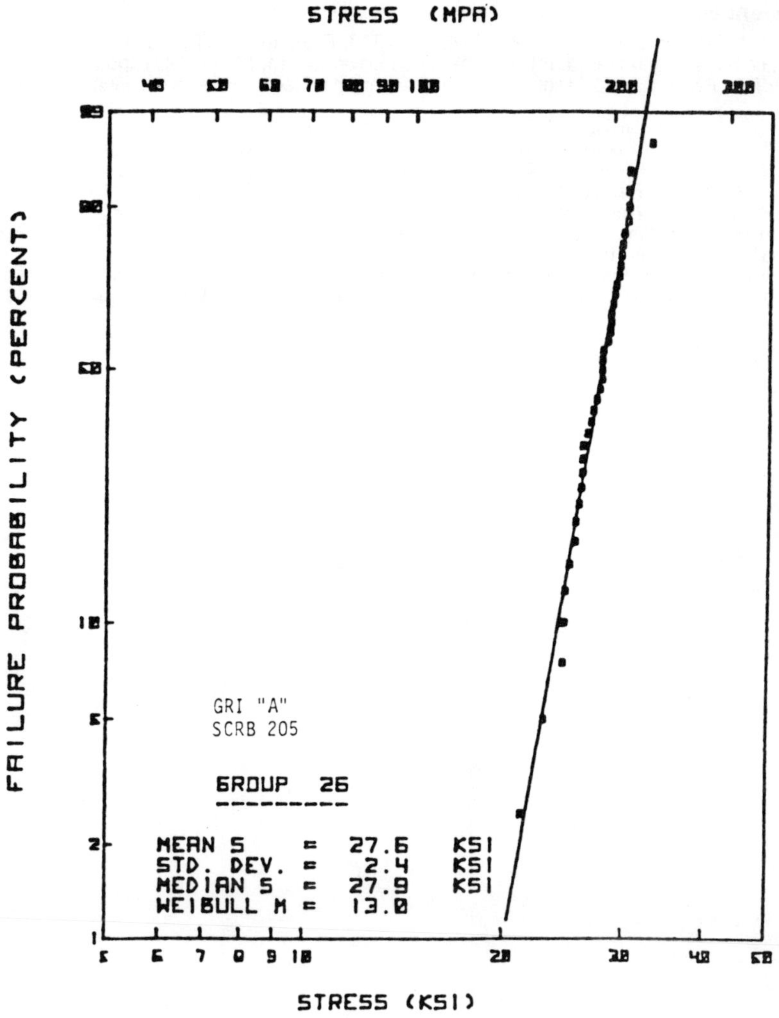


Fig. 4. Weibull plot for Coors C-ring data.

silicon carbide will perform well as a material for high-temperature heat recovery in very corrosive environments.

Finally, it is projected that, for quantities of several thousand feet, it will be possible to achieve the target selling price of $15 per linear foot for unfinned, one-inch-diameter tubes with an eighth-inch wall.

In summary, based on the test results to date and the current cost projections, we believe that the challenge to develop a cost-effective ceramic tube process for high-temperature heat-recovery systems has been successfully met. Long-term reliability can be determined only with further testing. The commercialization (Phase III) activity with the appropriate production scale-up will be contingent on the market development.

## References

[1]V. J. Tennery and G. C. Wei, ORNL/TM-7578, Evaluation of Tubular Ceramic Heat Exchanger Materials in Residual Oil Combustion Environment, March 1981, pp. 1, 76.

[2]EPRI Project 545-2, High Temperature Ceramic Heat Exchangers, Final Report, July 1979.

[3]V. J. Tennery, ORNL/TM–7580, Contract No. W–7405, p. 58–59.

[4]D. Kotchick, AiResearch Manufacturing Company; private communication.

[5]D. Kotchick, AiResearch Manufacturing Company; private communication.

[6]Preliminary AiResearch Report on Coors/GRI Ceramic Evaluation, April 1984.

[7]Preliminary AiResearch Report on Coors/GRI Ceramic Evaluation, April 1984.

[8]High Temperature Metallic Recuperator Contract GRI–5080–342–0303, Final Report GRI–81/0160, Solar Turbines Incorporated.

[9]Analyses and Testing of Materials in Support of Recuperation Projects, Waste Heat Recovery Contractor's Information Exchange Meeting, Torrance, CA, July 26–27, 1983.

# Ceramic Recuperator Tube Manufacturing Method Comparison

JOHN M. HALSTEAD

Sohio Engineered Materials Company
Niagara Falls, NY 14302

Current methods for manufacturing ceramic recuperator tubes, including plastic extrusion, water-based extrusion, and slip casting, are compared and recommendations are made. The effect of fins to enhance the heat-transfer surface is also discussed and evaluated in terms of cost effectiveness.

The Sohio Engineered Materials Company (Carborundum) is currently developing many different configurations of recuperator tubes: short tubes and long tubes; closed-end tubes; tubes with fins; on the OD, the ID, or both. Activities are underway to scale-up to make large quantities of tubes.

The following items are discussed in this paper: (1) outline of manufacturing methods, (2) advantages and disadvantages, (3) technical considerations, (4) cost analysis comparisons, and (5) the effect of dimensional tolerances.

### Manufacturing Methods

Three methods can be used to manufacture recuperator tubes: water-based extrusion, plastic extrusion, and slip casting.

*Water-based extrusion* is the least complex of the three methods (Fig. 1). SiC powders and binders are mixed and compounded with water and extruded with a properly sized and configured die. The extruded tubes are then dried in a low-temperature oven. The tubes must be fixtured prior to sintering to accommodate approximately 20% linear shrinkage while maintaining correct dimensional tolerances and geometry. Once sintering is complete, cutting to length or other post-sintering operations can be completed prior to final inspection.

*Plastic extrusion* (Fig. 2) is somewhat more complex as the components may have to undergo treatment to remove the plastic binder. However, secondary forming activities may be performed while the parts still contain the plastic binder. An example of secondary forming may be closing the ends of tubes or the attaching of collars or flanges.

*Slip casting* (Fig. 3) begins with a model of the component, from which a plaster mold is made. A bimodal mix of coarse and fine SiC grains is then ball-milled and made into a slip. The component is drain-casted; i.e., the slip is allowed to remain in the mold until a wall has built up to sufficient thickness, and the excess slip is drained off. The component is then demolded, air-dried, and baked. In the case of a tube, one end will be closed and must then be machined off.

A presintering treatment is required. Just as with extruded components, fixturing is required to accommodate shrinkage during sintering. The rest of the processing method is similar to the manufacture of extruded components.

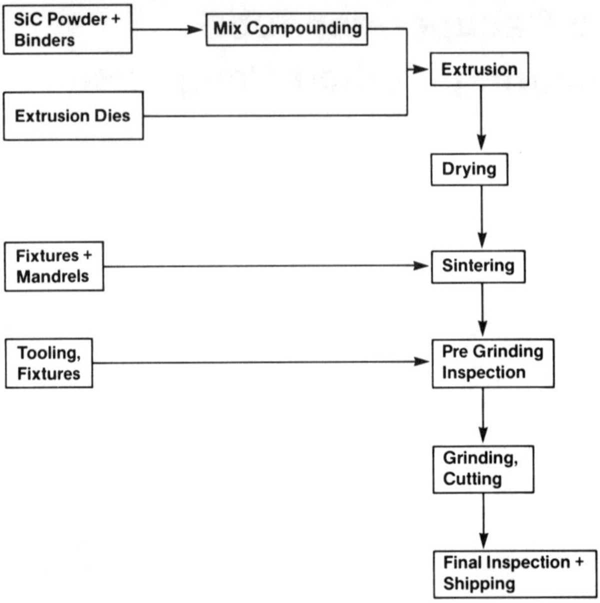


Fig. 1. Processing steps for water-based extrusion of tubes.

### Advantages and Disadvantages

Each method, water-based extrusion, plastic extrusion, and slip casting, has its advantages and disadvantages in terms of shape and size capability (Fig. 4).

#### Size Capability

Figure 5 shows the current size capability for each of the three manufacturing methods. Generally, they represent the current state-of-the-art at Carborundum and are subject to change with further development. Also, binder removal and sintering conditions are not fully developed for large-diameter, long tubes.

### Technical Considerations

From a technical standpoint, each manufacturing method has its own sensitivities. Both methods have a thickness limitation. However, with water-based extrusion, thicker walls are advantageous; with plastic extrusion, very thin walls are advantageous. Plastic extruded components also are very fragile after the binder has been removed and the sintered density is slightly lower than for water-based tubes. Slip-cast parts are sensitive to relative humidity during the air-drying step and to handling problems.

Among other technical considerations for aqueous extrusion are drying yields, thickness limits, and secondary forming. Plastic extrusion has problems with binder removal, fragility, density, and thickness limits. Slip casting considerations include powder consistency, environmental control, mold life, handling, and length and thickness limits.

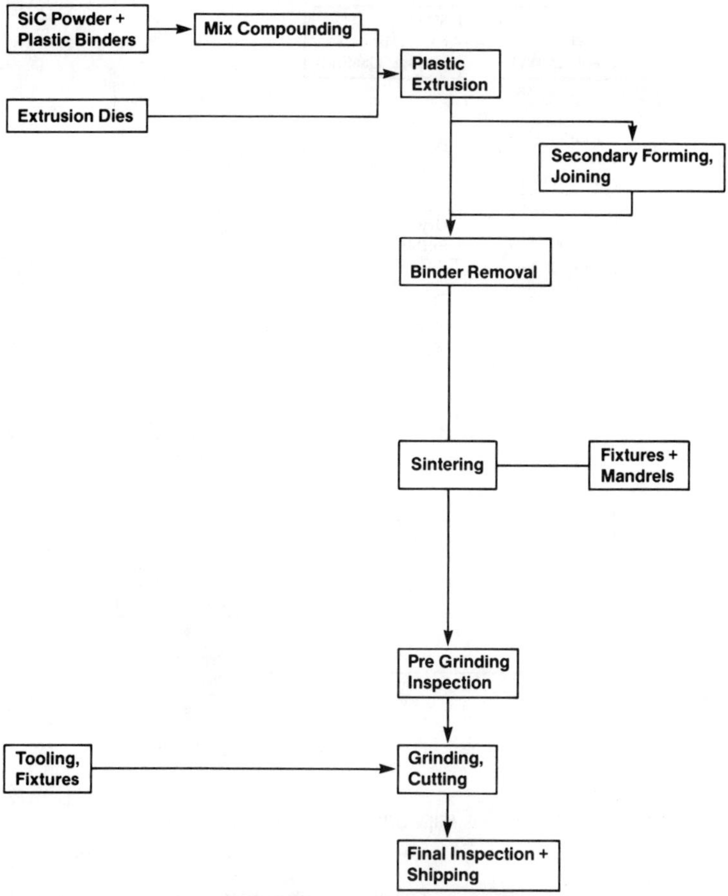


Fig. 2. Processing steps for plastic extrusion of tubes.

## Manufacturing Cost Comparisons

### Forming Cost vs Diameter

To compare the various manufacturing methods, three individual tube sizes were analyzed: ¼-in. OD, 1-in. OD and 6-in. OD. In the two smaller sizes, the two extrusion methods are compared. In the 6-in. diameter size, all three forming methods are compared.

*¼-in. OD:* In this size, plastic extrusion is slightly lower in cost due to a faster extrusion rate, primarily due to its being a continuous operation. A higher mix preparation cost is more than offset by the higher extrusion efficiency.

*1-in. OD:* In this instance, the extrusion speed on the plastic system does not overcome the added mix preparation cost, primarily due to the added mass of the tube in terms of raw material required. Water-based extrusion is lower cost in this size.

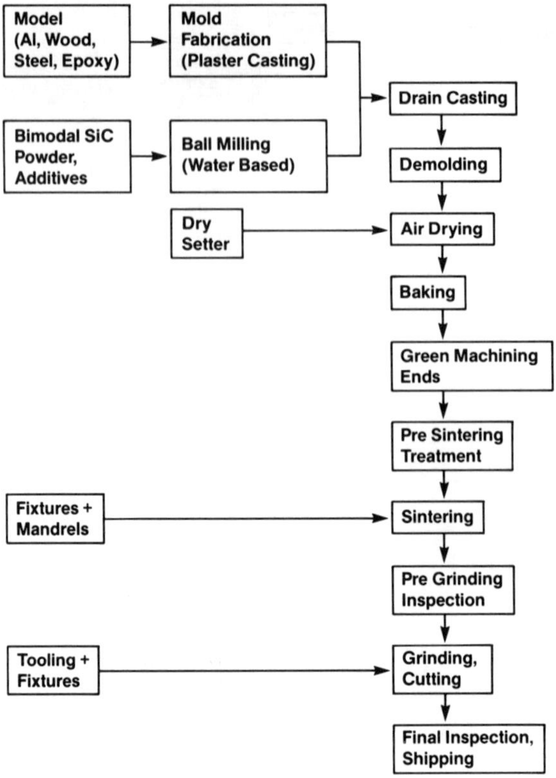


Fig. 3. Processing steps for slip casting of tubes.

| | Plastic Extrusion | Water Base Extrusion | Slip Casting |
|---|---|---|---|
| External Long Fins | Yes | Yes | Limited |
| Internal Long Fins | Yes | Yes | No |
| Flanges | Adds Cost | Adds Cost | Yes |
| Closed End | Adds Cost | Adds Cost | Yes |
| Int. Dia. Control | Good | Good | Poor |
| Bending, Loops Spirals | Good | Poor | No |
| Joining | Good | Poor | Poor |
| Non Parallel Walls | No | No | Yes |

Fig. 4. Shape capability comparison of plastic extrusion, water-based extrusion, and slip casting.

*6-in. OD:* For this size, water-based extrusion remains the most cost-effective. Mix preparation and binder removal and/or drying are the lowest for water-based extrusion.

Slip casting maintains an advantage only in material cost due to the coarse grain included in the bimodal mix.

| | Plastic Extrusion | Water Base Extrusion | Slip Casting |
|---|---|---|---|
| Length: | Continuous | 6′ | About 4 Feet |
| Diameter Max. | 6″ | 3″ | 18″ |
| Min. | — | ? | 1/2″ |
| Wall Thickness Max. | .5″ | ? | .30″ |
| Min. | .030″ | .050″ | .15″ |

Fig. 5. Present size capability of plastic extrusion, water-based extrusion, and slip casting.

**Sintering Cost vs Diameter**

The unique nature of tubular components affects sintering costs as a function of total volume (length times OD area). As the diameter of the tube increases, the volume of the furnace occupied increases faster than the weight of the tube.

Also, for the fixed diameter continuous furnace currently used to manufacture these tubes, cost effectiveness drops until a size is reached at which only one tube will fit in the furnace. Then, as the size of that one tube increases, cost effectiveness increases until the capacity limit (single-tube diameter) is reached. Figure 6 depicts this scenario. The "worst case" is the smallest single tube shown.

**Overall Manufacturing Cost vs Diameter**

Figures 7 through 9 combine the effects of material, forming, and firing to show relative costs as a function of diameter (OD) size in three ways: (1) total tube cost, (2) tube cost per pound (finished weight), and (3) tube cost per square foot surface area. In each scenario a 2-in.-OD tube is considered the "base case" and is assigned a value of 100.

Fig. 6. Schematic of decreasing tube size.

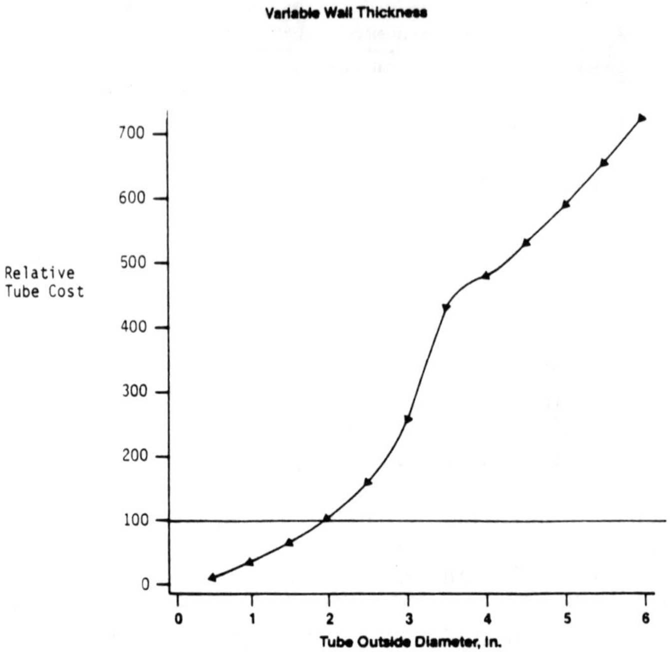


Fig. 7. Tube cost vs tube diameter.

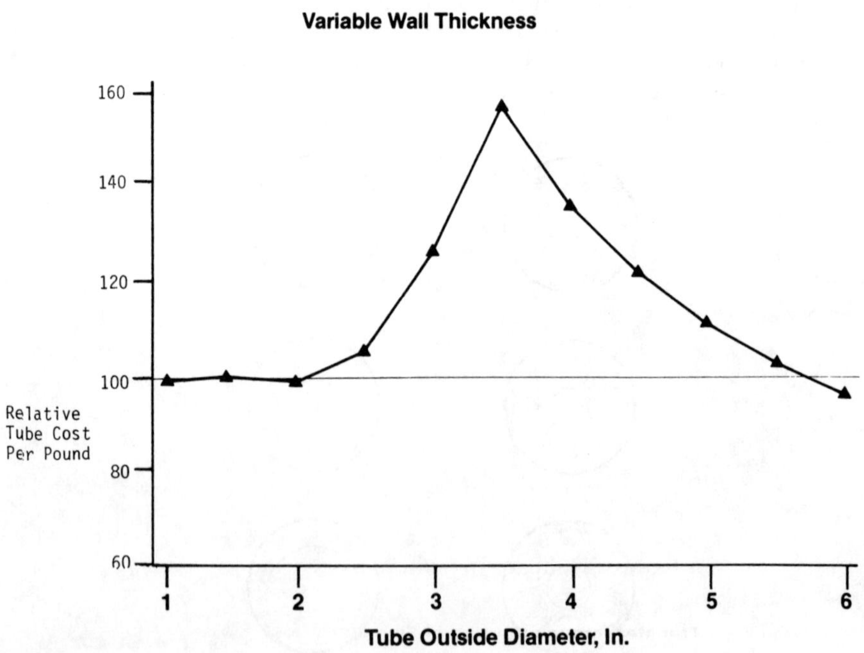


Fig. 8. Tube cost per pound vs tube diameter.

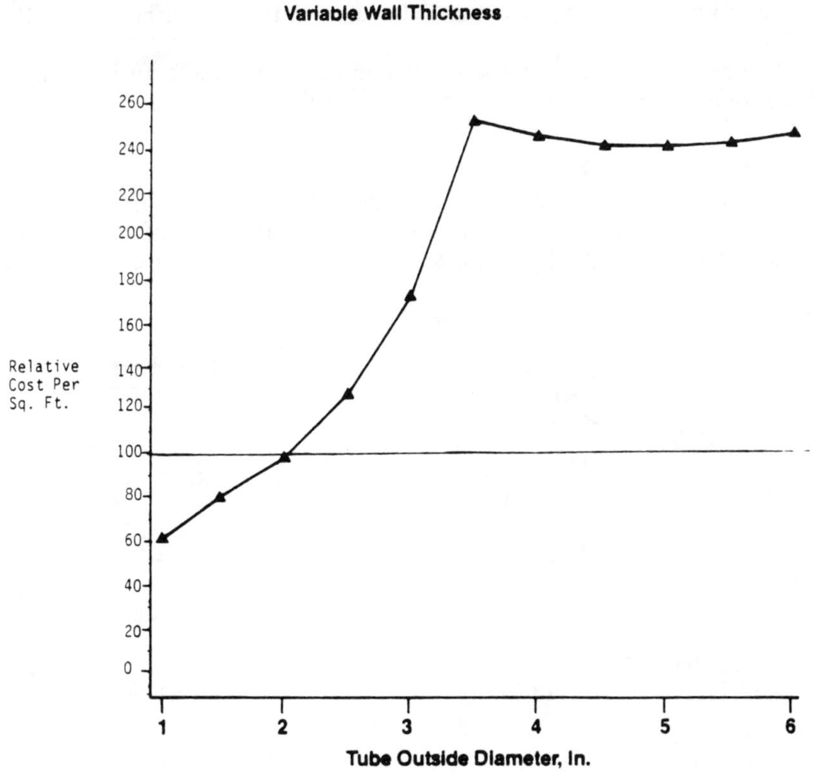


Fig. 9. Surface area cost vs tube diameter.

#### Total Tube Cost vs Tube Diameter

Cost per tube increases steadily with a "bump" in the curve at the lowest point where only one tube can be fired at once. This point can be shifted by changing the size of the furnace chamber and it can be mitigated by, perhaps, firing smaller diameter tubes in the same batch but the relative cost factor remains the same.

#### Tube Cost Per Pound vs Tube Diameter

Costs per pound are minimized at both the smallest and largest tube sizes due to more optimum furnace loading conditions. The highest cost point is at the smallest tube size where only one tube can be placed in the furnace.

#### Tube Cost Per Unit Surface Area vs Tube Diameter

Here, the most cost-effective tube is the smallest diameter tube up to the previously described "mid-range" tube and a leveling between that size and the largest tube size. This is a combined function of efficiency of furnace loading as well as lesser impact of material cost on total cost with smaller diameter tubes.

#### Enhanced Heat Transfer through the Use of Fins

Heat-transfer surface area determines, to a large extent, the effectiveness of any heat-transfer device. Heat-transfer surface area can be increased by the use of fins or other surface-area enhancement features either on the ID or the OD.

Figures 10 and 11 show the relative impacts of manufacturing cost both on the total cost of the tube and in terms of tube cost on square foot surface area.

**Internal Fins**

Internal fins would be used where the ID is the controlling surface for heat transfer. In the round tube shown in Fig. 10, the internal surface area is stated as 1.13 square feet and a cost per tube of "$X" and a cost per square

Example; 60 in. Long

| | **Round Tube**<br>1.0 Inch O.D.<br>.866 Inch I.D. | **Internal Fins**[I]<br>1.111 Inch O.D.<br>.977 Inch I.D.<br>0.2 In x .06 |
|---|---|---|

| | | |
|---|---|---|
| Cost per Tube | $X | $X+67% |
| Surface Area per Tube | 1.13 Sq. Ft. | 4.64 Sq. Ft. |
| Cost per Unit Surface Area | $Y | $Y-45% |

**Conclusion: Internal Fins are Probably Cost Effective**

Fig. 10. Effect of internal fins on tube cost. (The slight dimensional change is to equilibrate the internal gas-flow area of the tube with and without fins.)

.866 In. I.D. 60 In. Long

| | **Round Tube**<br>1.0 Inch O.D. | **Finned Tube**<br>1.5 Inch O.D., 0.25 x .067 Fins |
|---|---|---|

| | | |
|---|---|---|
| Cost per Tube | $X | $X+120% |
| Surface Area per Tube | 1.38 Sq. Ft. | 4.64 Sq. Ft. |
| Cost per Unit of Surface Area | $Z | $Z-35% |

**Conclusion: External Fins Might Be Cost Effective**

Fig. 11. Effect of external fins on tube cost.

foot surface area of "$Y" (where $Y=$X/1.13). Simply stated, while fins add approximately 67% to the cost of the example tube, heat-transfer surface is increased fourfold, resulting in a 45% *decrease* in cost per square foot surface area.

**External Fins**

External fins, used where the OD is the controlling surface area for heat transfer, can also be cost-effective. In the example shown in Fig. 11, our same tube with a cost of $X now contains 1.38 square feet of *external* surface area and a cost per square foot surface area of "$Z" (where $Z=$X/1.38).

When external fins are added, the cost of the tube is increased approximately 120%. However, the external surface area is increased by nearly 240%, resulting in a 35% decrease in cost per square foot surface area.

**Total Cost Effectiveness**

It has been clearly shown that fins can be cost-effective in terms of tube cost per square foot surface area with internal fins slightly more cost-effective than external fins. However, technical considerations such as the desirability of turbulence and the potential for fouling are equally important factors. In instances where fins are appropriate from a technical standpoint, they can also assist in improving the cost effectiveness of the heat-transfer system.

## Dimensional Tolerances

Another vitally important factor in the cost of ceramic recuperator tubes is dimensional tolerances. Due to the lack of ductility of ceramic components, system designers tend to specify very exact tolerances so that easy assembly can be achieved and undesirable mechanical stresses can be avoided. However, there is a direct correlation between the degree of dimensional tolerance and product yield which, in turn, affects the component cost.

There are two primary aspects to dimensional tolerances: (1) diameter (OD) and (2) straightness.

Figures 12 and 13 demonstrate that there is "some" normal distribution of dimensions which will be the outcome of the manufacturing process as

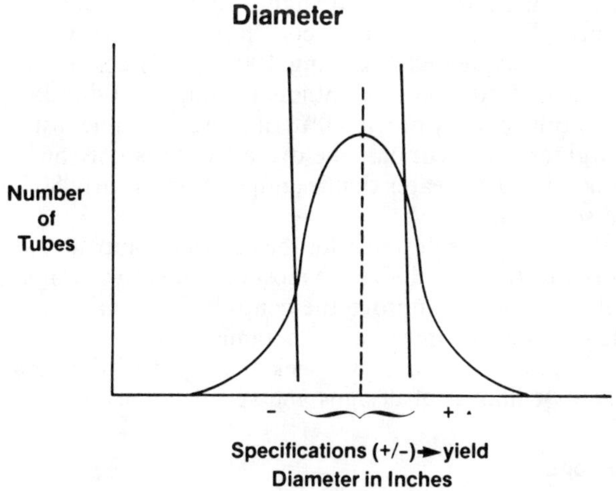


Fig. 12. Bell curve illustrating normal distribution of diameter.

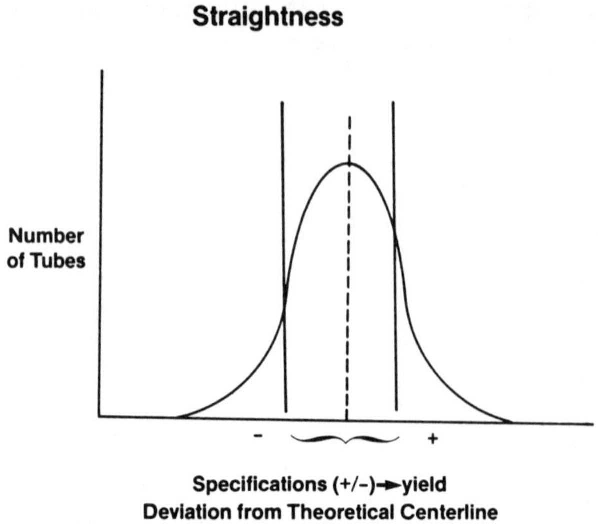


Fig. 13. Bell curve illustrating normal distribution of straightness.

signified by the "bell curve." Also, for each specification (i.e., diameter and straightness) there is a tolerance band, specified by the customer, which is signified by the vertical lines. The yield of "good" tubes, then, is the number of tubes on the bell curve that fall between the tolerance bands.

The ceramic manufacturer has some degree of control over the shape and steepness of the bell curve; the system manufacturer and/or end-use customer have some degree of control over the width and position of the tolerance band.

It should also be noted that individual tolerances may be independent events and the yield impacts of each may be combined to determine the overall yield achieved in the manufacturing process.

Figure 14 is an example of two slightly different yield scenarios which have significantly different total impacts. In a base case where an example tube might have an unyielded (meaning 100% yield) cost of $100, yields of 90% on diameter and 80% on straightness (example A) due to specifications can increase the tube cost by nearly 40%. In a less desirable instance (example B), implying tighter specifications, the overall yield is only 56% and the cost of the tube is nearly 30% greater than example A and nearly 80% greater than the unyielded scenario.

Clearly, the optimum solution is for the ceramic companies and the system builders to work together to achieve the most cost-effective design. The ceramic companies will continue to improve the control of the manufacturing process and the system designers must take "ceramic design" into account in the construction of systems and actively seek the input of the ceramic supplier concerning the cost impact of designs and specifications.

#### Recommendations

A number of important concepts have been reviewed and discussed, including alternate manufacturing methods, the cost impact of diameter size

**Tolerances may be independent events**

**Example:**

| | | |
|---|---|---|
| **diameter yield** | **90%** | **80%** |
| **straightness yield** | **80%** | **70%** |
| **overall yield** | **72%** | **56%** |
| Impact on cost:<br>Base case $100 unyielded<br>plus effect of another<br>yield affecting event | **$139** | **$179** |

Fig. 14. Yield scenarios which indicate different total impacts.

and dimensional tolerances, and the cost effectiveness of fins to enhance the heat-transfer surface.

The general conclusions are:

(1) Water-based extrusion tends to be the most cost-effective for all but very small diameter tubes, such as ½-in. OD or less.

(2) Both plastic extrusion and slip casting have significant usefulness where unique geometries are important.

(3) Small-diameter tubes are more cost-effective on a cost per square-foot surface area basis than large-diameter tubes.

(4) Both external and internal fins can be cost-effective methods of increasing surface area where the application is appropriate.

(5) Dimensional tolerances can dramatically affect the manufacturing cost of tubes.

In terms of recommendations, it is very clear that a close interaction in design of heat-transfer equipment between the ceramic supplier and the equipment manufacturer is the best way to achieve the optimum in performance at the most attractive cost.

General guidelines for design of heat transfer equipment should include:

(1) Use the smallest diameter tubes possible.

(2) Think in terms of cost per unit surface area, not per pound or per tube.

(3) Review the cost of a variety of sizes and/or configurations early in the design phase.

(4) Carefully consider the costs and benefits of dimensional tolerances.

# Materials and Process Development for Heat-Exchanger Tubes

M. L. Torti, N. I. Paille, and J. J. Litwinowich

Norton Co.
Worcester, MA 01606

The objective of this program was to develop a more cost-effective, long-life silicon carbide heat-exchanger tube. The work concentrated on two materials: nitride-bonded silicon carbide and siliconized silicon carbide. Based on previous work, the strength and oxidation resistance of established siliconized silicon carbides were judged adequate for the heat-exchanger tubing application, but costs were very high. To reduce the costs of siliconized silicon carbide, a simplified firing procedure and extrusion forming—as opposed to slipcasting—were evaluated. For the nitride-bonded silicon carbides it was felt necessary to improve the strength and oxidation resistance, so these concerns were addressed as well as the development of extrusion capability for this composition. This paper reviews the physical and mechanical properties obtained from this approach both as-manufactured and after exposure to a heat-exchanger test stand environment with hot products of gas combustion but without other corrosive elements.

### Siliconized Silicon Carbide

Optimization of the siliconized silicon carbide was reasonably successful on a laboratory scale based on density and room-temperature strength, as can be seen from the data in Table I. The strengths were measured on O rings cut from a 2 in. diameter tube and tested in compression, thus testing the as-fired

Table I. Property Summary of Siliconized Silicon Carbide 2 in. Diameter Tube

| Composition | Bulk Density ($g/cm^3$) | Free Silicon (wt%) | Strength O Ring (psi) |
|---|---|---|---|
| Slipcast Standard firing (NC-430) | 3.07 | 13 | 159.5 (23 116) |
| Extruded standard firing | 3.02 | 18 | 150.5 (21 814) |
| Extruded simplified firing | 2.97 | 23 | 149.2 (21 625) |
| Recrystallized (CS-101) | 2.69 | 0 | 88.4 (12 808) |

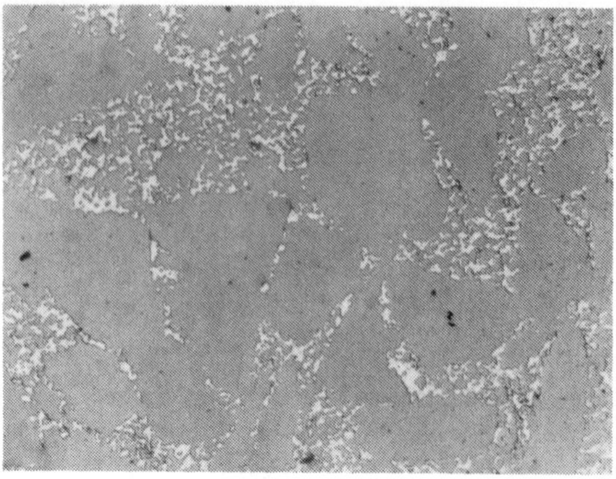

Fig. 1. NC–430 siliconized silicon carbide, ×240.

surfaces of the tubes. The first cost-reduction effort was the development of extrusion capabilities for these compositions. The results are shown in the second line of Table I. There is slight drop in overall density, which represents an increase in the amount of free silicon, as this is still a fully dense body. There is a very slight drop in O-ring strengths. The existing firing procedures for the NC–430 were not developed with cost as a major criterion but rather to minimize the free silicon content. This was not felt to be as much of a factor in heat-exchanger tubes as cost reduction. Combining extrusion with a simplified firing operation resulted in the properties on the third line. There is a slight drop in the bulk density and a corresponding increase in free silicon but a good retention of strength compared to the extruded material listed above. For comparison, properties of a recrystallized silicon carbide are included. This is a pure silicon carbide self-bonded but with no silicon impregnation and with approximately 15–20% open porosity. Note that the strength is almost one-half that of the siliconized tubes.

Figure 1 shows the microstructure at ×240 of the standard NC–430. The light areas are free silicon and the gray areas silicon carbide. The structure is bimodal, with large silicon carbide grains and a network of finer grains in-between. Figure 2 shows the results with the extruded but standard firing. There is very little difference, although the higher silicon content is discernible. Figure 3 shows the results of both extrusion and simplified firing. This particular structure shows better bonding between the large silicon carbide grains and no discernible increase in free silicon. The results of extrusion and simplified firing on a macro scale are shown in Fig. 4, which is a slice through a 2 in. diameter extruded and siliconized tube. Figure 5 is the same section at a higher magnification to show the uniformity of silicon carbide distribution and the completeness of the silicon impregnation.

The strengths of these tubes were measured in a variety of ways. Table II shows results for the NC–430 composition. The first column shows 4-point strengths on ground surface bars cut from the tubes. The average strength is 33 590 psi and the Weibull modulus is 12.25. The next series of samples was O rings sliced from the tube to test the as-fired surface with the highest stresses on the internal surface of the tube. In this case, the Weibull modulus is much

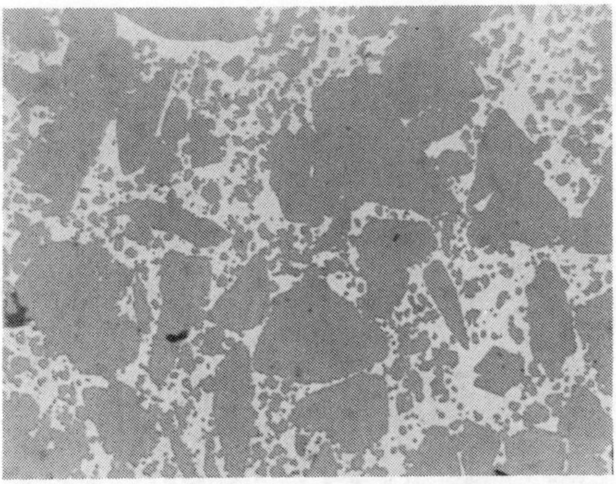

Fig. 2. Extruded, standard fired siliconized silicon carbide ×240.

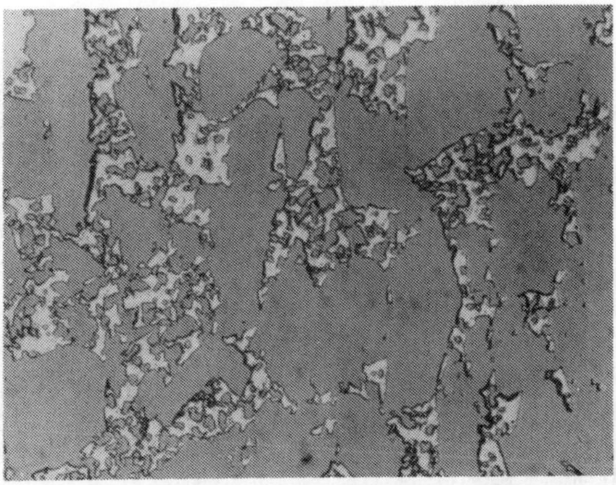

Fig. 3. Extruded, simplified fired siliconized silicon carbide ×240.

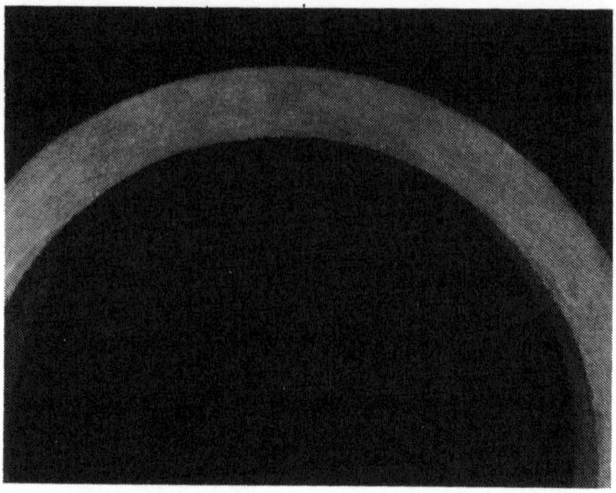

Fig. 4. Cross section of an extruded 2 in. diameter siliconized silicon carbide tube.

Fig. 5. Magnified cross section of an extruded 2 in. diameter siliconized silicon carbide tube.

Table II. Comparative Strengths of Cast 2 in. Diameter Siliconized Silicon Carbide Tubes (NC-430)

| | Ground Surfaces 4-Point Bend | As-Fired Surfaces O Ring |
|---|---|---|
| Average strength (MPa) | 231.8 | 159.5 |
| No. of specimens | 42 | 12 |
| Weibull modulus | 12.25 | 4.13 |
| $\sigma_c$ (MPa) | 240.9 | 175.6 |
| $\sigma_0$ (MPa) | 129 | 31.9 |

Table III. 2 in. Diameter Tube Flexural Strengths of Siliconized Silicon Carbide (Extruded, Standard Firing)*

| | (MPa) psi | |
|---|---|---|
| Average strength | 64.4 | (9 329) |
| Standard deviation | 14.4 | (2 087) |
| No. of specimens | 0.06 | (9) |

*4-point, 37 in. outer span, 29 in. inner span.

Table IV. Closed-End Heat-Exchanger Tubes of Cast Siliconized Silicon Carbide*

| | Room Temp. Unexposed | 1300°C† psi | | Room Temp. after Exposure at B&W |
|---|---|---|---|---|
| | Standard firing (NC-430) | | | |
| Average strength | 21 665 | 28 600 | 27 700 | 27 245 |
| Standard deviation | 4 278 | 4 000 | 7 200 | 6 485 |
| No. of samples | 6 | 10 | 12 | 12 |
| | Simplified firing (CS101K) | | | |
| Average strength | 29 230 | 30 300 | 30 200 | |
| Standard deviation | 5 280 | 3 900 | 2 600 | |
| No. of samples | 12 | 12 | 12 | |

*B&W design, 3½ in. OD; O-ring tests, 1 in. wide.
†Tests carried out at B&W.

lower than on the ground bars and the characteristic $\sigma_z$ and $\sigma_0$ strengths are also lower. Obviously, there is a different flaw population in these tests. However, the O-ring strength is felt to be more representative of the actual heat-exchanger tube. Finally, a procedure was developed to take the entire tube and test it in 4-point bending. The results are shown in Table III. At this writing there are not enough data to do a meaningful Weibull analysis. A test rig is being constructed, which will rotate a tube while it is being stressed so that the tensile stress is applied to the entire circumference and most of the length of the tube rather than just a single point. Internal pressurization is also planned as a routine quality control test on these tubes, but data from that equipment are not yet available.

In a related program, a comparison of standard vs simplified firing was made for slipcast tubes. Table IV presents results on siliconized silicon carbide tubes, 3½ in. OD with closed ends, which were evaluated in the Babcock & Wilcox design in their original test module.[2] Room-temperature strengths for the standard NC–430 material and a version of the simplified fired material prior to exposure are given in the first column. In this particular set of data, an unusually good strength for the simplified material is obtained, above that of the standard NC–430. Babcock & Wilcox took O-ring sections from these tubes and tested them in compression at 1300°C. The next two columns of data represent their results. In all cases, the strengths were slightly higher than the room-temperature values, which is typical for these materials where surface crack healing may occur during high-temperature exposures. The re-

Table V. Property Summary of Nitride-Bonded Silicon Carbide 2 in. Diameter Tube

| Composition | Bulk Density (g/cm³) | Open Porosity (%) | Strength O Ring (MPa) (psi) |
|---|---|---|---|
| Standard slipcast (CX-589) | 2.51 | 15.9 | 85 (12 322) |
| Improved slipcast | 2.57 | 15.3 | 86 (12 489) |
| Extruded | 2.28 | 28.2 | 72 (10 400) |
| Extruded modified | 2.50 | 20.0 | 49 (7 109) |

sults are very reassuring. After exposure at Babcock & Wilcox for approximately 600 h between 732° and 1287°C (1350°–2350°F), the tube shown in the last column was returned to Norton and strength tests carried out, showing no degradation in strength; if anything, there was some improvement over the unexposed strengths.

#### Nitride-Bonded Silicon Carbide

Results of the work on the nitride-bonded silicon carbides are shown in Table V. The standard material is a slipcast mix containing silicon carbide and free silicon, which is fired in a nitrogen atmosphere to develop a silicon nitride bond. Note that the strengths are considerably lower than those of the siliconized silicon carbide but approximately equal to the recrystallized silicon carbide. There is a substantial amount of open porosity. Initially, we attempted to improve the cast mix to increase the density and reduce the open porosity. We achieved some improvement in density but not much improvement in open porosity and not much improvement in strength. Extruding rather than casting caused a significant drop in density. Further developments enabled us to recover most of the drop in density and get back up to 2.5 $g/cm^3$. We had a large amount of open porosity in the initial extrusions but were able to reduce that significantly in the improved material. However, we lost strength as we went to higher density.

#### Permeability

Because of the open porosity, permeability of these tubes is a consideration for their use as heat-exchanger tubing. However, we are primarily looking at low differential pressure applications. Table VI shows the results of evaluations carried out on these materials with a simple gas permeability tester in which the tube was pressurized internally to 100 mm Hg and then the leakage measured in liters per minute per square meter. As would be expected, no leakage occurred with the siliconized silicon carbides. We measured high leakage with the recrystallized silicon carbide but relatively low leakage with the nitride-bonded silicon carbide, presumably because of the finer structure of the open porosity in the nitride-bonded material. These values for the nitride-bonded silicon carbide are low enough that leakage would be on the order of a percent or less in an actual heat-exchanger environment. However, the values for the recrystallized silicon carbide are high enough to affect the efficiency of the

Table VI. Air Permeability (Differential Pressure 100 mm Hg)

| Composition | Flow 1·min·m² |
|---|---|
| Siliconized SiC | |
| Standard | |
| (NC-430) | 0 |
| Extruded | 0 |
| Recrystallized SiC | |
| (CS-101) | 30 |
| Nitride-bonded SiC | |
| Standard | |
| (CX-589) | 2.40 |
| Improved cast | 0.79 |
| Extruded | 2.40 |
| Extruded-modified | 2.20 |

heat exchanger. That is the main reason why we did not continue optimization of recrystallized silicon carbide for heat-exchanger application.

**Test Stand Evaluation**

The most important result is the performance of these tubes in a heat-exchanger environment. In a companion paper,[1] R. G. Graham describes his heat-exchanger test stand. Table VII presents the results for a siliconized extruded tube and a nitride-bonded cast tube, which were exposed in that test stand for relatively long times. Because of the nature of the heat-exchanger test, with cool air flowing through the tube and the hot burner gases on the outside, there is a temperature differential over the length of the tube. The tests attempted to duplicate the environment of a heat exchanger rather than maintaining a specified temperature over a period of time. Therefore, the temperatures indicated in Table VII are necessarily approximate. Furthermore, one of the difficulties in these extended tests with frequent cycling is the failure of thermocouples to adhere to the tubes. Thus good tube-temperature data are not obtainable.

**Soot Blowing Stress**

The most severe conditions for ceramic heat-exchanger tubes may be the stresses the tube is subjected to when a jet of compressed air is used to blow soot or other accumulation off the tubes. In this case, a specific area of the siliconized silicon carbide was used repeatedly for soot blowing with an ambient-temperature compressed air jet approximately 5 or 6 in. from the surface of the tube. Blowing was carried out for approximately one minute, long enough to see a distinct cooling of the tube from the heat-exchanger test temperature down to a black heat. The O-ring strength values for the siliconized silicon carbide hot end of the tube show $\approx$15% decrease after the 2400-h exposure. The difference between 119 and 127.6 MPa (17 300 and 18 500 psi) for the soot blown and the higher temperature areas may not be significant. Essentially, no change in strength is seen for the cooler end of the tube. The nitride-bonded tube performed very well, except for an area at the temperature of maximum oxidation susceptibility for this composition. For temperatures in the 815° to 926°C region, the strength is only 32 MPa. That particular location showed

Table VII. Silicon Carbide Tube Strengths after Exposure*

| Siliconized, Extruded 2400-h Exposure | | | Nitride-Bonded, Cast 1800-h Exposure | |
|---|---|---|---|---|
| Average | No. of Tests | Hot End, °C (°F) | Average | No. of Tests |
| 18 504 | 6 | 982 (1800) | 14 765 | 1 |
| | | | 4 710 | 1 |
| 17 328 | 6 | | 17 706 | 6 |
| soot blown area | | | | |
| | | Cold End | | |
| 22 424 | 6 | 537 (1000) | 12 192 | 4 |

*2 in. Diameter, 4 ft Long Tubes, C&H Heat Exchanger Test Stand, O-Ring Strengths (psi).

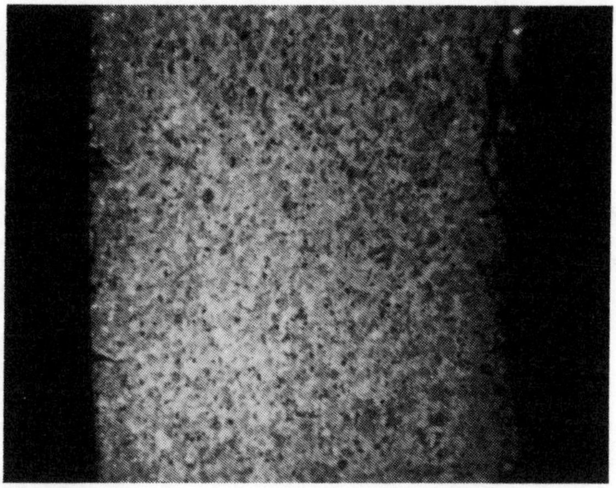

Fig. 6. Cross section of nitride-bonded silicon carbide tube after exposure at 815° to 926°C for 2400 h.

significant oxidation and some surface cracking in the removed tube, although it had not failed in service. Figure 6 shows a cross section of that tube showing the surface cracks penetrating 0.01 to 0.02 in. into the cross section of the tube. Unfortunately, in the heat-exchanger environment, some part of the tube will be at the critical temperature some time unless the whole tube is maintained below that temperature. These tubes are lower in cost because of the nature of the raw materials and processing and, if they are used in heat exchangers at temperatures below the 815°C region, they may be quite useful, but at the current state of the art, they should not be utilized in applications where the temperatures will be in the 815° to 982°C region for any extended period. During extensive testing, no tube breakage was encountered due to mounting or to thermal cycling of the heat-exchanger test stand. We had expected thermal stress failures in at least some of weaker compositions, but none occurred.

#### Summary

A more cost-effective route to a good-quality siliconized silicon carbide tube has been developed without significant reduction in properties. Several years ago, 2-in. diameter siliconized silicon carbide tubes sold for approximately $100/ft. At the current state of the art we are in the range of about one-third to one-sixth of that number, depending on quantities. Development was carried out on a laboratory extrusion facility and firing tests done in relatively small batches, although in a number of cases in production-sized furnaces. The next phase of this program should be to demonstrate production capability of extruded, simplified-firing tubes on pilot-plant scale equipment with products of that production put under evaluation in a heat-exchanger test module and in actual heat-exchanger industrial environments. The nitride-bonded material may represent a cost-effective tubing for the cooler portion of an all-ceramic heat exchanger, but needs further development work for exposure at temperatures over 815°C.

**Acknowledgment**

Portions of this work were done under contract to the Gas Research Institute and their support is gratefully acknowledged.

**References**

[1]R. G. Graham, "Development of an Air to Air Heat Exchanger with All-Ceramic Internals," C&H Combustion Company, Troy, MI.

[2]J. F. Snyder, D. R. Petrak, and P. L. Barnesburg, "High Temperature Burner Duct Recuperator System," Babcock & Wilcox, Lynchburg, VA.

# CVD Technology for Heat Exchangers and Heat Pipes

P. REAGAN, A. N. SCOVILLE, R. LEAF, W. COLE, AND F. HUFFMAN

Thermo Electron Corp.
Waltham, MA 02254

High-temperature ceramic heat exchangers and heat pipes can be fabricated using chemical vapor deposition (CVD) technology. The resulting components have superior oxidation resistance and can operate at temperatures up to 1600°C. Additionally, components can be fabricated in one piece, eliminating sealing and joining problems.

The energy contained in the exhaust gases from high-temperature, direct-fired processes represents a major source for potential heat recovery in the United States. In 1977, such processes represented approximately 11% of the national energy consumption of all types.[1] In addition, corrosive chemicals which inherently exist in the exhaust of direct-fired processes often preclude the use of conventional heat-recovery equipment. Recent developments in equipment design, utilizing ceramic materials and fluidized beds, are promising but suffer from limitations, such as range of operating conditions, reliability, and cost. By using chemical vapor deposition (CVD) technology developed at Thermo Electron, a new generation of advanced, ceramic-based heat exchangers can potentially be constructed. Particular features of these heat exchangers are: (1) no joints to fail or leak, (2) high strength and high resistance to thermal shock, (3) high effectiveness, (4) capability of high-temperature operation (up to 1700°C), (5) compactness and light weight, (6) corrosion resistance and chemical stability, and (7) ease of cleaning (because of their thermal shock resistance).

Two alternatives for fabricating such a heat exchanger have been explored. One utilizes heat-pipe technology to transfer the heat and the second draws on the emerging technology of advanced ceramic composites to fabricate heat exchangers in one piece. Both concepts take advantage of advances in CVD technology developed as part of the DOE program on thermionic energy conversion.[2]

Chemical vapor deposition is a technique for fabricating materials on a substrate immersed in a vapor. Gaseous precursors are introduced into a reactor containing a hot substrate. When the vapor contacts the hot surface, a thermally activated chemical reaction occurs, forming the solid product directly on the surface. As a result, a fine-grained, cohesive, dense layer forms which can be built up to the desired thickness. An additional benefit is that the deposit assumes the shape of the substrate, allowing the formation of near net shape components. In addition, the cost of manufacturing is considerably reduced because the material cost is lower and machining of the refractory material is eliminated. A typical CVD reactor for deposition of silicon carbide is shown in Fig. 1. Details of the process are reported elsewhere.[3]

### Heat-Pipe-Based Heat-Exchanger Design

The heat pipe (HP) is a high-performance heat-transfer device, able to transfer thermal energy at high rates over considerable distances with almost no temperature drop. In its basic form, the heat pipe consists of a sealed tube lined with a porous capillary wick that is saturated with a volatile fluid, as shown in Fig. 2. Heat absorbed by the evaporator zone vaporizes the fluid; vapor flows to the other end where it condenses, giving up the heat. The fluid cycle is then completed as the condensed fluid flows back to the evaporator in the wick. The axial heat transfer is equal to the mass circulation rate times the latent heat of vaporization of the working fluid. With proper design, heat pipes can exhibit effective thermal conductivities hundreds of times greater than those of copper.

This remarkable heat-transfer capability makes heat pipes well suited for high-temperature heat recovery. A design concept of a recuperator using CVD HPs is shown in Fig. 3. Flue gases and combustion air would be ducted in a counterflow manner, with heat being transferred through the heat pipe from one gas to the other. In the hottest stage, the heat pipe would use lithium working fluid. The outside would have a coating of CVD SiC to protect the device from the corrosive combustion gas. Since heat-pipe operating temperatures are controlled by the vapor pressure of the working fluid, sodium or potassium would be substituted for lithium as the combustion gas temperature is reduced. Bonding of the individual heat pipes at a single circumference allows free expansion, thus reducing the thermal stresses and eliminating the need for sliding seals.

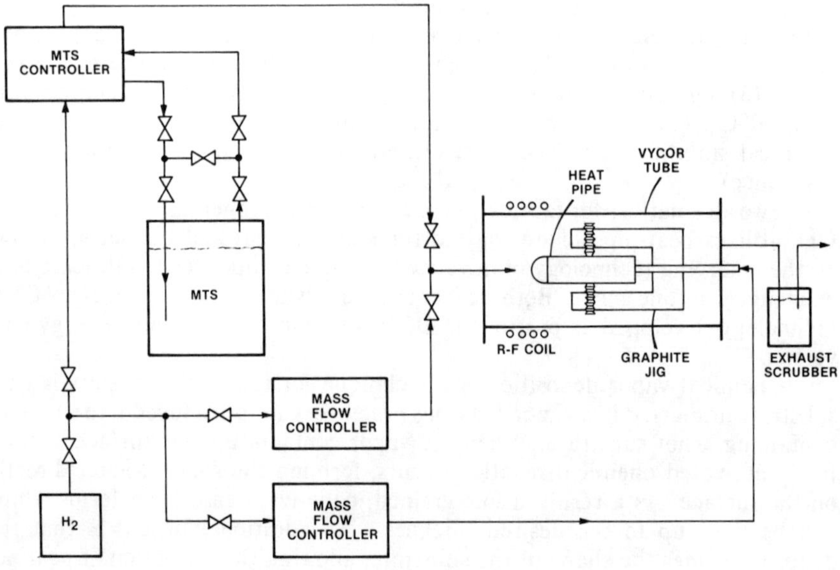


Fig. 1. Simplified flow diagram of a reactor for CVD deposition of SiC.

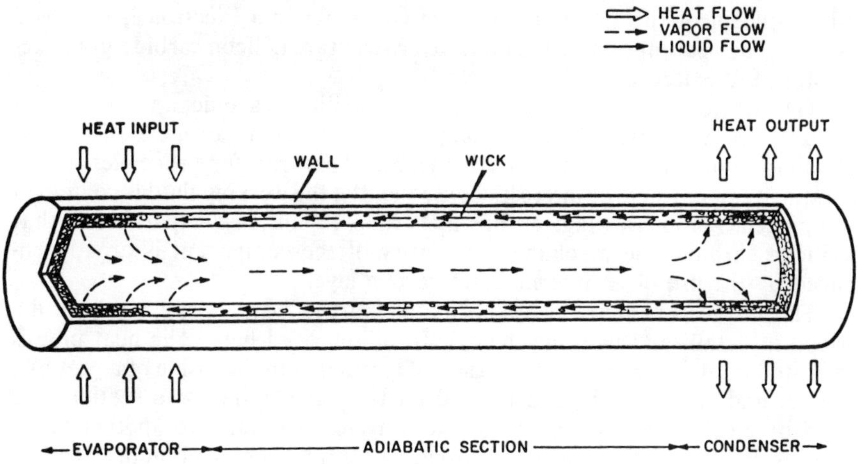


Fig. 2. Conventional heat pipe.

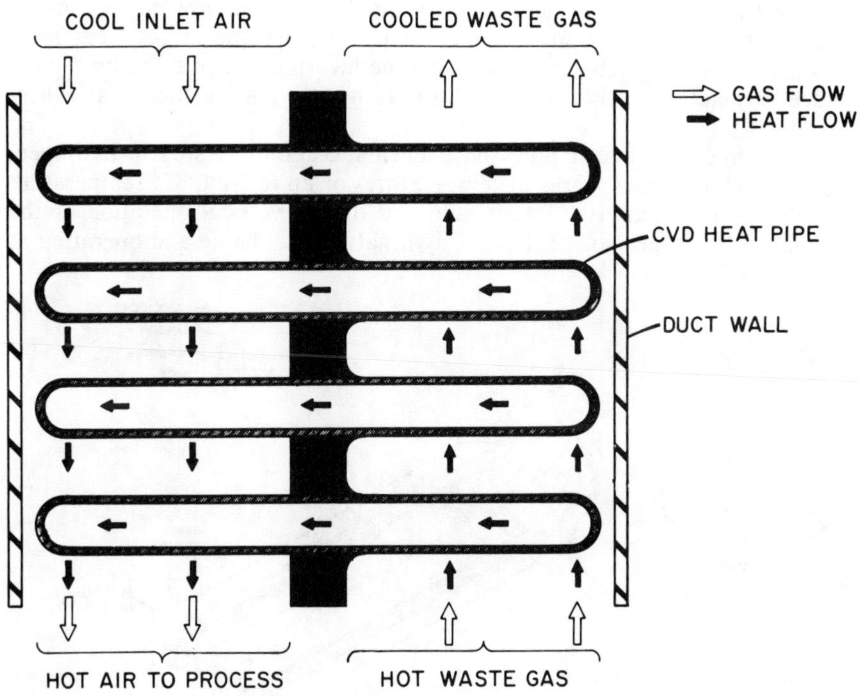


Fig. 3. Design concept of recuperator using high-temperature CVD heat pipes.

The development of a high-temperature (i.e., >1200°C) heat pipe that can operate in air or in a combustion atmosphere is an extremely challenging technical problem. Since conventional metals would be molten at some of the operational temperatures of interest, refractory materials must be used. The design concept of the CVD HP formulated by Thermo Electron is sketched in Fig. 4. The heat pipe consists of a trilayer structure (silicon carbide-graphite-tungsten) fabricated by CVD.

The trilayer structure is required because silicon carbide, an outstanding high-temperature material, is not compatible with the alkali metals (lithium, potassium, and sodium) which have the best heat-transfer properties at the operating temperatures of interest. Consequently, the working fluid is contained by tungsten. Although tungsten has a thermal expansion comparable to that of silicon carbide, the mechanical integrity of the component is greatly enhanced by the use of an intermediate carbon layer.

The fabrication steps for alkali metal-charged heat pipes have been described in detail[4] and will only be briefly summarized here. The heat pipe is fabricated in two sections, with tungsten deposited onto the inside of a graphite mandrel and silicon carbide deposited on the outside. The two sections are joined by electron beam welding of the tungsten. Lithium (or another alkali metal) is distilled in the heat pipe through a fill tube contained in one section. The fill tube is then sealed by electron bombardment-induced melting. A final layer of silicon carbide is deposited over the whole outside surface and the ends. Modifying the tungsten deposition parameters at the end of a run gives a microstructural surface, such as shown in Fig. 5; this provides a wicking surface for return of the molten lithium to the hot zone of the heat pipe. Alternatively, a tungsten screen wick can be incorporated into the heat pipe, bonded in place by a thin layer of CVD tungsten. This surface is shown in Fig. 6.

Lithium-charged heat pipes have been successfully tested in hydrogen, vacuum, and air atmospheres at temperatures of up to 1400°C. Temperature gradients of less than 10°C were achieved during vertical operation of the heat pipe. Heat pipe No. 12, heated by a natural gas flame and operating at

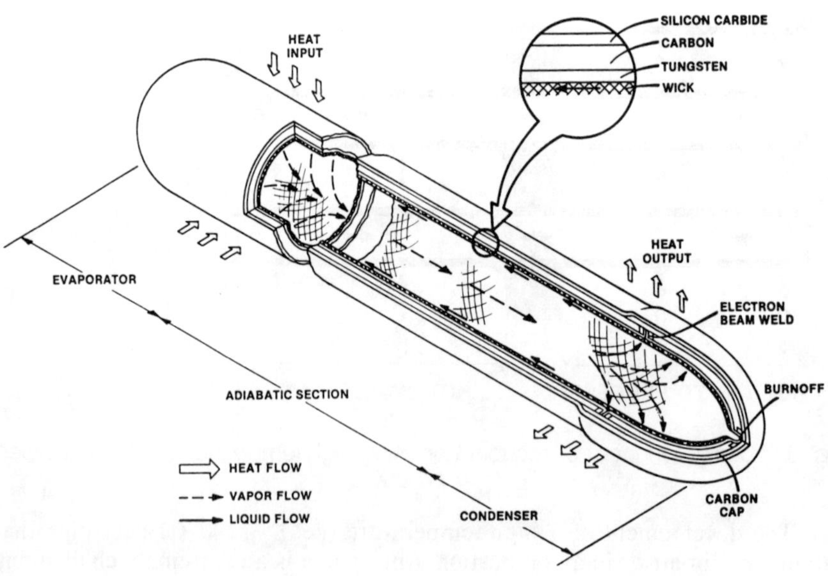


Fig. 4. High-temperature CVD heat pipe.

Fig. 5. Microstructural tungsten wick.

Fig. 6. Tungsten screen bonded to tungsten substrate by CVD tungsten.

1200°C, is shown in Fig. 7. Testing of this heat pipe was terminated after 40 h, with the heat pipe still functioning. These CVD heat pipes have operated in air for extended periods at temperatures greater than those reported for any other heat pipe.

Fig. 7. Lithium-charged heat pipe operating in air at 1200°C. Material of construction: SiC-graphite-tungsten composite.

**Ceramic Composite Recuperator Design**

This concept involves utilizing CVD technology to impregnate a fibrous ceramic preform with silicon carbide to form a one-piece ceramic recuperator. The recuperator would incorporate four flanges for air and flue gas to enter and leave—everything between these flanges would be of one-piece integral construction.

This approach would also permit flexibility in system design to match the needs of specific heat-recovery applications. Integral recuperator modules could also be fabricated in several configurations—shell and tube, matrix, or baffled passages. Moreover, the silicon carbide modules could be coupled in series to provide high effectiveness, or coupled in parallel for high flow rates.

Ceramic-fiber composites are currently being developed for a variety of high-value-added applications, such as heat engines. In these cases, the addition of the fibers significantly reduces the risk of catastrophic failure caused by the brittle nature of ceramics. In high-temperature recuperators, however, the major benefit is that very thin-walled and complicated structures can be fabricated in one piece. The resulting component is lightweight and more resistant to thermal shock than thick-walled monolithic ceramics.

Fig. 8. Graphite-cloth tube impregnated with SiC.

The technology required for this recuperator design is currently being developed at Thermo Electron as a modification to the standard CVD process. This modification, commonly known as chemical vapor infiltration (CVI), involves adapting the process parameters to favor infiltration of a fibrous preform rather than deposition on the surface.

When applied to the integral ceramic recuperator, a preform would be fabricated from one layer of SiC cloth. In effect, separate components of the recuperator would be woven and then sewn together to achieve the desired recuperator configuration. This preform would then be suspended in a CVD reactor and densified with silicon carbide. The process parameters can be modified at the end of the run to deposit a thin, coherent layer of silicon carbide on the inside and outside of the recuperator, thereby ensuring that the structure is leak-tight.

Several test pieces that show the feasibility of the technique have been fabricated. In particular, several small tubes and flat sheets of woven carbon or silicon carbide fiber* have been infiltrated with CVD SiC. A typical structure is shown in Fig. 8. A photomicrograph of the cross section is shown in Fig. 9. This photomicrograph clearly shows how readily the CVI process can occur, especially in thin sections.

The integrity of these structures has been tested in combustion atmospheres. Minimal degradation occurs over 200 h at 1200°C, presumably due to the formation of a protective silicon dioxide coating. More research needs to be done in this area since the long-term stability of composite materials has not been established.

---

*Nicalon, Nippon Carbon Co., Tokyo, Japan.

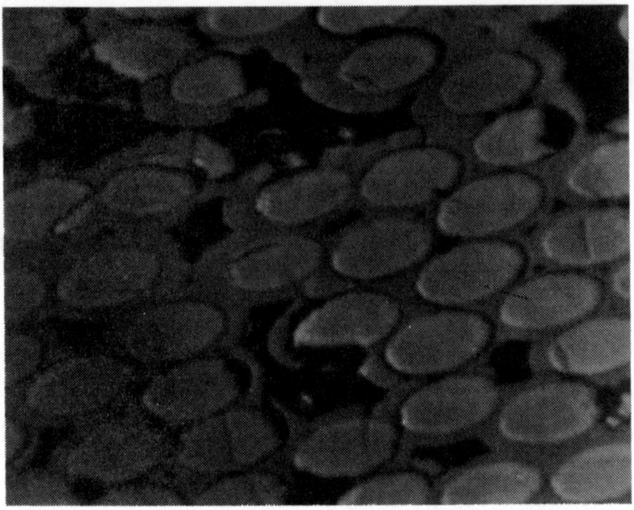

Fig. 9. Optical micrograph of carbon braid infiltrated with SiC.

## Conclusions

Two novel approaches for recovering waste heat have shown feasibility in initial studies. Both processes rely on utilizing CVD to fabricate stronger, lighter, and more corrosion-resistant components than waste heat-recovery equipment currently available. Thus, the largely untapped source of high-temperature waste heat as well as heat from highly corrosive industrial furnaces will be recoverable.

## References

[1]M. Gerstner and R. Stake, "Survey of Potential Energy Savings Using High Effectiveness Recuperators for Waste Heat Recovery from Industrial Flue Gases," TID–28954, prepared for U.S. Department of Energy by AiResearch Manufacturing Company, Torrance, California, October 15, 1977.

[2]D. B. Goodale, P. Reagan, G. Miskolczy, D. Lieb, and F. N. Huffman, "Characteristics of CVD Silicon Carbide Thermionic Converters," Proc. 16th IECEC, p. 1974 (1981).

[3]P. Reagan, D. Lieb, G. Miskolczy, D. Goodale, and F. Huffman, "CVD Fabrication of Thermionic Converter and Heat Pipe," *Ceram. Eng. Sci. Proc.*, 4, 520 (1983).

[4]P. Reagan, A. N. Scoville, R. Leaf, F. Huffman, and R. Irwin, "Chemically Vapor Deposited Heat Pipes for Application in High Temperature Heat Recovery," Proc. 19th IECEC (1984).

# Olivine: A Cost-Effective Heat-Exchange Medium

O. J. WHITTEMORE
University of Washington
Seattle, WA 98195

HAYNE PALMOUR, III
North Carolina State University
Raleigh, NC 27607

**There are vast quantities of the mineral olivine in North Carolina and Washington. This mineral is refractory, has a high volume heat capacity, and has no polymorphic transformations. Uses as heat-exchange media currently involve off-peak power heat storage; in the past it involved glass furnace regenerators. Recent studies of olivine-based heat-exchange refractories are discussed.**

Olivine is a ferromagnesium silicate mineral occurring in a number of locations in the world. In the United States, three companies mine olivine from several large deposits in the state of North Carolina, while two mine olivine in the state of Washington from the Twin Sisters mountain range, which consists principally of olivine. Commercial deposits are also mined in Norway, Japan, Sweden, Yugoslavia, Spain, and India.[1]

## Properties

In mineralogy, the term "olivine" means a member of the chrysolite group of minerals which conform to the orthosilicate formula, $2RO \cdot SiO_2$, where "R" can be Fe, Mg, Ca, Mn, Zn, and Pb. However, the term usually refers to the isomorphous series of solid solutions of forsterite [$2MgO \cdot SiO_2$] and fayalite [$2FeO \cdot SiO_2$]. The large mined deposits usually consist of about 90% forsterite and 10% fayalite. The mineral is green—hence its name. The rock "dunite" consists almost entirely of olivine and also owes its name to the dun color of the oxidized rock. Structurally, olivine is in the nesosilicate class where single $SiO_4$ tetrahedrons are bonded to two Mg or Fe ions each.

Goldschmidt[2] first advocated the use of olivine as a refractory in 1926 and obtained several patents. They were licensed to the Harbison-Walker Refractories Co. who, in 1933, first offered sale of olivine refractories under the trade name "Forsterite." Their properties were described by Birch and Harvey.[3] Research was also conducted by Stone on North Carolina olivine as a refractory.[4] At the University of Washington, the U. S. Bureau of Mines sponsored six consecutive masters theses on olivine from 1936 to 1940.[5,6]

The principal uses of olivine in heat exchangers are in glass furnace regenerator bricks, in incinerators, and in electric heat storage furnaces.[1] It is also used as a flux and slag, for foundry molding sand, as a steel tundish backing filler, and in the bottom of steel mill soaking pits.

Olivine is an attractive refractory for heat storage for the following reasons:

(1) *Low cost.* The mineral occurs in massive concentrated form and requires only mining and crushing and little beneficiation before processing into refractories. No calcination is required to obtain coarse grain, as is the case for alumina, magnesia, and fireclay refractories.

(2) *High heat-storage capacity.* Magnetite ($Fe_3O_4$) has a higher heat capacity owing to magnetic properties but is limited in temperature. Magnesia and alumina have slightly higher capacities but are more costly.

(3) *Refractoriness.* Olivine melts at about 1800° and can be used as a refractory up to 1400°C.

(4) *No crystallographic inversions.* Since heat exchange involves frequent cycling of a refractory from high to low temperatures, inversions such as in silica would cause rapid deterioration; olivine has no inversion problem. However, it is true that olivine may have problems in some oxidizing-reducing environments because of the iron content.

(5) *High electrical resistivity.* Sheathing of electrical heating elements from olivine would not be required.

**Processing Additives**

In the development of olivine refractories, several additives have been studied; these additives improve the properties of the products. The additives are:

(1) *Magnesia.* Goldschmidt[2] recognized that, in furnace use, olivine might contain low-melting phases due to the oxidation of the ferrous iron in fayalite and also from low-melting serpentine, an alteration product of olivine. Thus, he proposed the addition of magnesia, which would react to form forsterite and magnesioferrite spinel, as in Eqs. (1) and (2).

$$2FeO \cdot SiO_2 + 3MgO = 2MgO \cdot SiO_2 + MgO \cdot Fe_2O_3 \tag{1}$$

$$MgSiO_3 + MgO = 2MgO \cdot SiO_2 \tag{2}$$

(2) *Alumina.* In oxidizing-reducing conditions, the change in iron from the ferrous to the ferric state and back can cause disintegration. Therefore, alumina additions were used in the "Forsterite A" brick of Harbison-Walker Co. (proposed by E. F. Osborn) to stabilize the spinel phase.

(3) *Kaolin.* Palmour et al.[7] demonstrated that a small addition of a plastic kaolin would not only serve as a binder for heat-storage media but also result in a thermal-shock-resistant product. Although some liquid sintering resulted, there was substantial grain-to-grain solid state bonding and microcracking through the olivine grains. One might speculate that some cordierite might be formed from the reaction between forsterite and kaolin, as in Eq. (3).

$$2MgO \cdot SiO_2 + 2(Al_2O_3 \cdot 2SiO_2) = 2MgO \cdot 2Al_2O_3 \cdot 5SiO_2 \tag{3}$$

(4) *Calcium aluminate.* Smith, of the Olivine Corp., Bellingham, Washington, has developed an olivine castable composition using calcium aluminate cement which he uses to manufacture precast slabs, 6 ft by 2½ ft by 6 in., which are assembled into waste incinerators.[8] At this writing, these slabs cost $160 each, illustrating the low cost of olivine.

(5) *Sodium phosphates.* Whittemore developed olivine compositions using sodium polyphosphate or sodium hexametaphosphate as the bond for heat-storage brick compositions which do not require firing.[9] Compositions containing only 1% of the phosphate were found to be as strong after drying as

Table I. Properties of Olivine vs Other Heat Storage Materials

| Brick | Density (% theor.) | Volume Heat Capacity (cal/cm²·°C) | Thermal Conductivities (W/m·K) | Thermal Expansion Coefficient $\alpha \times 10^6$ (per °C) |
|---|---|---|---|---|
| Olivine: | | | | |
| NCSU | 82 | 0.68 | 1.3–2.1 | |
| UW | 86 | 0.70 | 2.7 | 7.2 |
| English | 73–79 | 0.62–0.66 | 1.8 | |
| German | 82 | 0.63 | 1.3–2.0 | |
| Magnesite | 83 | 0.79 | 8.0 | 14 |
| Magnetite | 77 | 0.87 | 1.6–2.1 | 11 |

NCSU = North Carolina State University.
UW = University of Washington.

heat-storage bricks which had been fired. The strength after drying did not diminish when the product was fired at several temperatures up to 1400°C.

### Olivine vs Other Heat-Storage Materials

Table I lists properties of various olivine bricks, together with magnesite and magnetite, all of which have been used for electric heat storage furnaces.

Densities of olivine brick vary considerably, as can be seen in the table. As the volume heat capacity varies directly with density, it is desirable to obtain as dense a brick as possible which yet would retain heat-shock resistance. The volume heat capacities follow the Dulong and Pettit rule of about 6 cal·g·atom·°C, except for magnetite, which owes its higher heat capacity to its magnetic properties. A differential thermal analysis of magnetite gave high endothermy compared with alumina up to 600°C but at that temperature the curve reversed and was quite exothermic. The Curie point of magnetite is 585°C. Magnetite thus is a superior heat-storage material up to its Curie point. However, it also is a poor electrical insulator, so sheathed elements must be used in electric storage units.

The thermal conductivities of the various materials are similar, except for magnesite which is considerably higher. For electric heat storage units this can be a disadvantage as higher temperatures can be stored in olivine-based units. The thermal expansion coefficient of olivine is lower than the other materials, which means better thermal shock resistance.

### Use of Olivine in Glass Furnace Regenerators

Olivine checker brick have been used in glass furnace regenerators for years because of their low cost and the excellent heat-storage properties. Some problems have occurred with carry-over from the batch materials. Also, if oxidizing-reducing conditions occur, swelling can occur. This problem was alleviated by the addition of alumina, as mentioned previously.

### Use of Olivine in Refractory Incinerators

As mentioned above, calcium aluminate-bonded olivine slabs are being used in the construction of waste incinerators. An adjunct to these furnaces has been rock towers where lump olivine is used as heat-storage media.[8]

**Use of Olivine in Electric Heat-Storage Heaters**

For the last 30 years, electrical power load-leveling of thermal generating units has been assisted by the use of off-peak heat storage units. Lower rates are granted to customers using these units at night. Because of the low cost, olivine has been a popular material for the construction of these units in Europe and, in the United States, heat-storage units using olivine are also beginning to be used. In England, domestic units are now being built using magnetite bricks because of the higher heat-storage capacity. These do have the drawbacks of temperature limitations and also must be shielded from the heaters because of low electrical resistivity. They do serve the purpose in that domestic units are more compact.

Studies have been carried out in the United States on improvement of olivine brick for thermal heat-storage units. As previously mentioned, at the North Carolina State University, kaolin-bonded bricks were developed with excellent thermal shock resistance. Tests also were conducted at Purdue University[10] in a project associated with the American Society of Heating, Refrigerating and Air-Conditioning Engineers to establish standards for electrically heated thermal energy storage units. In these tests in a commercial unit, North Carolina olivine bricks equaled or slightly surpassed the performance of European magnesite. This was surprising since magnesite has higher specific heat. However, the lower thermal conductivity of the olivine allowed a temperature of 850°C in the core where the magnesite bricks reached only 750°C, thus accounting for the higher heat storage.

A project was carried out at the University of Washington to develop a lower-cost olivine brick which did not require firing.[11] Additions of only 1% sodium polyphosphate or sodium hexametaphosphate gave dried strengths as high as those in the commercial English olivine bricks. The strength was retained after heatings at 300°, 500°, 725°, and 1400°C.

Over three tons of bricks were molded in a commercial friction press* and cured at 260°C. The bricks had a density of 2.8 g/cm$^3$, much higher than that of commercial olivine bricks. The higher density was attributed to the massive olivine,† to the superior molding power of the friction press, and to the simple yet "gap-graded" mixture of 80% 6-mesh and finer olivine with 20% olivine fines, 180-mesh and finer.[11]

Some of these bricks were tested at Purdue University[12] in a 30 kW central storage unit.‡ These unfired dense bricks were found to both store and deliver up to 14% more heat than would fired olivine. The higher capacity is directly related to the higher density.

**Conclusions**

Olivine is a good heat-storage medium possessing sufficient refractoriness for most applications. The mineral requires no calcination, and possesses no destructive crystal inversions. Large deposits in North Carolina and Washington can supply more olivine than any conceivable requirement.

---

*North American Refractories Co., Renton, WA
†Olivine Corp., Bellingham, WA
‡TPI Corp., Johnson City, TN

## Acknowledgment

The developments at North Carolina State University, at the University of Washington, and at Purdue University were supported by the Department of Energy. The North Carolina State University project was initiated with a grant from the North Carolina Energy Institute. The Washington project was started with support from the Washington Mining and Mineral Resources Research Institute.

## References

[1]J. Griffiths, "Olivine-Exchanging New Uses for Old," *Ind. Mineral.*, **9**, 65–79 (1984).

[2]V. M. Goldschmidt, "Olivine and Forsterite Refractories in Europe," *Ind. Eng. Chem.*, **30** [1] 32–34 (1938).

[3]R. E. Birch and P. A. Harvey, "Forsterite and Other Magnesian Silicates as Refractories," *J. Am. Ceram. Soc.*, **18** [6] 176–92 (1935).

[4]R. L. Stone, "Thermochemistry of North Carolina Olivine in the Manufacture of Forsterite Refractories," *J. Am. Ceram. Soc.*, **22** [10] 342–48 (1939).

[5]H. Wilson and K. G. Skinner, "Review of Four Year's Research on the Refractory Properties of Pacific Northwest Olivine," *J. Am. Ceram. Soc.*, **23** [5] 136–38 (1940).

[6]O. J. Whittemore, "Basic Refractories from Washington Olivine"; M. S. Thesis, University of Washington, Seattle, 1940.

[7]H. Palmour, B. M. Gay, and R. L. Cochrane, "Olivine Refractory Bricks for Heat Storage Applications," U. S. Pat. No. 4 303 448, Dec. 1, 1981.

[8]C. Smith, Olivine Corp., Bellingham, WA; personal communication.

[9]O. J. Whittemore, "Unfired Olivine Heat Storage Media," Oak Ridge National Lab. Report ORNL/Sub/81/9063/1, Oct. 1983.

[10]R. J. Schoenhals, W. R. Laster, and M. R. Elter, "Thermal Energy Storage Testing Facility"; pp. 394–98 in Proceedings of D.O.E. Phys. & Chem. Energy Storage Annual Contractors Review Meeting, U. S. Dept. of Energy, Washington, D. C., Dec. 1982.

[11]O. J. Whittemore, "Unfired Washington Olivine Heat Storage Media"; in Opportunities in Thermal Storage R & D, Electric Power Research Institute, Palo Alto, CA, July 1983.

[12]R. J. Schoenhals, "Opportunities and Challenges Suggested by Performance Tests of Electrically Charged Thermal Energy Storage Units," ibid.

# Analysis of Candidate Silicon Carbide Recuperator Materials Exposed to Industrial Furnace Environments

J. I. FEDERER and T. N. TIEGS

Oak Ridge National Lab
Oak Ridge, TN 37831

Several SiC ceramics were exposed to the combustion environments in six industrial furnaces to determine their corrosion resistance. The materials were sintered $\alpha$-(Hexoloy SA), Sintride, recrystallized (NC-400), CVD SiC-coated NC-400, siliconized (NC-430), reaction-sintered (SC-X and KT), and $Si_3N_4$-bonded (C/75 and CN-178). Tubes of these materials were exposed in two aluminum remelt furnaces, a forge furnace, a steel reheat furnace, and two steel soaking pits at 925° to 1250°C for 530 to 5545 h. Significant corrosion occurred in specimens exposed to aluminum remelt furnaces and one of the steel soaking pits, whereas corrosion in the other furnaces was substantially less or negligible. The average C-ring fracture strength of Hexoloy SA (the only material tested) was substantially decreased by exposure to an aluminum remelt and a steel reheat furnace. These results show that SiC ceramics are susceptible to both corrosion and strength degradation when exposed to certain furnace environments.

In recent years the Department of Energy (DOE) has conducted research and development through industrial contractors to advance waste heat recovery technology. In 1982, DOE contracted with Garrett AiResearch Manufacturing Company* and Babcock & Wilcox Company to design, develop, build, and field-test advanced high-temperature burner-duct-recuperator (HTBDR) systems for preheating combustion air to 1090°C (2000°F).[1] This performance goal required that durable ceramic heat-exchanger elements be fabricated, and that energy losses by leakage through seals be minimized. Both contractors selected SiC ceramics as candidate materials for recuperator tubes on the basis of refractoriness and physical and mechanical properties; however, the corrosion resistance of the materials in various industrial furnace environments had not been determined.

The intended recuperator applications would involve recovery of waste heat from fouling, corrosive, flue gases in industrial furnaces such as steel reheat, steel soaking pit, and aluminum remelt furnaces. In these applications recuperator materials are exposed to flue gases containing numerous chemical species at temperatures up to 1300°C. Significant corrosion of recuperator tubes would adversely affect the economics of recuperation for a particular furnace because of tube replacement expense and furnace downtime. Both contractors used exposure tests, therefore, to assess the corrosion resistance of candidate materials. Subsequently, the specimens were analyzed and selected

*Funding for certain tasks was provided by the Gas Research Institute (GRI) in a cooperative program with DOE.

materials were subjected to flexure tests at high temperatures. The objective of these analyses and measurements was to correlate corrosion and changes in flexure strength with exposure conditions.

### Materials

Silicon carbide ceramics were selected as candidate materials for heat-exchanger tubes and structural components in preference to oxide ceramics such as alumina and mullite. Although the latter materials, especially alumina, are also strong and oxidation-resistant, SiC is significantly less susceptible to thermal shock because of inherently higher thermal conductivity and lower coefficient of thermal expansion. The properties of SiC ceramics depend strongly on composition and density; therefore, several types were included in these exposure tests for comparison.

The exposed materials, shown in Table I, represent most of the commercially available SiC ceramics that might be considered for heat-exchanger applications. Alumina and mullite were exposed in one furnace for comparison with the SiC ceramics. The specimens were mostly tubes measuring nominally 25 or 89 mm OD (1 or 3.5 in.) and 25 to 30 cm long (10 to 12 in.) except that the specimens in aluminum remelt No. 2 were only 5 cm long (2 in.). One $Si_3N_4$-bonded SiC (CN-178) was in the form of 15 mm by 15 mm by 25 cm bars. Not all materials were included in each test.

### Exposure Conditions

Corrosion testing involved installation of specimens in industrial furnaces for exposure to high-temperature combustion environments. Although all furnaces burned natural gas, the composition of the flue gases depended strongly on the type of furnace operation (as will be shown) and, possibly, on specific location within the furnace or flue. Table II shows the exposure conditions in six industrial furnaces. Maximum estimated specimen temperatures occurred only during periods of normal use. Normal use, however, involved numerous heating and cooling cycles for loading and unloading the materials being processed. The maximum temperatures shown in Table II, therefore, did not occur continuously during the normal use periods.

The exposures occurred during typical furnace operations involving feed materials, heating and cooling cycles, maximum temperatures, fluxes, and air-fuel mixtures. The aluminum remelt furnaces were charged with NaCl-KCl-$Na_3AlF_6$ flux along with metal to be melted. In addition, the molten aluminum in furnace No. 2 was sparged with chlorine gas during a portion of each normal cycle. The use of chlorine gas and chloride fluxes is normal industry practice for purifying scrap aluminum and for minimizing oxidation. The presence of chloride compounds in the flue gases, therefore, is not unusual. Steel billets in the soaking pits were covered with topping compounds, which is standard industry practice to minimize oxidation. Although flue gases in the furnaces were not analyzed, analysis of material that deposited on some specimens revealed significant quantities of metals (to be shown). Gaseous species in the fuel combustion products no doubt included $N_2$, $O_2$, $CO_2$, and $H_2O$.

Specimens were mostly exposed in crossflow configuration, that is, with the long axis perpendicular to the direction of flue gas flow. None of the specimens was exposed in the heat-exchanger mode with air flowing through the inside. The combination of high temperature and medium length exposure

Table I. Candidate Materials Exposed to Industrial Furnace Environments

| Type | Designation | Nominal Composition (wt%) | Manufacturer* |
|---|---|---|---|
| Sintered $\alpha$-SiC | Hexoloy SA | >99 SiC | A |
| Sintered SiC | Sintride | >99 SiC | B |
| Recrystallized SiC | NC-400 | >99 SiC | C |
| Recrystallized SiC coated with SiC | NC-400/CVD SiC | >99 SiC | C, D |
| Siliconized SiC | NC-430 | 90 SiC-10 Si | C |
| Siliconized SiC coated with zircon | NC-430/zircon | 90 SiC-10 Si | C |
| Reaction-sintered SiC | SC-X | 88 SiC-12 Si | E |
| Reaction-sintered SiC | KT | 84 SiC-16 Si | A |
| $Si_3N_4$-bonded SiC | C/75 | 75 SiC-23 $Si_3N_4$-2 oxide | F |
| $Si_3N_4$-bonded SiC | CN-178 | 75 SiC-23 $Si_3N_4$-2 oxide | C |
| Alumina | AD-998 | 99.8 $Al_2O_3$ | E |
| Mullite | Mullite | 72 $Al_2O_3$-28 $SiO_2$ | E |

*A—The Carborundum Company
B—General Electric Company
C—Norton Company
D—Garrett AiRearch
E—Coors Porcelain Company
F—NGK Locke, Inc.

Table II. Exposure Conditions in Industrial Furnaces

| Furnace | Maximum Estimated Specimen Temperature* °C (°F) | Exposure Period† h (days) |
|---|---|---|
| Aluminum remelt No. 1‡ | 1040 (1900) | 5545 (231) |
| Aluminum remelt No. 2‡ | 1150 (2100) | 530 (22) |
| Steel forge‡ | 1175 (2150) | 4440 (185) |
| Steel reheat‡ | 1100 (2010) | 1455 (61) |
| Steel soaking pit No. 1‡ | 925 (1700) | 2280 (95) |
| Steel soaking pit No. 2§ | 1250 (2300) | 3460 (144) |

*Occurring during normal furnace operation.
†Period of normal furnace operation involving heating and cooling for loading and unloading.
‡Conducted by Garrett AiResearch.
§Conducted by Babcock & Wilcox.

time in steel soaking pit No. 2 represents one of the most severe thermal exposures among the group shown in Table II. A lower temperature location was not available in steel soaking pit No. 2.

## Results

### Appearance of Specimens

The exposed specimens are shown in Figs. 1–4. Specimens exposed in the aluminum remelt furnaces were covered with deposits (Figs. 1 and 2), whereas specimens in the steel reheat furnaces and soaking pit No. 1 were relatively clean, as shown in Fig. 3. The deposit on specimens in aluminum remelt furnace No. 1 (Fig. 1) was mostly porous and loosely adhering; however, a thin glassy layer was detected at the surfaces during microstructural examination. By contrast, the deposit on specimens in aluminum remelt furnace No. 2 (Fig. 2) was a glassy material, relatively rough in some areas and somewhat smoother at 180° locations. Specimens exposed in steel soaking pit No. 2 (Fig. 4) had a slag layer which was thickest on the downstream side. Slag on the upstream side was dull, but on the downstream side was glassy in appearance. High exhaust-gas flow rates apparently forced molten slag, which deposited on the upstream side, to flow to the other side. Post-exposure heating confirmed that the slag melted in the 1250° to 1300°C range.

A few specimens broke during the exposures. One each Hexoloy SA and C/75 specimen broke in aluminum remelt No. 1, as did the only AD-998 specimen. One NC-400 specimen broke in aluminum remelt No. 2. Cracking of $Si_3N_4$-bonded SiC specimens occurred in all exposures. The C/75 material, in particular, cracked extensively in most cases and exhibited severe wall thinning in steel soaking pit No. 2.

### Characterization of Deposits

The characteristics of deposits on specimens in three furnaces are shown in Table III (deposits did not form in the other three furnaces). The sources of deposits include the materials being heated, fluxes and topping compounds, and furnace refractories. The K and Na content of the deposits on specimens in the aluminum remelt furnaces evidently derived from the NaCl-KCl flux used to purify and protect the molten aluminum from oxidation. Both deposits consisted largely of silicate glass, and several crystalline compounds were de-

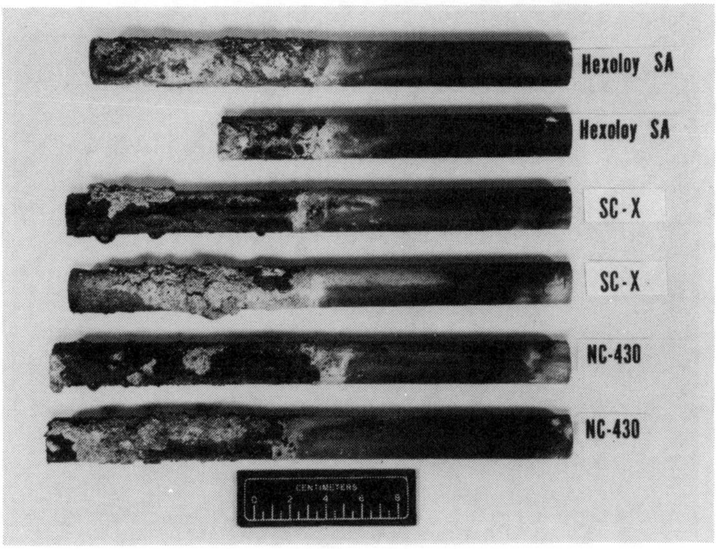


Fig. 1. Materials exposed in aluminum remelt furnace No. 1.

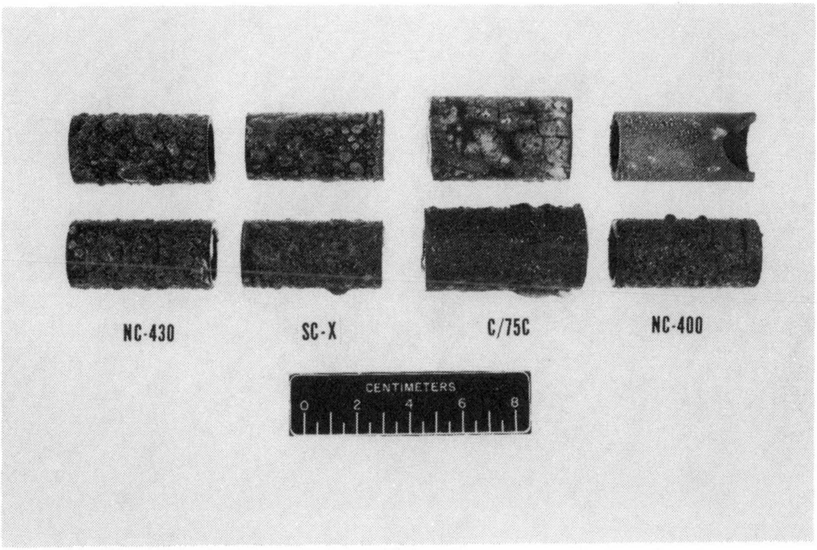


Fig. 2. SiC ceramics exposed in aluminum remelt furnace No. 2.

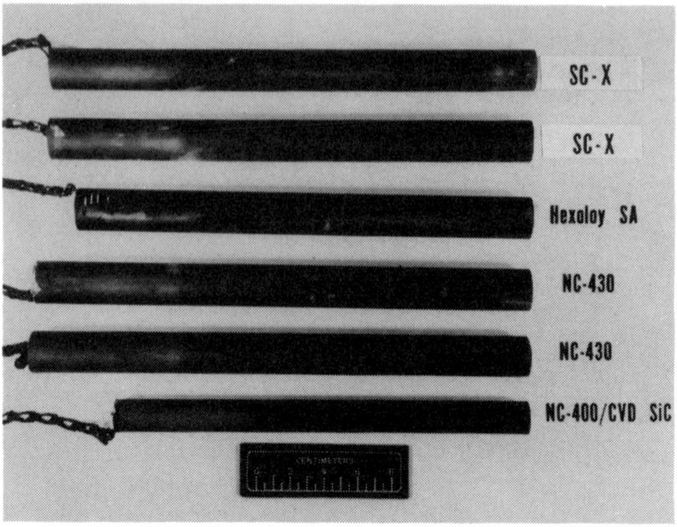


Fig. 3. SiC ceramics exposed in steel soaking pit No. 1.

Fig. 4. Materials exposed in steel soaking pit No. 2.

tected by X-ray diffraction, as shown in Table III. The deposits melted over the range 1150° to 1300°C. Specimens in aluminum remelt No. 2 had evidently been exposed to such temperatures because the deposit had obviously been molten. Many elements in the deposit on specimens in steel soaking pit No. 2 were probably derived from steel billets being heated, in particular Cr, Fe, Si, Mn, and Ni, and from topping compounds. This deposit, which had obviously been molten, melted over the range 1250° to 1300°C.

Table III. Characteristics of Deposits on Specimens Exposed to Aluminum Remelt Furnace Environments

| Content* (ppm) | Aluminum Remelt No. 1 | Aluminum Remelt No. 2 | Steel Soaking Pit No. 2 |
|---|---|---|---|
| | | Composition | |
| Major† | Al, Si | Al, K, Na, Si | Cr, Fe, Si |
| 1000–3000 | Fe, K, Na, Zn | Ba, Ca, Cu, Fe, Zn | Cu, K, Mn |
| 500–1000 | Ca, Ti | Cl, Mg | Ca, Na, Ni |
| 100–500 | B, Ba, Bi, Cr, Cu, Mg, Mn, Ni, P, Pb, S, Sn | Mn, S, Sr | P, S, Ti, Zn |
| <100 | Ce, Cl, Co, Mo, Sc, V, Zr | B, Co, Cr, F, La, Mo, Ni, P, Rb, Ti, V, Zr | As, B, Cl, Co, Ga, Ge, Mg, Mo, Rb, V, Zr |
| | | Crystalline compounds | |
| | $Al_2O_3$ | (Ca, Mg, Fe, Al) silicate | ‡ |
| | $Na_2O \cdot Al_2O_3 \cdot 2SiO_2$ | | |
| | | Melting range (°C) | |
| | 1200–1300 | 1150–1300 | 1250–1300 |

*Determined by semiquantitative spectrographic analysis.
†5 to 20%.
‡Not determined.

#### Weight Changes

Weight changes occurring in specimens in the forge and steel reheat furnaces are presented in Table IV. Specimens in these furnaces apparently were exposed to relatively clean atmospheres compared to those in the aluminum remelt furnaces because no deposits were evident on the surfaces. The weight

Table IV. Weight Changes of SiC Ceramics Exposed to Industrial Furnace Environments

| Material | | Weight Change (%) | |
|---|---|---|---|
| Type | Designation | Forge Furnace (1175°C, 2150 h) | Steel Reheat Furnace (1100°C, 2010 h) |
| Sintered-$\alpha$ | Hexoloy SA | 0 | 0.2 |
| Recrystallized | NC-400 | −0.1 | 0.2*<br>0.1* |
| Recrystallized/CVD coated | NC-400/CVD SiC | | 0.6 |
| Siliconized | NC-430 | 0 | 0.4*<br>0.5* |
| Reaction-sintered | SC-X | 0 | 0.5 |
| $Si_3N_4$-bonded | C/75 | 17.5 | 4.2*<br>1.9* |
| | CN-178 | 3.8 | 1.4 |

*Two specimens in exposure test.

changes in these specimens, therefore, were probably caused by oxidation. Table IV shows that the $Si_3N_4$-bonded materials had significantly higher weight changes (gains) than the other specimens. Even the porous NC-400 and the materials containing silicon as a primary phase (NC-430 and SC-X) had smaller weight gains than the $Si_3N_4$-bonded materials. The weight gains exhibited by C/75 and CN-178 are tentatively attributed to oxidation of the $Si_3N_4$ bonding phase in these ceramics. This oxidation might have caused the previously mentioned extensive cracking of some specimens of these materials.

#### Dimensional Changes

The outer diameters of exposed specimens were measured to determine changes indicative of corrosion. Measurements were made with a micrometer on apparently clean specimens, i.e., those from the forge, steel reheat, and steel soaking pit No. 1. Specimens from the aluminum remelt furnaces and steel soaking pit No. 2, as previously mentioned, had deposits on the surfaces. The diameters of these specimens were measured with a microscope on samples prepared for ceramographic examination. In either case three or four measurements were made at approximately equal intervals around the circumference. The corrosion rates shown in Table V reveal that significantly more corrosion occurred in specimens exposed in both aluminum remelt furnaces and steel soaking pit No. 2 than in the other furnaces. Corrosion rates were significantly higher in aluminum remelt No. 2 than in aluminum remelt No. 1. The higher rates in aluminum remelt No. 2 might be associated either with a much shorter exposure time (since corrosion rates sometimes decrease with increasing time) or with a higher temperature. Specimens in the forge and steel reheat furnaces and in steel soaking pit No. 1 had relatively small corrosion rates, which probably represent measuring uncertainty. A possible exception is NC-430, which exhibited surface recession in each exposure test.

The specimens in the aluminum remelt furnaces were covered with adherent glass deposits (also loosely adherent deposits in No. 1). The surface deposits might have participated in corrosion in these specimens. The data also indicate that the siliconized SiC ceramics (NC-430 and SC-X) corroded more than pure SiC (Hexoloy SA and NC-400) in the aluminum remelt furnaces.

Specimens in steel soaking pit No. 2 were covered with slag, which was thickest on the downstream side. The largest dimensional changes (reported in Table V) occurred on the upstream side. The specimens in this test exhibited much larger dimensional changes than those in the other tests, especially steel soaking pit No. 1. In the latter case the specimens were exposed to substantially lower temperatures and had no significant surface deposits, apparently because of a more protected location in the flue.

No dimensional changes were determined for several specimen types: NC-400/CVD SiC because the coated surfaces were too rough to obtain accurate pretest diameters; C/75 (except steel soaking pit No. 2) because this material either corroded extensively or cracked and swelled; CN-178 because of difficulty in determining pretest dimensions and because of cracking. The alumina and mullite exhibited no surface recession during exposure in aluminum remelt furnace No. 1.

The projected annual corrosion rates in aluminum remelt No. 1 shown in Table V ranged from 0.1 to 0.6 mm/y (0.006 to 0.022 in./y). Corrosion rates of this magnitude could probably be accommodated by design. The much

Table V. Corrosion Rates in SiC Ceramics Exposed to Industrial Furnace Environments*

| Furnace | Maximum temperature (°C) | Time (h) | Corrosion Rate, pm/s | | | | | | |
|---|---|---|---|---|---|---|---|---|---|
| | | | Hexoloy SA | Sintride | NC-400 | NC-430 | SC-X | KT | C/75 |
| Aluminum remelt No. 1 | 1040 | 5545 | −8.9 (−0.3)§ | † | −4.5 (−0.1) | −13.4 (−0.4) | −17.5 (−0.6) | † | ‡ |
| Aluminum remelt No. 2 | 1150 | 530 | † | † | −120 | −60 | −73 | † | ‡ |
| Forge | 1175 | 4440 | +1.6 | † | +9.7 | −4.0 | +2.4 | † | ‡ |
| Steel reheat | 1100 | 1455 | −1.6 | † | 0 | −22 | +7.3 | † | ‡ |
| Steel soaking pit No. 1 | 925 | 2280 | −1.5 | † | −0.8 | −5.4 | +3.1 | † | ‡ |
| Steel soaking pit No. 2 | 1250 | 3460 | −112 (−3.5) | −112 (−3.5) | † | −137¶ (−4.3) | † | −129 (−4.1) | −217 (−6.8) |

*Positive values indicate surface deposits.
†Not included in this test.
‡Not determined.
§Corrosion rate in mm/y.
¶Corrosion rate for NC-430 coated with zircon was −112 pm/s.

higher rates in steel soaking pit No. 2, which ranged from 3.5 to 6.8 mm/y (0.14 to 0.27 in./y), would probably be unacceptable. The exposure conditions for materials in this test, however, were much more severe than those anticipated for an actual recuperator. Corrosion rates in the other furnaces were not calculated because of relatively short exposure times or because of inconsistency in the data.

#### Microstructures of Exposed Materials

Typical microstructures of exposed materials are shown in Figs. 5 to 8. Sintered $\alpha$-SiC (Hexoloy SA) exposed to aluminum remelt No. 1 had a rougher surface than specimens exposed in some of the other furnaces, as illustrated in Fig. 5. The roughness indicates more extensive corrosion, which was consistent with the corrosion rate presented in Table V. Cracks oriented approximately parallel to the surfaces of the specimens in Fig. 5 occurred while cutting the samples with a diamond-impregnated cutoff wheel.

Corrosion of siliconized SiC (NC-430) apparently occurred in a general manner rather than selectively on either the Si or SiC phases (Fig. 6). The only indication of attack was slightly rougher surfaces on specimens exposed in aluminum remelt furnaces. Electron microprobe analyses, in agreement with previously discussed chemical analyses, indicated the presence of Al, Ca, Fe, K, Na, Si, and Zn in the glassy deposits on specimens exposed to aluminum remelt furnace atmospheres. No evidence for penetration of these or other elements into the silicon phase was detected.

As previously discussed, the $Si_3N_4$-bonded SiC ceramics cracked extensively during the exposure tests (Fig. 7). In addition to cracking, surface recession evidently occurred in specimens exposed to aluminum remelt furnaces. Figures 7(*a*) and (*b*) show glassy deposits on these materials. Large bubbles in the glassy deposit originate at the bonding phase, which appears to be recessed leaving SiC particles in relief. Significant corrosion of the bonding phase did not occur in specimens exposed to the steel reheat furnaces; however, cracks of the type shown in Fig. 7(*c*) occurred frequently. The indicated corrosion in aluminum remelt furnaces and the extensive cracking that occurred in all furnaces strongly suggest that these $Si_3N_4$-bonded SiC ceramics are not suitable for heat-exchanger tubes or other applications requiring leak tightness and strength.

The aluminum remelt furnace environments apparently were quite corrosive to the materials in these tests. Another study of corrosion of SiC ceramics in an aluminum remelt furnace revealed even higher recession rates than in the present work.[2] The investigators suggested that the dominant corrosion mechanism in their work was reaction of SiC, Si, and $SiO_2$ with halide species in the flue gases, such as NaCl, HCl, NaF, etc., resulting in formation of volatile silicon halides. Although unproven, this hypothesis seems to be the most likely explanation for the observed corrosion in both the previous and present work.

Figure 8 shows microstructures of several materials exposed to the steel soaking pit No. 2 environment. As mentioned previously, corrosion was highest on the upstream side. The slag coating evidently was molten at the exposure temperature, and, under the influence of high flue-gas flow rates, flowed from the upstream to the downstream side of the specimens. The slag also flowed from top to bottom of the specimens, which were oriented vertically in this test. This dynamic condition might have contributed to corrosion/erosion. The

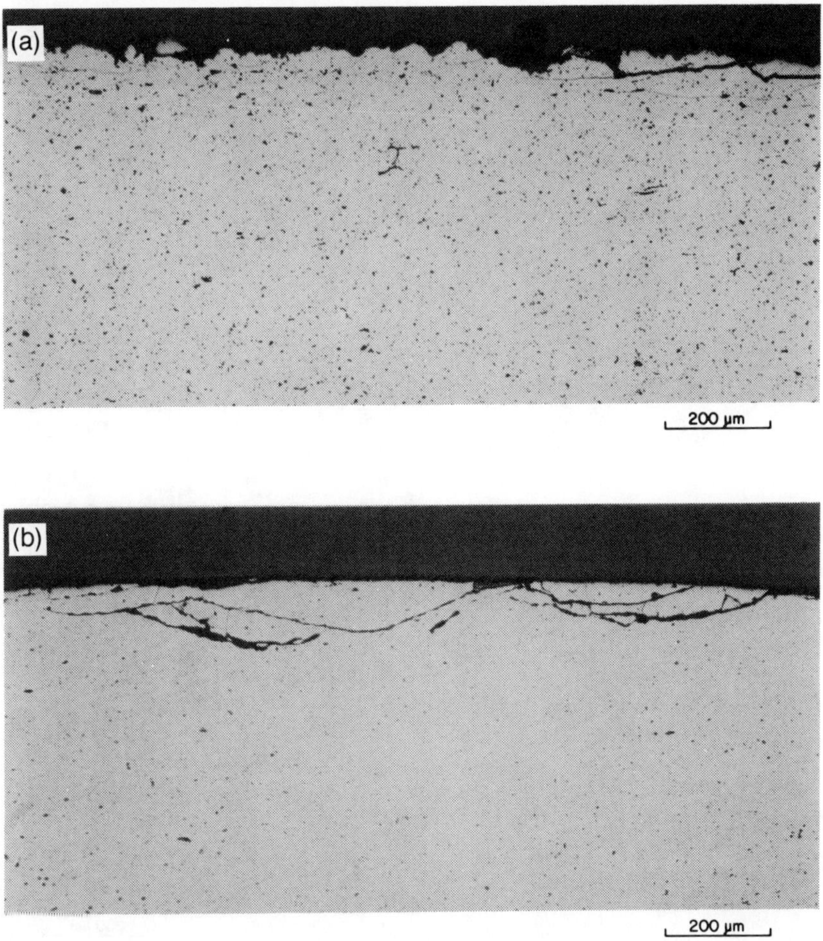


Fig. 5. Sintered $\alpha$-SiC (Hexoloy SA) exposed to industrial furnace environments. (*a*) Aluminum remelt No. 1; (*b*) steel forge; as-polished.

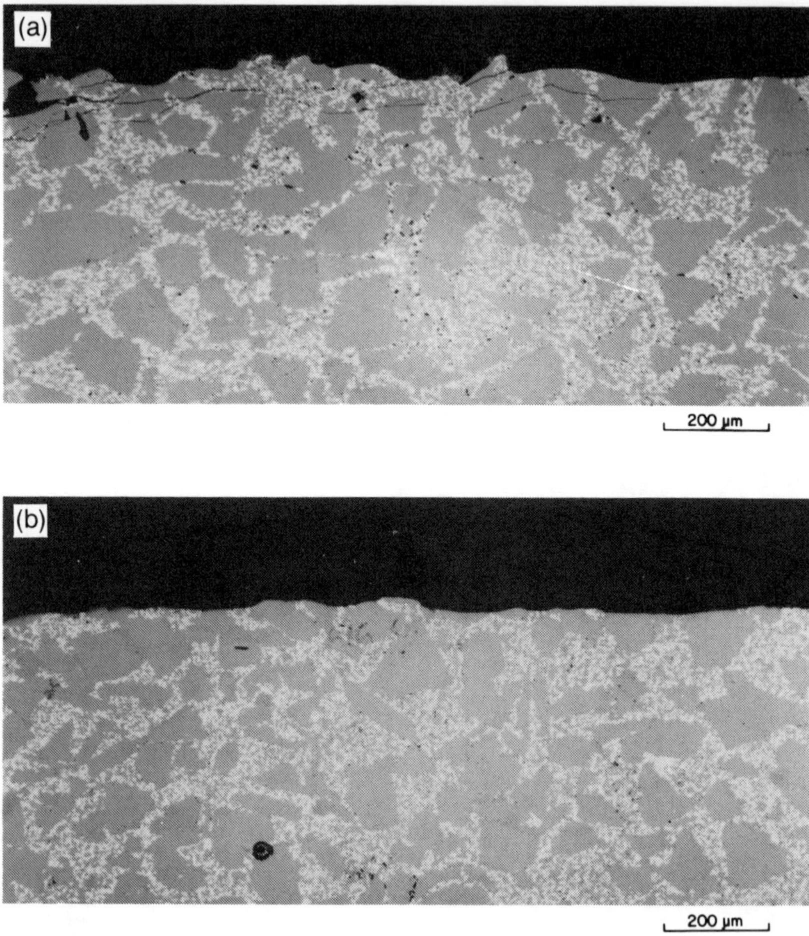
(a)
200 µm
(b)
200 µm

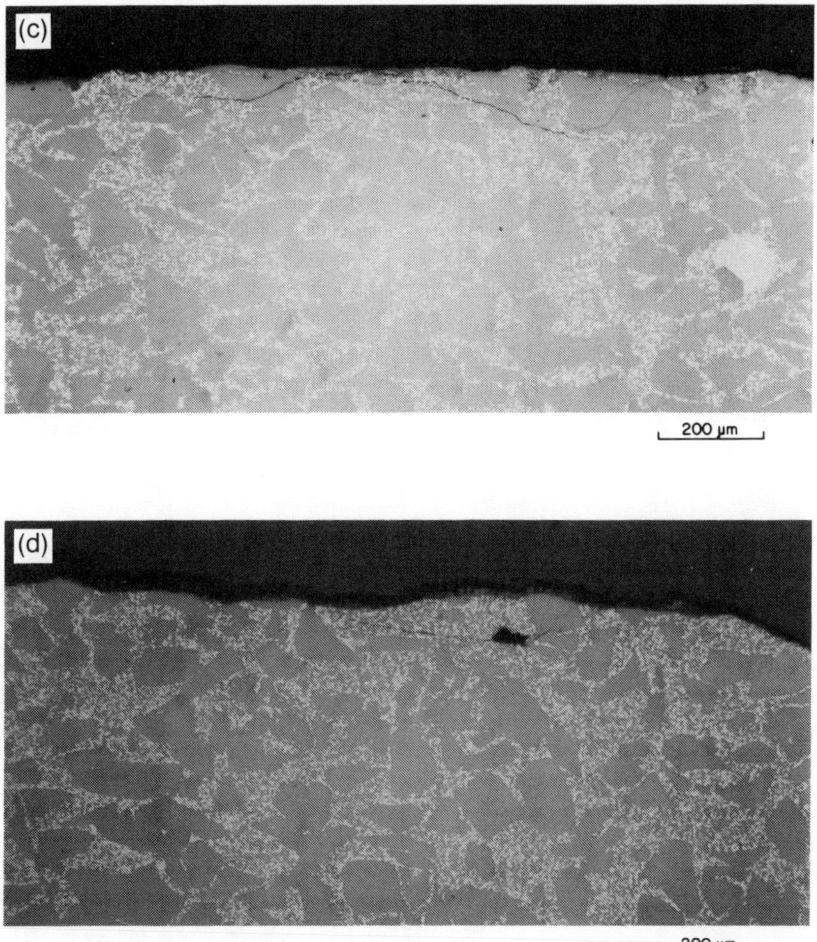


Fig. 6. Siliconized SiC (NC-430) exposed to industrial furnace environments. (*a*) Aluminum remelt No. 1; (*b*) aluminum remelt No. 2; (*c*) steel forge; (*d*) steel reheat; as-polished.

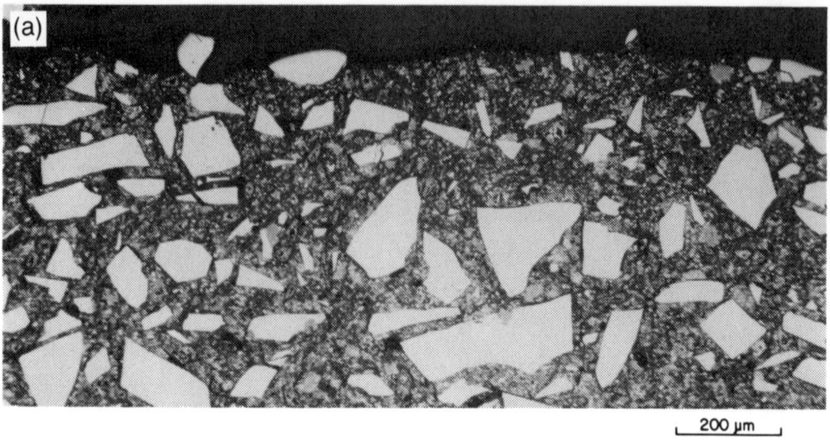
(a)
200 μm

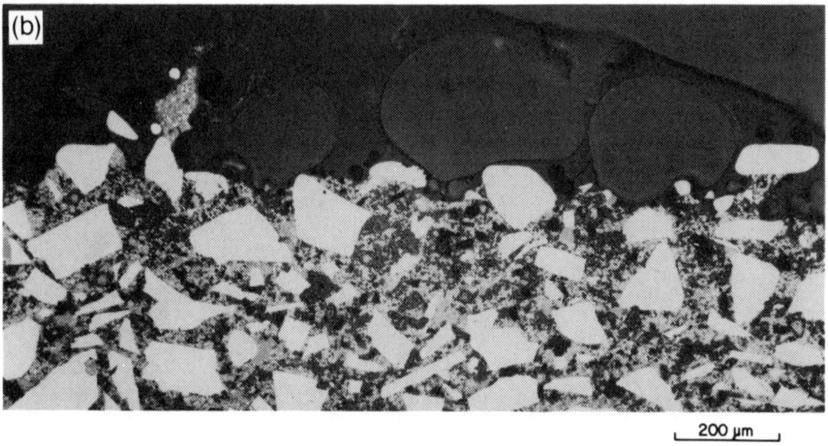
(b)
200 μm

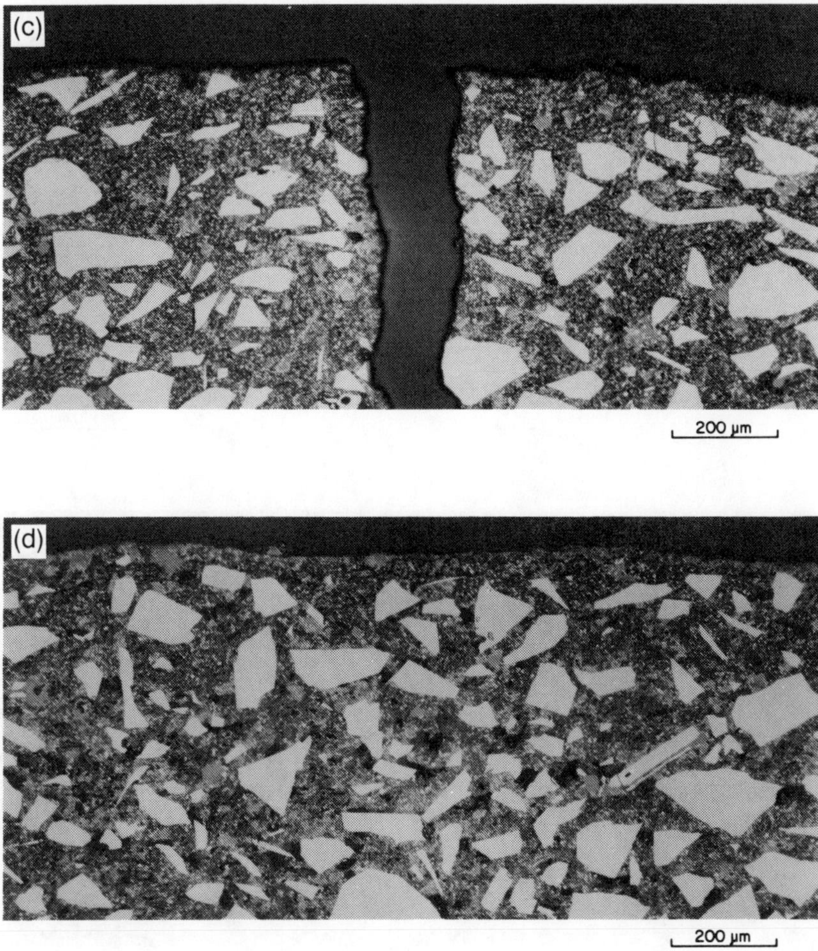


Fig. 7. $Si_3N_4$-bonded SiC (C/75) exposed to industrial furnace environments. (*a*) Aluminum remelt No. 1; (*b*) aluminum remelt No. 2; (*c*) steel forge; (*d*) steel soaking pit; as-polished.

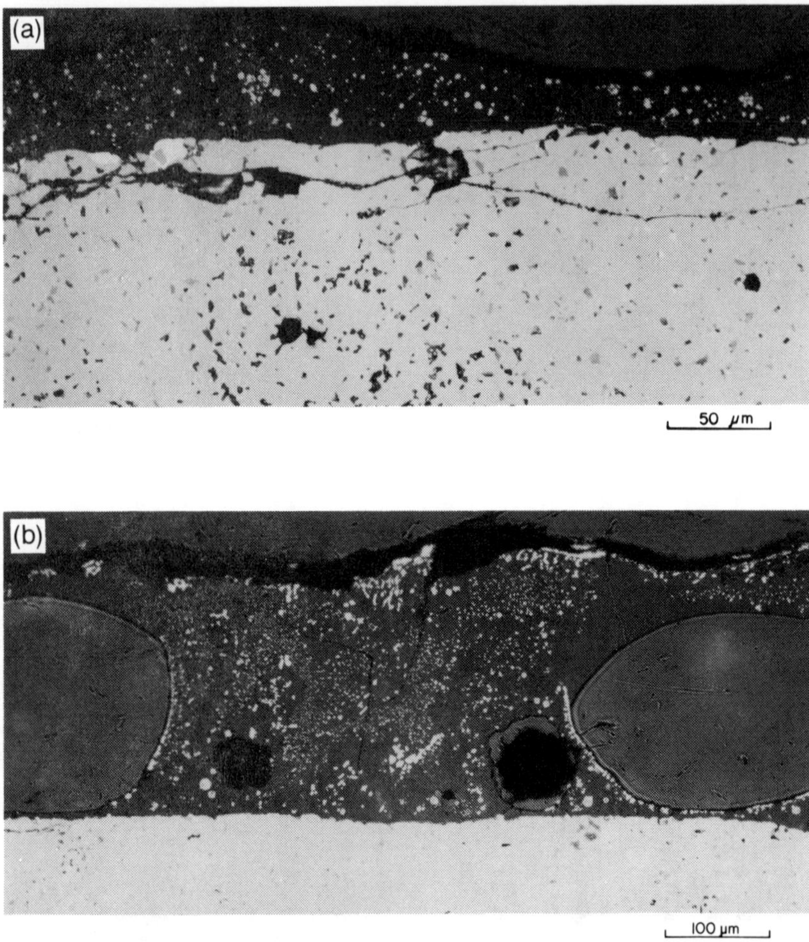
(a)
50 μm
(b)
100 μm

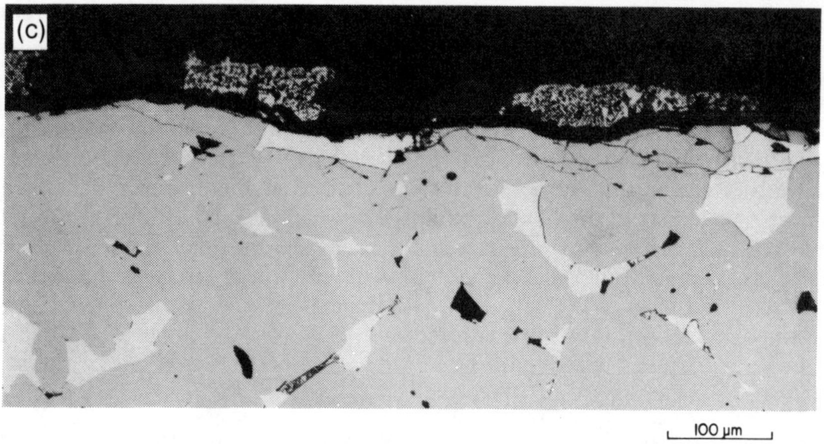

Fig. 8. SiC ceramics exposed to steel soaking pit No. 2 environment. (*a*) Hexoloy SA, upstream region; (*b*) Hexoloy SA, downstream region; (*c*) NC-430, upstream region; (*d*) C/75, cross section; as-polished.

upstream side might have eroded because the slag layer was thin or corroded faster because the slag was continually refreshed. Using chemical and electron microprobe analyses, the slag was determined to be mostly a silicate glass.

Typical microstructures of Hexoloy SA from upstream and downstream regions are shown in Figs. 8(*a*) and 8(*b*), respectively. Corrosion is indicated by surface roughness in both regions. Numerous particles in the slag were shown by electron microprobe analysis to contain high concentrations of iron, probably as hematite. Inclusions similar in appearance and located at the SiC-slag interface contained both iron and nickel. The particles might have precipitated from solution during cooling, as observed in a study of iron-bearing slags resulting from coal gasification.[3] At the exposure temperature of this test the dissolved iron might have reacted with SiC in the manner described by other investigators,[4,5] producing Fe-Si alloys and gas by reactions of the type

$$Fe_2O_3 + 3SiC \rightarrow Fe_2Si_3 + 3CO \tag{1}$$

Fe-Si alloys containing 30 to 70 at.% Si would be partially molten at the exposure temperature except for pure, stoichiometric FeSi (m.p. 1410°C).[6] Oxidation of the Fe-Si reaction product in the continually replenished slag would regenerate iron oxide and contribute silica to the slag. Voids in the slag might have been gas bubbles formed in reactions of the type mentioned above.

Figure 8(*c*) shows an upstream region of NC-430 where the slag layer was discontinuous. An erosion mechanism might have contributed to surface recession in regions where the slag was thin or absent. The corrosion rate of NC-430 coated with zircon was about 20% lower than that of uncoated NC-430 in steel soaking pit No. 2, which suggests that the coating provided some protection (see Table V). The coating apparently eventually dissolved in the slag because none was visible after the exposure. Figure 8(*d*) shows the $Si_3N_4$-bonded SiC (C/75) in a region of maximum surface recession. The slag appears to dissolve the matrix phase, which would eventually cause disintegration of the material.

#### Fracture Strength

The flexure strength of sintered $\alpha$-SiC was determined to assess degradation of mechanical properties caused by exposure to the furnace environments. C-ring specimens cut from tubes provided a convenient method of subjecting the outer surfaces to tensile stresses. The outer surfaces of the specimens were not deliberately treated or altered during specimen preparation or testing. In the case of specimens that had been exposed to aluminum remelt furnace No. 1 the loosely adhering deposits mostly detached during handling; however, the thin glass coating remained intact. Tubes were cut into 5-mm-wide rings by using a 75 $\mu$m diamond-impregnated cutoff wheel. C rings were made by cutting approximately 5-mm-long arcs from the rings. The final thickness of the rings was obtained by grinding both sides with 75 $\mu$m diamond abrasive. The outer edges of specimens were beveled with a 30 $\mu$m diamond lapping wheel.

The specimens were loaded so that a tensile stress was applied to the outer surface at a point opposite the open part of the ring. Specimens were heated to 1100°C in air, then the loading ram was advanced at 8.5 $\mu$m/s (0.02 in./min) until fracture occurred. The fracture stress ($\sigma$) was calculated from the load and specimen dimensions using the following equation applicable to the C-ring geometry as described by Ferber and Tennery.[7]

$$\sigma=\frac{P}{(r_o-r_i)w}\left[1+\left(\frac{r_o+r_i}{2\,r_o}\right)\left(\frac{R-r_o}{\bar{r}-R}\right)\right] \qquad (2)$$

where $P$ is load, $r_o$ outer radius, $r_i$ inner radius, $w$ width, $\bar{r}=(r_o+r_i)/2$, and $R=(r_o-r_i)/\ln\,(r_o/r_i)$.

The specimen dimensions and parameters used to calculate fracture strength from fracture loads for sintered $\alpha$-SiC are shown in Table VI. The fracture loads and strengths shown in Table VII reveal considerable variation around mean values. In the present case the conditions of the tensile surfaces might have significantly affected strength values. The weakest material, which had been exposed to aluminum remelt No. 1, had corroded more and had a rougher surface than material exposed to the other furnace environments.

Table VI. Dimensions and Parameters for Calculating C-Ring Fracture Strength of Sintered $\alpha$-SiC (Hexoloy SA)

| Dimensions and Parameters | Aluminum Remelt No. 1 | Forge | Steel Reheat | Steel Soaking Pit |
|---|---|---|---|---|
| $r_o$ (mm) | 12.8 | 12.8 | 12.6 | 12.8 |
| $r_i$ (mm) | 6.43 | 6.40 | 6.32 | 6.43 |
| $w$ (mm) | 5.10 | 5.10 | 5.10 | 5.10 |
| $\bar{r}$ (mm) | 9.62 | 9.60 | 9.46 | 9.62 |
| $R$ (mm) | 9.25 | 9.23 | 9.10 | 9.25 |
| $\sigma$ (MPa) | 1.87P* | 1.88P* | 1.93P* | 1.87P* |

*$P$ is breaking load in kg.

Table VII. C-Ring Fracture Strength of Sintered $\alpha$-SiC (Hexoloy SA) after Exposure to Industrial Furnace Environments

| | Aluminum Remelt No. 1 | Steel Reheat | Steel Soaking Pit | Forge |
|---|---|---|---|---|
| Number of samples | 12 | 12 | 11 | 9 |
| Minimum strength (MPa) | 148 | 174 | 164 | 181 |
| Maximum strength (MPa) | 263 | 290 | 329 | 316 |
| Median strength (MPa) | 206 | 232 | 247 | 249 |
| Mean strength (MPa) | 213 | 221 | 246 | 286 |
| Standard deviation (MPa) | 39 | 34 | 56 | 43 |

## Conclusions

The results described above allow the following principal conclusions concerning corrosion of SiC ceramics exposed to industrial furnace environments:

(1) Active corrosion processes, probably involving halide species, occur in aluminum remelt furnaces.

(2) The molten silicate slag on specimens in a steel soaking pit was extremely corrosive, possibly due to the presence of high concentrations of iron;

surface recession was probably augmented by erosion in the high-velocity flue gases.

(3) The C-ring fracture strength of Hexoloy SA was degraded by exposure, probably due to surface roughness associated with corrosion.

## References

[1]M. Coombs, D. Kotchick, and H. Strumpf, "High-Temperature Burner-Duct-Recuperator (HTBDR) Program"; Annual Report, October 1981 through September 1982, DOE/ID/12170, May 1983.

[2]A. D. Russell, C. E. Smeltzer, and M. E. Ward, "Waste Heat Recuperation for Aluminum Furnaces," GRI–81/0160, April 1983.

[3]R. J. Lauf, J. I. Federer, and V. J. Tennery, "Stabilization of Coal Wastes by Crystallization," ORNL/TM–7071, March 1980.

[4]M. K. Ferber and V. J. Tennery, "Evaluation of Tubular Ceramic Heat Exchanger Materials in Basic Coal Ash from Coal-Oil-Mixture Combustion," ORNL/TM–8385, October 1982.

[5]M. K. Ferber, J. Ogle, V. J. Tennery, and T. Henson, "Characterization of Corrosion Mechanisms Occurring in a Sintered SiC Exposed to Basic Coal Slags"; to be published in *Journal of the American Ceramic Society*, New York, 1985.

[6]M. Hansen, p. 713 in Constitution of Binary Alloys. McGraw-Hill, New York, 1958.

[7]M. K. Ferber and V. J. Tennery, "Evaluation of Tubular Ceramic Heat Exchanger Materials in Acidic Coal Ash from Coal-Oil-Mixture Combustion," ORNL/TM–7958, December 1981; pp. 28–29.

# The Slit-Ring Test for Evaluating Fracture in Tubular Cross Sections

D. J. LANDINI, J. E. FLINN, AND P. V. KELSEY, JR.

EG&G Idaho, Inc.
Idaho Falls, ID 83415

Slit-ring specimens were used to evaluate the fracture properties of several ceramic materials produced in tubular cross sections. The utility of the specimen and its applicability to heat-exchanger components is discussed. Data obtained from tests using slit-ring specimens are presented and compared to data obtained using other test geometries.

Structural ceramics with tubular geometries are being used increasingly as intricate components of heat-exchanger systems. These components experience stresses under operating conditions which may include high temperatures, extensive fouling (particulate deposits), and extremely corrosive industrial waste-heat streams. Effective design of heat-recovery systems requires adequate characterization of candidate tubular ceramics. Fundamental mechanical and material properties such as strength, oxidation/corrosion resistance, and thermal shock resistance need to be determined as well as fracture toughness and time-dependent failure.

To date, published properties of structural ceramics consist largely of flexural strength data from tests conducted on small bend bar specimens, usually fabricated from pressed billets, of well-defined dimensions and surface finish.[1-5] Although these results provide some valuable insights into the materials, they do not necessarily represent the behavior of larger "as-received" components fabricated in specific geometries such as tubes. Generally, tubular shapes are produced by slip-casting or extrusion processes which impart different microstructural features than does hot-pressing. Tests conducted on specimens of near net shapes and sizes would therefore yield pertinent, microstructurally specific information.

Typically, mechanical properties of materials in tubular geometries have been determined by diametral compression, pressurized burst, conventional "C"-ring, or expanded-ring tests. Although each of these tests lends itself to testing tubes, the test procedures typically require sophisticated loading systems or meticulous sample preparation and can involve complex analyses to obtain the results. A simpler technique for determining basic properties data would be beneficial.

The slit-ring specimen illustrated in Fig. 1 has a convenient geometry and can be used in conventional loading modes (tension or compression) with a minimum of fixturing. The test mode flexibility provides the option to stress in tension either the inner or outer surface of the ring. This is an important feature should in-service conditions dictate a specific source of concern for failure at one tube surface vs the other.

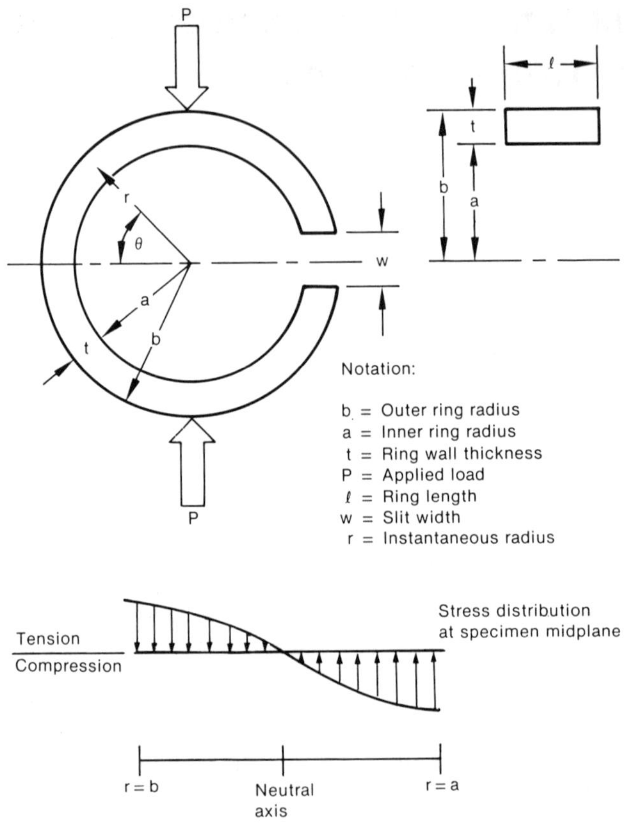


Fig. 1. Schematic of slit-ring specimen illustrating load arrangement and specimen dimensions. Stress distribution through the thickness is also shown.

This paper reports a study on use of the slit-ring specimen to generate mechanical properties data on tubular materials.

## Experimental Procedure

### Materials and Test Specimen Preparation

Two grades of siliconized SiC (NC-430* and SC-AXI†) and one grade of $Al_2O_3$ (998‡) were received as tubes. The silicon carbide tubes had been grit-blasted to remove surface deposits of silicon. This changes the surface flaw population from that of the "as-fired" tubes; however, the process is typical of the treatment used on production tubes. Therefore, tests on "as-received" tubes were considered representative.

Rings were sliced from tubes of each material using a diamond cut-off blade. A slow cutting speed was used to minimize chipping and machining damage (circumferential cracks) to the rings. The outer surface edges were

*Norton Co., Worcester, MA.
†Coors Porcelain Co., Golden, CO.
‡McDanel Refractory Porcelain Co., Beaver Falls, PA.

then beveled to ≈45° using a series of SiC grinding papers (120 to 600 grit) to further reduce edge flaw failures. A slit, approximately 0.5 cm wide, was placed in the rings using a thin diamond cut-off blade. The nominal dimensions of the rings are listed in Table I.

Twenty slit rings of SC-AXI were further prepared for indentation studies. The outer surfaces (tensile surface) were hand-ground and wet-polished to a 1 μm diamond paste finish. The specimens were then indented using a Knoop microhardness indenter[§] which introduced surface cracks of reproducible geometries and sizes. Indentation loads of 4.9, 10.8, 21.6, and 43.2 N were used to provide a variety of precrack depths. All indents were positioned at the specimen midplane, opposite the slit, with the major axis normal to the direction of the applied stress. Immersion oil was placed on the indent to minimize any changes in flaw size due to slow crack growth from residual stresses.

Table I. Results of Room-Temperature Slit-Ring Strength Tests

| Material* | Crosshead Speed (cm/min) | $\sigma_f$(avg) (MPa) | Weibull Modulus, $m$ | No. of Samples |
|---|---|---|---|---|
| NC-430 | 0.05 | 193(187–199)[†] | 16.2 | 20 |
| OD=7.6 cm | 0.005 | 189(182–196) | 14.3 | 10 |
| ID=6.4 cm | | | | |
| $l$=1.5 cm | | | | |
| SC-AXI | 0.05 | 174(162–186) | 5.8 | 30 |
| OD=2.5 cm | 0.005 | 179(164–186) | 5.9 | 10 |
| ID=1.9 cm | | | | |
| $l$=1.0 cm | | | | |
| 998 (alumina) | 0.05 | 305(294–316) | 11.7 | 30 |
| OD=2.5 cm | 0.005 | 298(281–315) | 13.1 | 10 |
| ID=2.2 cm | | | | |
| $l$=1.0 cm | | | | |

*Nominal ring dimensions.
[†]95% confidence interval is shown in parentheses.

#### Slit-Ring Fracture Tests

All slit-ring specimens were loaded in a compression mode on an Instron mechanical tester[¶] using flat loading platens at the positions indicated in Fig. 1. The resulting stress distribution across the thickness of the slit ring is also depicted in Fig. 1. Specimens were tested in ambient conditions and were loaded at crosshead speeds of 0.05 and 0.005 cm/min.

The fracture stress was determined from the analytical expression:

$$\sigma_f = \frac{P}{l(b-a)}\left[\frac{b+a}{r}\left(\frac{(b-a)-r\ln(b/a)}{(b+a)\ln(b/a)-2(b-a)}+1\right)\right]\cos\theta \qquad (1)$$

where $P$ is the applied force (negative for compressive loading) and $l$, $b$, $a$, $r$,

[§]Wilson Instrument Div., ACCO, Bridgeport, CT.
[¶]Instron Corp., Canton, MA.

and $\theta$ are as defined in Fig. 1. This expression was derived from simple curved beam theory with slit-ring responses confirmed through strain gage measurements from gages positioned at specific locations on the rings.[6,7] A typical profile of strain as a function of gage location is illustrated in Fig. 2. Immediately apparent from the profile is that fiber strains were nearly equivalent over a region described by $\theta = \pm\ 30°$. This equivalency indicates that maximum stresses are not limited to the exact midplane of the specimen. Further analyses are the subject of a future publication and are not discussed in detail here.

## Results and Discussion

### Slit-Ring Fracture Tests

The results of strength tests on slit-ring specimens of the three materials are summarized in Table I. Included in the table are the test condition, calculated mean strength, number of samples tested, and the statistical parameter Weibull's modulus ($m$).

The statistical distributions of the fracture strengths were evaluated using a two-parameter Weibull analysis[8]:

$$F = 1 - \exp\left[-(\sigma_f/\sigma_0)\right]^m \tag{2}$$

where $F$ is the probability of failure, $\sigma_f$ the fracture stress, $\sigma_0$ the characteristic stress, and $m$ the Weibull modulus. Values of $F$ were assigned using the ranking statistics method, where the strengths are ordered and assigned a rank, $n$, such that

$$F = n/(N+1) \tag{3}$$

where $N$ is the total number of specimens. Equation (2) can be linearized by a double logarithmic transformation to obtain:

$$\ln \ln \frac{1}{1-F} = m \ln \sigma_f - m \ln \sigma_0 \tag{4}$$

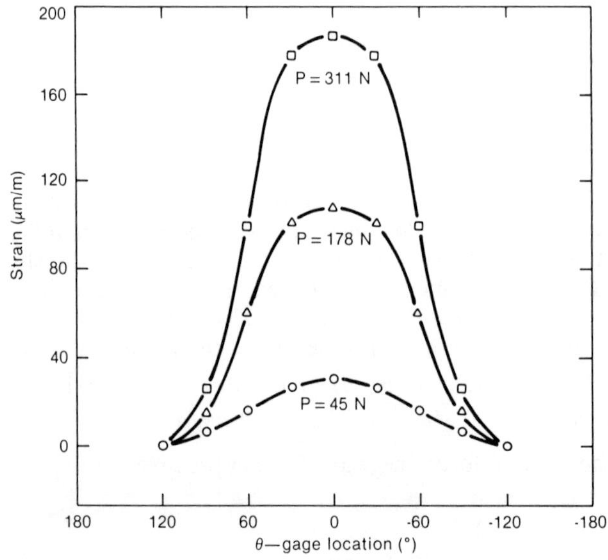


Fig. 2. Outer fiber strain profile for SiC slit rings under compressive loads.

A plot of ln ln $[1/(1-F)]$ vs ln $\sigma_f$ yields a line with a slope of $m$, the Weibull modulus, which can be obtained using a least-squares linear regression.[9] The Weibull modulus is a measure of the flaw-size distribution in the material and indicates the scatter of the material strength. The larger the value of $m$, the more uniform or consistent is the material.

Figures 3 through 5 illustrate the strength distributions for "as-received" slit-ring specimens tested at two loading rates. The data exhibit linear behavior

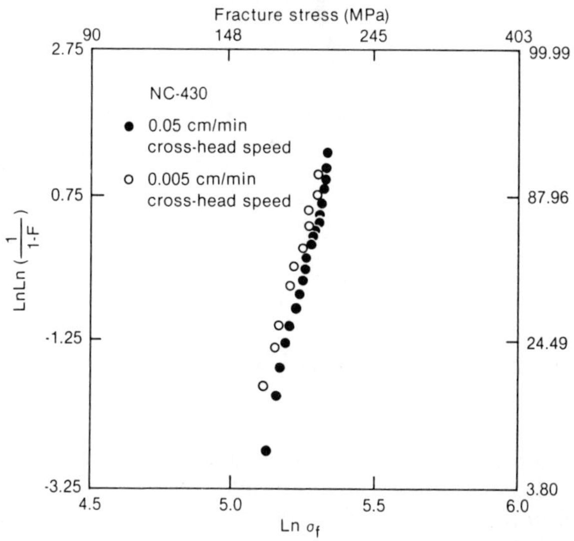


Fig. 3. Strength distribution of NC-430 under ambient test conditions.

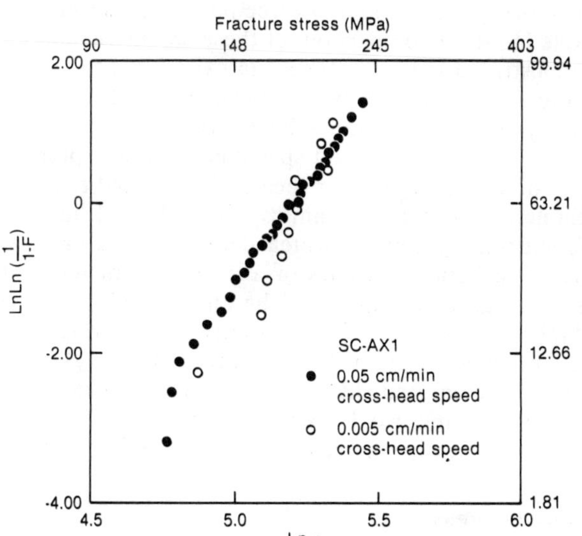


Fig. 4. Strength distribution for SC-AXI under ambient test conditions.

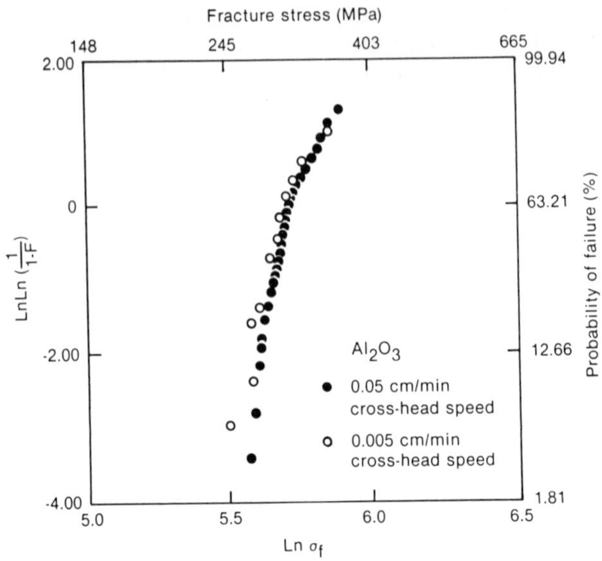


Fig. 5. Strength distribution of $Al_2O_3$ (998) under ambient test conditions.

when plotted according to Eq. (4). The Weibull moduli, as listed in Table I, indicate various degrees of material uniformity. The SC-AXI material displays a lower value of *m* than NC-430. The difference may be due in part to surface variations as "grit" blasting procedures may vary for the two types of tubes. Nonuniformity in the surface flaw populations could then occur and be detected as a variation in *m*.

Data from slit-ring specimens and three-point bend specimens of NC-430 are listed in Table II. Strict comparisons of these results cannot be made, since it has been demonstrated that variations can exist for the same grade of SiC produced in different years. However, several observations appear to be in order. Immediately obvious is the much higher value of strength resulting from the surface finish on three-point bend specimens. Sample preparation of this nature is typical for a vast majority of the specimens used to produce the data currently available on structural ceramics with potential heat-exchanger applications. The slit-ring specimens, which demonstrate a lower strength, are more indicative of realistic properties of a candidate material. The Weibull moduli were similar, although, as would be anticipated, the ground samples exhibited the highest value due to the uniformity of the surface flaw population created during sample preparation. The difference in $\sigma_f$ (avg) for "as-received" three-point bend specimens and slit-ring specimens is probably due to a specimen size effect. The slit rings would have a larger effective volume and surface area under stress, statistically increasing the probability of finding a flaw of critical size.

#### Crack Size vs Fracture Stress

The variation of fracture stress as a function of indentation load and crack depth is given in Table III. (All specimens listed in the table failed at the precrack site.) Crack depths were measured by microscopy of the fracture

Table II. Strength and Strength Distributions for NC-430 SiC Tested by Various Loading Techniques

| Test Method | Test Vol. ($cm^3$) | $\sigma_f$(avg) (MPa) | $m$ | No. of Samples |
|---|---|---|---|---|
| 3-point bend* "as-received" tensile surface | 0.14 | 213 | 14.2 | 12 |
| 3-point bend† 600-grit surface finish | 0.38 | 265 | 19.0 | 15 |
| Slit ring "as-received" tensile surface | 21.50 | 193 | 16.2 | 20 |

*Ref. 10.
†Ref. 11.

Table III. Fracture Stress (Slit-Ring Specimens) as a Function of Knoop Indentation Load at 25°C for SC-AXI

| Specimen Number | Indentation Load (N) | Fracture Stress (MPa) | Crack Depth ($\approx\mu$m) |
|---|---|---|---|
| 1 | 4.9 | 193.8 | 38 |
| 2 | 4.9 | 205.9 | 41 |
| 3 | 4.9 | 184.2 | 35 |
| 4 | 10.8 | 176.7 | 56 |
| 5 | 10.8 | 174.0 | 57 |
| 6 | 10.8 | 171.8 | 57 |
| 7 | 10.8 | 168.9 | 61 |
| 8 | 21.6 | 127.3 | 79 |
| 9 | 21.6 | 144.6 | 69 |
| 10 | 21.6 | 156.7 | 75 |
| 11 | 21.6 | 137.6 | 86 |
| 12 | 43.2 | 116.6 | 111 |
| 13 | 43.2 | 124.5 | 113 |
| 14 | 43.2 | 89.5 | 125 |
| 15 | 43.2 | 105.8 | 116 |

surface. An SEM photomicrograph of the fracture face is shown in Fig. 6; the semicircular flaw from a 43.2 N indent load is apparent on the fracture surface. The change in fracture stress, $\sigma_f$, with the inverse square root of crack depth for SC-AXI tested at room temperature is shown in Fig. 7. A reasonably good linear relationship is observed to exist, correlating with Griffith's criterion for brittle fractures. The "as-received" strengths are also indicated on the plot with the range covering the highest and lowest strengths obtained. The position of these strengths would appear to indicate the size range of the inherent flaws. The range of flaw sizes from Fig. 7 would be $\approx$30 to 90 $\mu$m. Flaws in this range can stem from microstructural features in this material or surface machining damage from the grit-blasting process used in tube preparation.

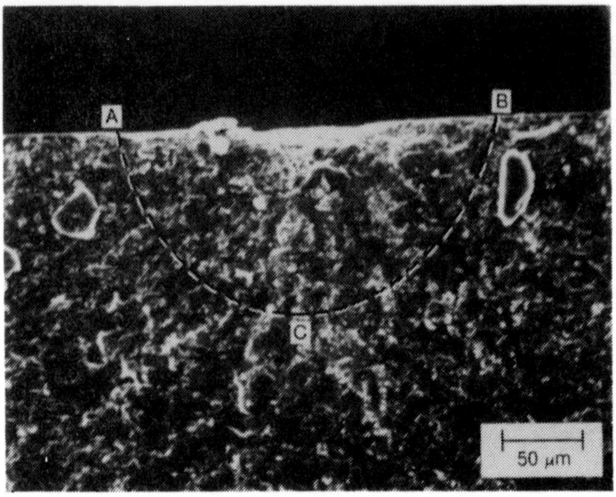


Fig. 6. Typical SEM photomicrograph of the fracture surface of a precracked specimen of SC-AXI tested at room temperature. The specimen was precracked using a 43.2 N indentation load. The arrows indicate the boundary of the precracked region, ABC.

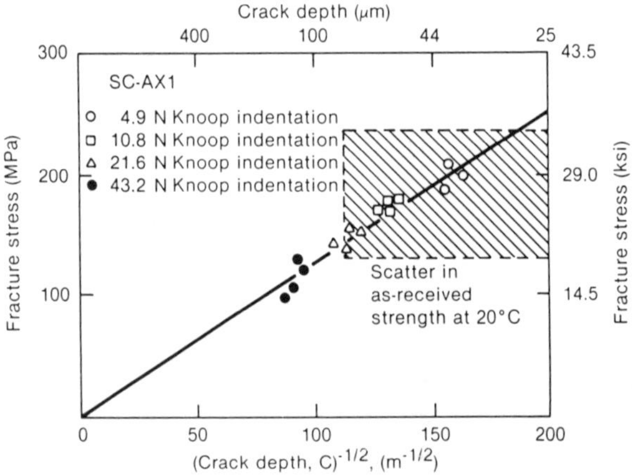


Fig. 7. Variation of fracture stress as a function of inverse square root of crack depth at room temperature for SC-AXI.

**Conclusions**

Slit-ring specimens were used to evaluate several basic mechanical properties of candidate heat-exchanger materials of tubular geometries. The slit-ring test offers the same simplicity in sample preparation and test fixtures as the three-point bend test, but the specimen is more representative of the near-net shapes and sizes of the manufactured component. Therefore, data obtained from slit rings would provide designers with more accurate information on which to base reliability and lifetime predictions for potential heat-exchanger materials.

**Acknowledgment**

This work was supported by the U.S. Department of Energy Assistant Secretary for Conservation and Renewable Energy, Office of Industrial Programs under DOE Contract No. DE–ACO7–76IDO1570.

**NOTE:** Since this paper was written, P. V. Kelsey, Jr. has relocated. The new address is Alcoa Technical Center, Ceramics Section–Building "B," Alcoa Center, PA 15069.

# Durability of SiC Heat-Exchanger Tubes in a High-Temperature Fluidized Bed

H. W. Carpenter

Rockwell International/Rocketdyne Div.
Canoga Park, CA 91304

The durability of SiC tube samples was evaluated in a simulated atmospheric fluidized bed environment at 982°C (1800°F) under both oxidizing and reducing conditions. No erosion, corrosion, or loss in strength was found. Hoop tensile strength was measured by a burst test method.

The shortage and high price of petroleum fuels make it desirable to develop more efficient methods for using coal in the generation of power. Ceramic heat-exchanger tubing materials afford higher operating temperatures with concomitant higher-power generating efficiencies than are presently available from metal tubing materials. The concept of high-temperature atmospheric fluidized bed (AFB) heat exchangers offers a promising method for utilizing ceramic heat-exchanger tubes.[1] Of the candidate ceramic materials for coal-fired AFB heat-exchanger applications, silicon carbide (SiC) is the most promising.

SiC has a number of attractive properties, including mechanical properties that are essentially unaffected to the maximum operation temperature of coal-fired fluidized beds, around 1093°C (2000°F); thermal conductivity higher than that of high-alloy metals; the best hardness and resistance to erosion of commercial materials; resistance to oxidation to 1648°C (3000°F); competitive cost compared with high-alloy metals; and, finally, it is made from nonstrategic materials, coke and silica.

A key consideration of heat-exchanger materials is that they should have a long life in the service environment. Since SiC is one of the most erosion-resistant commercial materials available and the oxidation rate of SiC is negligible to 982°C, a long life in the AFB environment would be predicted. However, SiC material has not been evaluated in the AFB environment and information in the literature is not adequate to make an accurate prediction of SiC tube durability. Under certain AFB conditions, it is possible that the oxidation rate of SiC might be accelerated due to continual removal of the protective layer of $SiO_2$, for example, by erosion, by dissolution in a glassy coating formed by impinging bed materials, or by the formation of SiO vapor rather than $SiO_2$.

For these reasons, it was necessary to assess the durability of SiC as a viable heat-exchanger surface by exposing it to the AFB environment. This effort was performed as a screening test for evaluating small samples of different classes of SiC materials in a small test facility where a large number of samples could be evaluated under controlled conditions. Objectives included:

(1) determining whether there is any surface wastage of SiC in the AFB environment and, if so, what the surface recession rate is so that the life of the heat-exchanger tubes can be predicted, (2) determining whether there is any degradation of material strength, (3) comparing types of SiC, (4) evaluating durability of SiC/SiC joints, and (5) determining how AFB operational parameters affect tube performance.

## Test Procedure

### Test Facility Design and Operation

A small AFB test unit was designed and built specifically for the SiC tube sample exposures. A schematic of the test facility is shown in Fig. 1 and a schematic of the fluidized bed section is shown in Fig. 2.

The equipment is best described in sequence by following the air flow through test facility (Fig. 1). Air was supplied by a plant compressor and the air velocity was controlled by a sonic nozzle downstream of the compressor. The air was then heated to 648°C (1200°F) in an electric tube heater.

The 648°C air entered a chamber below the AFB and flowed upward through the air distribution plate. The air distribution plate was made of Incoloy 800H* and contained 77 3/32-in. diameter holes on 0.500-in. centers, and each hole was counterbored from the bottom to preclude plugging by the bed material.

The 982°C test temperature was achieved by burning methane gas supplied through two burner pipes (similar to a fireplace log burner) located on the top of the air distribution plate. The gas was ignited by the hot bed. The heating and cooling rate of the bed was typically limited to 204°C/h (400°F/h) to avoid excessive thermal strains, although much faster rates were possible. Bed temperature, measured by the six thermocouples, was uniform to within ±5°F, at a superficial air velocity of 6 ft/s, but variations increased as air velocity increased. The bed temperature varied ±10°F at an air velocity of 12 ft/s and ±20°F at an air velocity of 18 ft/s. These variations were observed from top to bottom, side to side, and they were caused by poor distribution of the methane gas in the high-velocity air. This poor distribution of methane was the result of too few holes (there were only three) in the methane distributor pipes. Peak temperatures in the bed over the 100-h runs were 1010°C (1850°F) at an air velocity of 6 ft/s, 998°C (1830°F) at 12 ft/s, and 1010°C (with one exclusion to (1026°C (1880°F)) at 18 ft/s.

The bed was 18 inches high when fluidized, and the height of the fluidized bed was maintained three ways: manually, with visual observation through a port located in the hood on top; automatically by a bed overflow tube (excess bed material fed into the bed overflow tank where it was monitored); and indirectly by measuring bed pressure at four levels bracketing the top of the bed. The fluidized bed pressure was plotted vs bed height and the top of the bed was, by definition, where the bed pressure was equal to atmospheric pressure.

The bed was composed of spent material from a large coal-fired AFB test facility. Spent bed material was a mixture of limestone (calcium carbonate), calcium sulfate (limestone reacted with sulfur from coal combustion), and coal ash.

---

*Huntington Alloy Products Div., International Nickel Co., Inc., Huntington, WV.

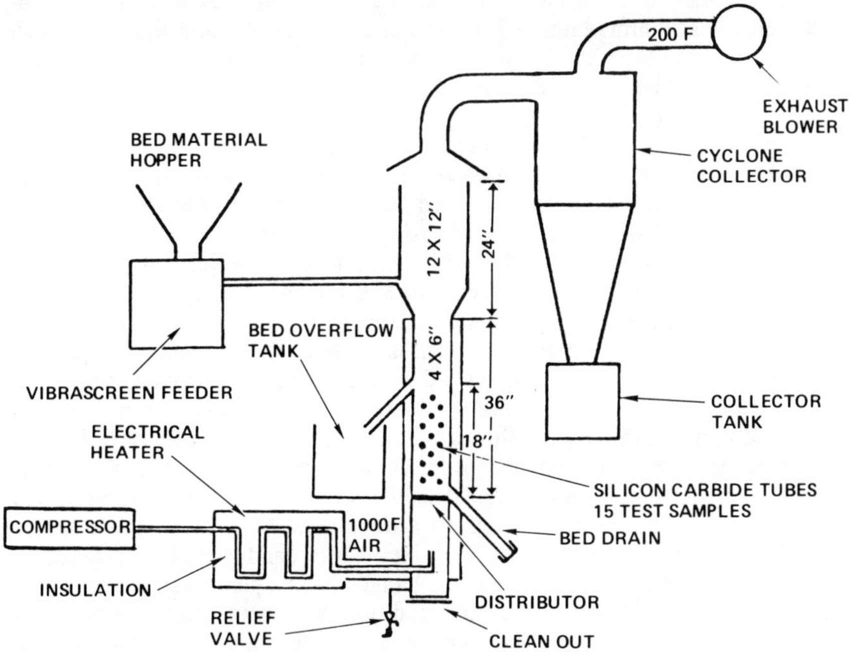


Fig. 1. Unfired air fluidized bed SiC material corrosion/erosion test apparatus.

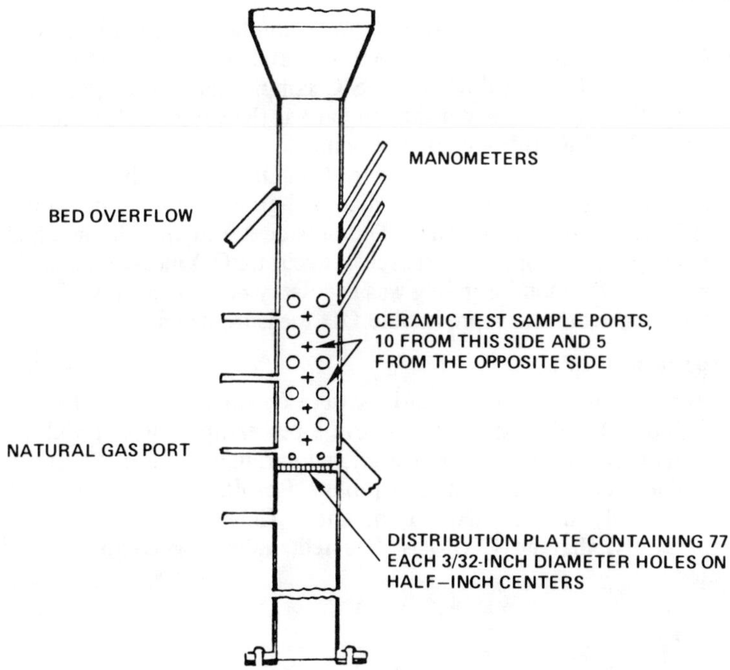


Fig. 2. Unfired air fluidized bed erosion/corrosion test bed.

The horizontal cross section of the bed measured 4 by 6 inches. The 4-in. dimension accommodated 3.7-in. long tubular samples and the 6-in. width accommodated a tube array with 1.5-in. centers. The freeboard section was 42 inches high. The lower 18 inches had the same cross-sectional dimensions as the AFB section, but the upper 24 inches of the freeboard opened to a 12- by 12-in. cross section. The purpose of the large cross section was to entrap the particles blown out of the fluidized bed and to allow them to fall back. New bed material was continually fed into the large freeboard zone by a vibrating screw feeder. The test bed casing was made of ¼-in. thick Incoloy 800 plate insulated on the outside with a fibrous ceramic blanket.

Exhaust from the freeboard was sucked into a hood by a large fan. A controlled amount of ambient air was mixed with the exhaust to reduce the temperature to 93°C (200°F) to prevent damage to the ducting, cyclone, and fan. Fines blown out of the fluidized bed were collected by a cyclone separator.

Four runs of 100 h each were conducted. Run numbers 1, 2, and 3 were conducted under oxidizing conditions at superficial bed velocities of 6, 12, and 18 ft/s, respectively. Current designs for commercial AFB plants indicate that optimal superficial bed velocity is between 3 and 6 ft/s. Thus, superficial bed velocities of 12 and 18 ft/s represent erosive conditions much more severe than they will be in large plants. Run No. 4 was conducted under reducing conditions. Commercial AFB heat exchangers will operate under oxidizing conditions, but local, highly reducing conditions may occur, for example, near the coal-feed ports. Thus, it was necessary to determine if a reducing environment would have a detrimental effect on the SiC heat-exchanger tubes. The superficial air velocity was 6 ft/s.

#### Test Samples

Three types of SiC tube materials were employed: siliconized SiC (Si/SiC) tubes,[†] sintered alpha SiC (SASC) tubes,[‡] and chemically vapor deposited (CVD) SiC tubes.[§] One-third of the Si/SiC samples had a butt joint and one-third of the SASC samples had a shiplap joint at the center of the tube length. None of the CVD SiC tubes contained joints.

The SiC tube sample dimensions were 1-in. OD with a ⅛-in. wall thickness by 3.7 in. long (Fig. 3). A notch was diamond-ground in one end so it could be locked in place during exposure. The outside surface at the notched end was diamond-ground to provide a source of accurate OD measurements before and after testing. Diamond grinding was necessary because most surfaces were too irregular for accurate, reproducible OD measurements.

#### Gas Composition

All runs under oxidizing conditions were conducted using preheated air (648°C (1200°F)) and methane to achieve a bed temperature of 982°C. Gas composition analyses taken at the operational parameters for 6, 12, and 18 ft/s superficial air velocities are listed in Table I. Results show that the fluidizing gas was essentially air plus a small amount of $CO_2$ ($H_2O$ was not analyzed) and, importantly, that no $CH_4$ was detected, indicating complete methane combustion.

---

[†]NC-430, Norton Co., Worcester, MA.
[‡]Carborundum Corp., Niagara Falls, NY.
[§]Materials Technology, Inc., Dallas, TX.

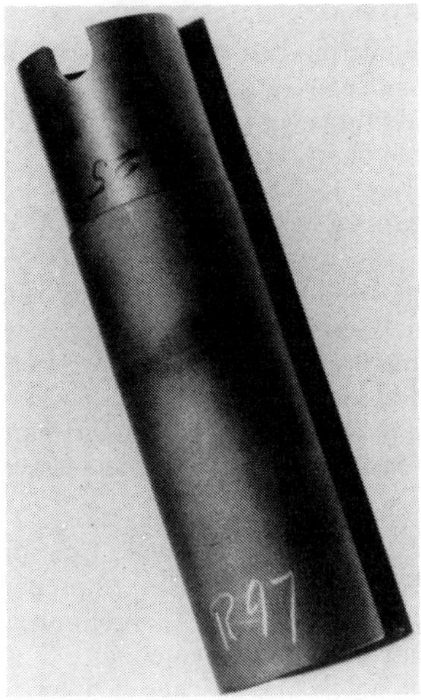

Fig. 3. SiC tube sample.

## Table I. Combustion Gas Analyses

| CONDITION | SUPERFICIAL AIR VELOCITY (FT/SEC) | COMPOSITION (% BY VOLUME)[1] [2] | | | | |
|---|---|---|---|---|---|---|
| | | $N_2$ | $O_2$/Ar | $CO_2$ | CO | $CH_4$ |
| OXIDIZING | 6 | 81.5 | 16.6 | 1.8 | ND[3] | ND |
| | 12 | 80.8 | 16.9 | 1.7 | ND | ND |
| | 18 | 80.7 | 18.2 | 1.4 | ND | ND |
| REDUCING[4] | 6 | 89.0 | 0.7[5] | 4.0 | 0.6 | 5.6 |
| | 6 | 88.0 | 0.9[5] | 4.3 | 0.8 | 5.8 |
| | 6 | 89.2 | 0.6[5] | 3.7 | 0.7 | 5.8 |
| | 6 | 89.4 | 0.8[5] | 3.5 | 0.9 | 5.0 |
| | 6 | AVG. 88.4 | 0.7 | 3.9 | 0.7 | 5.6 |

NOTES: 1. THE ANALYSES WERE MADE BY GAS CHROMATOGRAPHY ON A WATER-FREE BASIS
2. RESULTS DO NOT NECESSARILY ADD UP TO 100% BECAUSE EACH GAS WAS ANALYZED INDEPENDENTLY
3. ND = NONE DETECTED; LESS THAN 0.5%
4. THESE SAMPLES WERE TAKEN AT 24 HOUR INTERVALS
5. AT LEAST 0.5% (i.e., MORE THAN HALF OF THE TOTAL AMOUNT) IS ESTIMATED TO BE ARGON

The test run under reducing conditions and a superficial air velocity of 6 ft/s was conducted using nitrogen gas, an excess of methane, and only enough air to achieve a bed temperature of 982°C. The compositions of four gas samples taken at 24-h intervals are also listed in Table I. These results show excess unburned methane in the bed and low oxygen content. The difference between oxygen and argon cannot be distinguished in this analytical method, so most of the reported quantity of $O_2$/Ar in this case is actually argon.

**Posttest Sample Evaluation**

The effect of the fluidizing bed environment on the SiC tubes was evaluated after every 100 h of exposure as follows:

1. A visual inspection was conducted to record changes in color, texture, and any other observable characteristic.
2. Samples were weighed.
3. The OD was measured every 30 degrees on the diamond-ground surface portion of the sample.
4. Surface coatings were analyzed by X-ray diffraction, X-ray fluorescence, and chemical analyses.
5. Samples were burst-tested to determine whether the strength of the material increased or decreased.
6. Fragments from the burst test were mounted and polished to determine if the microstructural features at and near the surface had changed.

**Burst Tests**

All samples were burst-tested at the end of the program to provide hoop tensile strength data. The tube specimens were prepared for burst testing by sealing the ends with epoxy plugs (Fig. 4). Preliminary tests were conducted on SiC tube samples that were instrumented with 21 strain gauges in critical locations (Fig. 5), at the notch, the lip caused by grinding the surface at one end, and the beginning of the epoxy plug. Results demonstrated that (1) strains decreased linearly along the length of tube buried in the epoxy plug to a nominal value at the end of the SiC tube, (2) strains at stress risers (the notch and the lip) were well below the level in the center of the tube between the epoxy plugs, and (3) strains in the center of the tube were uniform.

Pressure was increased at 6.9 MPa (1000 psi) per minute to failure. The burst strength of each tube, which corresponds to the hoop stress at failure, was calculated using the thick wall cylinder formula:

$$\sigma_H = P_b \left( \frac{D^2 + d^2}{D^2 - d^2} \right) \tag{1}$$

where $\sigma_H$ is hoop tensile strength (burst strength), $P_b$ burst pressure, $D$ average outside diameter, and $d$ inside diameter calculated using "average" wall thick-

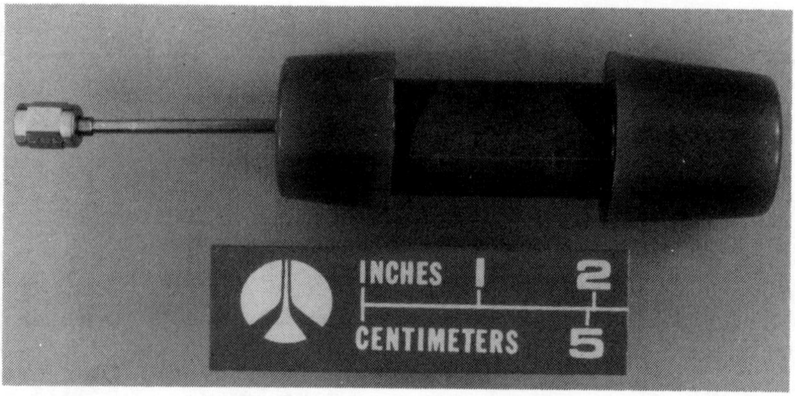


Fig. 4. NC-430 SiC burst sample.

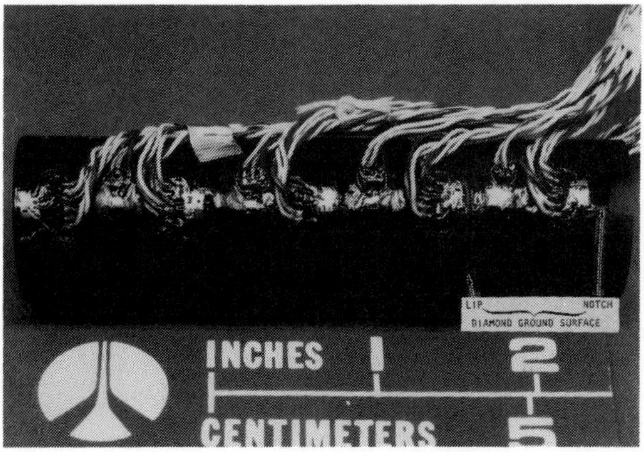


Fig. 5. Strain gage location on burst test specimen front side (before epoxy plugs were cast on the ends).

ness, or using "minimum" wall thickness, as noted. Unless otherwise stated, the strength referred to in this report is based on the "average" wall thickness.

## Results

### Oxidizing Condition Tests

SiC tubes were exposed in three separate tests conducted under oxidizing conditions at superficial air velocities of 6, 12, and 18 ft/s. Each run was conducted at a nominal bed and sample temperature of 982°C, for 100 h. Typical appearance of the three types of SiC tubes before and after exposure is shown in Figs. 6 and 7.

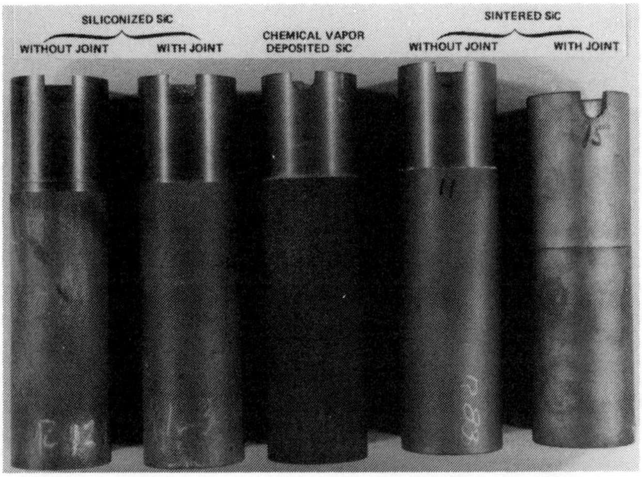


Fig. 6. Ceramic tube samples, as-fabricated condition.

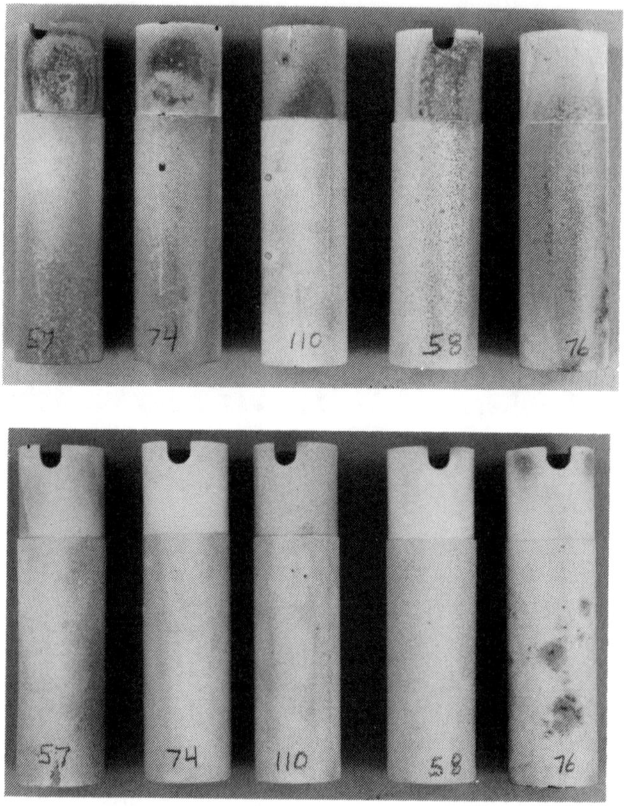


Fig. 7. SiC samples exposed in the 4- by 6-in. AFB under oxidizing conditions (*A*) bottom side; (*B*) top side.

**Surface Coating**

All of the SiC tubes were covered with a thin coating that varied in thickness between 0.1 and 1 mil. The weight gain averaged 0.21 g per tube. No apparent differences were noted as to duration in the AFB, location on the tube (except for color and hardness), location in the bed, bed velocity, or surface condition (viz., diamond-ground vs as-fabricated, and siliconized vs sintered vs CVD). The coating was beige on the top sides of the tubes and rust-colored on the bottom sides. The rust-colored areas were harder and more tenaciously bonded to the SiC surface.

Analyses showed that the primary composition was CaO, secondary compositions were $CaCO_3$, $CaSO_4$, and $Ca(OH)_2$,[¶] and iron oxide was responsible for the rust coloring. A significant result of these analyses was the absence of alkalies and alkaline earths, which are found in coal slag, and which would readily react with any $SiO_2$ that formed on the SiC surface.

---

[¶]$Ca(OH)_2$ probably formed after the test because CaO readily hydrates when exposed to ambient air.

**Burst Strength**

Burst-test results for the SiC tubes are listed in Tables II through VI. The upper data set in each table is that of the control samples, which were tested in the as-received condition. The next three sets of data are the results of the specimens exposed in Runs 1, 2, and 3. The last specimens listed were exposed to all three test runs.

The data for Si/SiC samples show two important results. First, the scatter in the data (the average strength was 16.5 ksi with a standard deviation of 1.6 ksi) is notably lower than that of the other SiC materials. This agrees with other burst-test results for Si/SiC tubes.[2,3] The low scatter is noteworthy because the tubes exhibited considerable variation in OD and wall thickness from tube to tube and within individual tubes. It is this variation in dimensions that accounts for the differences in calculated strength based on either the average or minimum wall thickness. Second, there is no evidence that strength decreased due to exposure in the 4- by 6-in. AFB. The average strength of the control samples was 16.5 ksi, whereas that of exposed samples was 15.9, 15.8, and 16.2 ksi, from Runs 1, 2, and 3, respectively, all within the standard deviation.

The average strength of the as-received Si/SiC specimens with joints, 10.1 ksi (Table III), was lower than that of the specimens exposed in the AFB, 12.8 to 15.9 ksi. Again, the results show no evidence of strength degradation due to exposure in the fluidized bed. Significantly, no fracture paths followed

Table II. Burst Test Results: AFB Tube Samples: Si/SiC

| TUBE NUMBER | AFB EXPOSURE | | TENSILE STRENGTH, KSI BASED ON: | | AVERAGE TENSILE STRENGTH |
|---|---|---|---|---|---|
| | HOURS AT 1800 F | AIR VELOCITY FT/SEC | MINIMUM WALL THICKNESS | AVERAGE WALL THICKNESS | |
| 121 | NONE | NONE | 11.4 | 10.9 | |
| 118 | NONE | NONE | 18.3 | 15.7 | |
| 117 | NONE | NONE | 17.5 | 15.9 | |
| 120 | NONE | NONE | 17.5 | 16.1 | |
| 119 | NONE | NONE | 17.0 | 16.2 | |
| 116 | NONE | NONE | 17.0 | 16.3 | |
| 121 | NONE | NONE | 20.6 | 18.9 | |
| | | | 18.0 | 16.5 | AVERAGE |
| | | | 1.4 | 1.6 | STANDARD DEVIATION |
| 103 | 100 | 6 | 16.7 | 15.2 | |
| 108 | 100 | 6 | 18.1 | 15.9 | |
| 104 | 100 | 6 | 19.3 | 16.6 | |
| | | | 18.0 | 15.9 | AVERAGE |
| | | | 1.3 | 0.7 | STANDARD DEVIATION |
| 109 | 100 | 12 | 13.0 | 12.1 | |
| 110 | 100 | 12 | 20.9 | 19.5 | |
| | | | 17.0 | 15.8 | AVERAGE |
| 114 | 100 | 18 | 12.8 | 11.8 | |
| 115 | 100 | 18 | 21.2 | 20.5 | |
| | | | 17.0 | 16.2 | AVERAGE |
| 102 | 300 | 6,12,18 | BROKE DURING REMOVAL DUE TO FAULTY HOLDERS | | |
| 105 | 300 | 6,12,18 | | | |

Table III. Burst Test Results: AFB Tube Samples: Si/SiC with Joint

| TUBE NUMBER | AFB EXPOSURE | | TENSILE STRENGTH, KSI BASED ON: | | AVERAGE TENSILE STRENGTH |
|---|---|---|---|---|---|
| | HOURS AT 1800 F | AIR VELOCITY FT/SEC | MINIMUM WALL THICKNESS | AVERAGE WALL THICKNESS | |
| 136 | NONE | NONE | 4.6 | 4.4 | |
| 140 | NONE | NONE | 10.4 | 9.9 | |
| 138 | NONE | NONE | 11.0 | 10.6 | |
| 134 | NONE | NONE | 12.1 | 11.2 | |
| 135 | NONE | NONE | 13.5 | 11.3 | |
| 141 | NONE | NONE | 12.2 | 11.6 | |
| 137 | NONE | NONE | 13.2 | 11.8 | |
| | | | 11.0 | 10.1 | AVERAGE |
| | | | 3.0 | 2.6 | STANDARD DEVIATION |
| 129 | 100 | 6 | 14.2 | 13.8 | |
| 123 | 100 | 12 | 13.3 | 12.8 | |
| 130 | 100 | 18 | 13.9 | 12.9 | |
| 128 | 300 | 6,12,18 | 17.2 | 15.9 | |

Table IV. Burst Test Results: AFB Tube Samples: SASC

| TUBE NUMBER | AFB EXPOSURE | | TENSILE STRENGTH, KSI BASED ON: | | AVERAGE TENSILE STRENGTH |
|---|---|---|---|---|---|
| | HOURS AT 1800 F | AIR VELOCITY FT/SEC | MINIMUM WALL THICKNESS | AVERAGE WALL THICKNESS | |
| 87 | NONE | NONE | 21.3 | 20.9 | |
| 83 | NONE | NONE | 30.1 | 29.7 | |
| 90 | NONE | NONE | 30.0 | 30.0 | |
| 86 | NONE | NONE | 30.9 | 30.4 | |
| 88 | NONE | NONE | 31.3 | 30.8 | |
| 91 | NONE | NONE | 31.8+ | 31.8+ | |
| 84 | NONE | NONE | 35.7 | 35.4 | |
| 78 | NONE | NONE | 37.9 | 37.5 | |
| | | | 31.1 | 30.8 | AVERAGE |
| | | | 4.9 | 4.9 | STANDARD DEVIATION |
| 73 | 100 | 6 | 36.6 | 36.2 | |
| 76 | 100 | 12 | 38.3 | 38.0 | |
| 77 | 100 | 18 | 16.9 | 16.8 | |
| 74 | 300 | 6,12,18 | 19.8 | 19.4 | |
| 75 | 300 | 6,12,18 | 32.9 | 32.0 | |

the join line (Fig. 8), which indicates that the joints were not a load-limiting feature. The lower strengths, compared to samples without joints, were probably due to the discontinuities caused by butting together tubes of different OD and ID dimensions.

The average strength of the as-received SASC tubes was much higher, 30.8 ksi (Table IV). The differences in strengths calculated using average and minimum wall thicknesses were small because these tubes had very uniform dimensions. The strength levels of the five tubes exposed in the fluidized beds were mixed, one being near the control average, two lower, and two higher. One of the high values was the highest strength measured, while the other

Table V. Burst Test Results: AFB Tube Samples: SASC with Joint

| TUBE NUMBER | AFB EXPOSURE | | TENSILE STRENGTH, KSI BASED ON: | | AVERAGE TENSILE STRENGTH |
|---|---|---|---|---|---|
| | HOURS AT 1800 F | AIR VELOCITY FT/SEC | MINIMUM WALL THICKNESS | AVERAGE WALL THICKNESS | |
| 160 | NONE | NONE | 7.3 | 7.1 | |
| 163 | NONE | NONE | 9.2 | 8.8 | |
| 159 | NONE | NONE | 10.5 | 10.5 | |
| 162 | NONE | NONE | 10.6 | 10.5 | |
| 158 | NONE | NONE | 11.1 | 10.8 | |
| 164 | NONE | NONE | 11.3 | 11.2 | |
| 161 | NONE | NONE | 11.9 | 11.6 | |
| | | | 10.3 | 10.2 | AVERAGE |
| | | | 1.6 | 1.6 | STANDARD DEVIATION |
| 152 | 100 | 12 | 5.3 | 5.1 | |
| 153 | 100 | 18 | 10.2 | 9.8 | |
| 97 | 300 | 6,12,18 | 8.0 | 7.8 | |

Table VI. Burst Test Results: AFB Tube Samples: CVD SiC

| TUBE NUMBER | AFB EXPOSURE | | TENSILE STRENGTH, KSI BASED ON: | | AVERAGE TENSILE STRENGTH |
|---|---|---|---|---|---|
| | HOURS AT 1800 F | AIR VELOCITY FT/SEC | MINIMUM WALL THICKNESS | AVERAGE WALL THICKNESS | |
| 59 | NONE | NONE | 7.4 | 6.3 | |
| 58 | NONE | NONE | 13.1 | 11.4 | |
| 69 | NONE | NONE | 19.0 | 17.5 | |
| 72 | NONE | NONE | 27.5 | 23.1 | |
| 61 | NONE | NONE | 29.2 | 26.7 | |
| 71 | NONE | NONE | 35.2 | 28.7 | |
| 62 | NONE | NONE | 33.5 | 29.2 | |
| 70 | NONE | NONE | 36.3 | 34.0 | |
| 67 | NONE | NONE | 41.7 | 37.3 | |
| | | | 27.0 | 23.8 | AVERAGE |
| | | | 11.5 | 10.3 | STANDARD DEVIATION |
| 55 | 100 | 6 | 37.6 | 37.1 | |
| 58 | 100 | 12 | 26.4 | 24.7 | |
| 60 | 100 | 18 | 39.7 | 37.9 | |
| 56 | 300 | 6,12,18 | 20.0 | 16.6 | |
| 57 | 300 | 6,12,18 | 25.6 | 22.9 | |
| | | | 22.8 | 19.8 | AVERAGE |

was the third highest. Thus, it was concluded that exposure in the fluidized bed did not degrade strength.

The strength of the SASC tubes with joints (Table V) was only one-third that of the tubes without joints, an average strength of 10.2 ksi compared to 30.8 ksi strength. Examination showed that the low strength was due to incomplete bonding of the joined surfaces. All of the jointed specimens fractured, either entirely or partially, around the joint. The strength values of the joined tubes exposed to the environment of the fluidized bed were slightly lower than the average strengths of the as-received tubes, but only one of the exposed

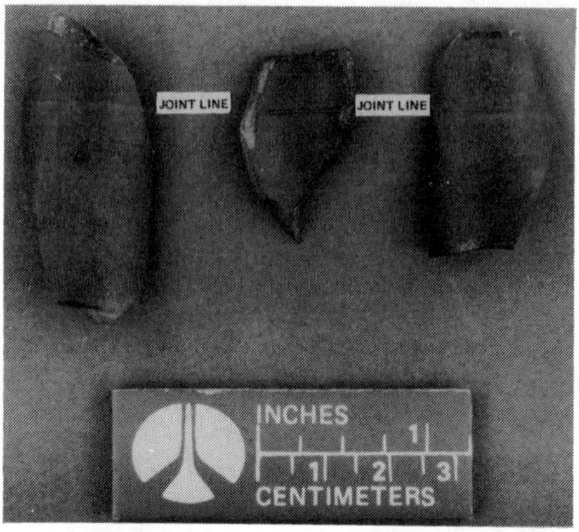


Fig. 8. Fragments of fractured Si/SiC tubes with butt joints.

tubes was below the range of values for as-received tubes. No significance can be drawn from these results due to the presence of large unbonded surfaces in the shiplap joints.

CVD SiC tubes exhibited a large variation in strength, 6.3 to 37.3 ksi (Table VI). The standard deviation of 10.3 ksi was almost one-half that of the average strength, 23.8 ksi. The strength values based on average and minimum wall thicknesses also showed large disparities, which were due to the extreme variation of diameter, wall thickness, and eccentricity in the tubes. The strength values of the CVD tubes exposed in the fluidized bed included the highest and the third highest values measured. Thus, there was no evidence that strength was degraded by exposure in the fluidized bed.

**Reducing-Condition Test**

Seven trial runs were necessary to determine the optimum operating parameters under reducing conditions. Consequently, some of these SiC tube samples were exposed to seven thermal cycles and 4 h at 982°C under both oxidizing and reducing conditions. Since spare samples were not available to replace all of these tubes prior to run 4, certain samples were exposed to (1) only the pretest runs, (2) both the pretest runs plus run 4, and (3) only run 4. Tubes exposed to the preliminary test runs were coated with a thin (1 mil) layer of CaO, similar to the tubes from runs 1, 2, and 3. The tubes exposed only to the reducing conditions of test run 4 were uncoated. Thus, under oxidizing conditions, a thin layer of CaO formed on the tubes almost immediately and the thickness did not change appreciably with further testing. Under reducing conditions, however, no coating formed on the SiC tubes. There were, however, minute specks of a light-colored material (presumably CaO) visible on the surface under ×10 magnification. These specks of coating were probably deposited during the short heatup period that was conducted under oxidizing conditions. The methane fuel was not introduced into the 4- by 6-in. AFB until the bed temperature was 593°C (1100°F), 3 h after the test started.

Excessive metal corrosion of the Incoloy 800H tube holders in the 4- by 6-in. AFB test unit caused difficulty in removal of the samples after testing, and several SiC tubes were broken in the process. The excessive metal corrosion was not due to the reducing conditions in run 4, but rather to the long-term exposure, over 400 h, at 982°C, mostly under highly oxidizing conditions.

Results of the burst tests are listed in Table VII in three groups of samples. Group A samples were exposed only to the seven preliminary test runs for a total of 4 h at 982°C under both oxidizing and reducing conditions. Group B samples were exposed to both the preliminary runs and run 4, which was 100 h at 1800°F under reducing conditions. Group C samples were exposed only to the conditions of run 4.

The data used for determining whether the strength of SiC tubes was affected by exposure to the reducing fluidized bed conditions in run 4 are summed up in Table VIII. The strength of all samples, excluding samples with joints, are below the average strength of control samples but they are within the range of values measured on control samples. Thus, there is no clear evidence indicating that the strength of SiC was graded under reducing conditions in the fluidized bed.

**Microstructure**

Ceramographic examinations were made on the three types of SiC tubes in both the as-received and exposed conditions. No significant differences were noted either at the surface or in the bulk of the material.

Table VII. Burst Test Results for SiC Heat-Exchanger Tubes[1] Run No. 4

| GROUP | TUBE NO. | MATERIAL (2) | AFB EXPOSURE 800 F DURATION, (HRS) | AFB EXPOSURE 800 F CONDITION (3) | HOOP TENSILE STRENGTH, KSI | COMMENTS |
|---|---|---|---|---|---|---|
| A | 63 | CVD | 4 | OX AND RED | 23.5 | SAMPLES THAT WERE EXPOSED IN THE PRELIMINARY RUNS TO RUN NO. 4. THESE PRELIMINARY RUNS WERE USED TO LEARN HOW TO OPERATE THE BED UNDER REDUCING CONDITIONS. |
| | 64 | CVD | 4 | OX AND RED | 28.1 | |
| | 82 | SSC | 4 | OX AND RED | 19.7 | |
| | 155 | SSC+J | 4 | OX AND RED | 10.5 | |
| | 124 | Si/SC+J | 4 | OX AND RED | 13.2 | |
| | 125 | Si/SC+J | 4 | OX AND RED | 11.6 | |
| B | 79 | SSC | 4 + 100 | 4 HOURS OX 100 HOURS RED | NOTE (4) | THESE SAMPLES WERE EXPOSED IN THE AFB DURING BOTH THE PRELIMINARY TESTS AND RUN NO. 4. |
| | 80 | SSC | 4 + 100 | | 17.1 | |
| | 81 | SSC | 4 + 100 | | NOTE (4) | |
| | 111 | Si/SC | 4 + 100 | | 14.0 | |
| | 112 | Si/SC | 4 + 100 | | NOTE (4) | |
| | 113 | Si/SC | 4 + 100 | | 16.4 | |
| C | 65 | CVD | 100 | RED | 19.2 | THESE SAMPLES WERE ONLY EXPOSED IN THE AFB DURING RUN NO. 4. |
| | 66 | CVD | 100 | RED | 8.5 | |
| | 95 | SSC | 100 | RED | 18.5 | |
| | 154 | SSC+J | 100 | RED | NOTE (4) | |
| | 127 | Si/SC+J | 100 | RED | 14.9 | |
| | 132 | Si/SC+J | 100 | RED | 15.5 | |

NOTES: (1) TUBE DIMENSIONS WERE NOMINALLY 1 INCH OD X 3.7 INCHES LONG X 1/8 INCH WALL

(2) CVD = CHEMICAL VAPOR DEPOSITED
SSC = SINTERED SiC
SSC+J = SINTERED SiC PLUS JOINT
Si/SC = SILICONIZED SiC, TYPE NC430
Si/SC+J = SILICONIZED SiC, TYPE NC430 PLUS JOINT

(3) OX = OXIDIZING
RED = REDUCING

(4) BROKE OR CRACKED DURING REMOVAL FROM TEST BED DUE TO CORRODED METAL HOLDERS

Table VIII. Comparison of the Strengths of Samples Exposed in Run No. 4 to that of Control Samples

| SAMPLE NO. | SiC MATERIAL(1) | CONDITION | HOOP TENSILE STRENGTH KSI | |
|---|---|---|---|---|
| | | | VALUE | RANGE |
| 80 | SSC | RUN NO. 4 + PRELIMINARY RUNS | 17.1 | – |
| -- | SSC | CONTROL (REF. 1), NOTE (2) | 29.9 | 16.8 TO 37.5 |
| 111 | Si/SC | RUN NO. 4 + PRELIMINARY RUNS | 14.0 | – |
| 113 | Si/SC | RUN NO. 4 + PRELIMINARY RUNS | 16.4 | – |
| – | Si/SC | CONTROL (REF. 1), NOTE (2) | 16.0 | 10.9 TO 20.5 |
| 65 | CVD | RUN NO. 4 ONLY | 19.2 | – |
| 66 | CVD | RUN NO. 4 ONLY | 8.5 | – |
| – | CVD | CONTROL (REF. 1), NOTE (2) | 25.2 | 6.3 TO 37.3 |
| 95 | SSC | RUN NO. 4 ONLY | 18.5 | – |
| | | CONTROL (REF. 1), NOTE (2) | 29.9 | 16.8 TO 37.5 |
| 127 | Si/SC+J | RUN NO. 4 ONLY | 14.9 | – |
| 132 | Si/SC+J | RUN NO. 4 ONLY | 15.5 | – |
| – | Si/SC+J | CONTROL (REF. 1), NOTE (3) | 10.1 | 4.4 TO 11.8 |

REF. (1) ASR82-32, "COAL-FIRED CCGT R&D CONTRACT DE-AC01-80-ET15020," QUARTERLY PROGRESS REPORT, 1 NOV 1981 - 29 JAN 1982

NOTES: (1) SSC = SINTERED SiC
Si/SC = SILICONIZED SiC
CVD = CHEMICAL VAPOR DEPOSITED
+J = WITH SiC/SiC JOINT

(2) THE AVERAGE OF ALL SAMPLES TESTED (AS RECEIVED AND EXPOSED TO TEST RUNS 1, 2 AND 3). ALL SAMPLES WERE USED TO CALCULATE AN AVERAGE BECAUSE THE CONCLUSION WAS THAT THE EXPOSURE IN THE AFB HAD NO EFFECT ON STRENGTH.

(3) THE AVERAGE OF AS-RECEIVED SAMPLES ONLY.

## Summary of Results

The combined results of all tests on SiC tubes exposed in the 4- by 6-in. AFB environment did not show any detrimental effects on the tubes or on the strengths of the tubes: viz., (1) no evidence of corrosion was observed by microstructural or compositional analyses, (2) no measurable erosion occurred, and (3) differences in strength between exposed and unexposed tubes were within the normal scatter in data and no trend in decreasing strength was observed. All ceramic tubes were coated with a thin layer of CaO when exposed to oxidizing conditions, while all tubes were uncoated when exposed to reducing conditions.

## Acknowledgment

This work was funded in part by the U.S. Department of Energy under Contract DE&AC21–80–ET15020. The DOE Program Manager was Mr. L. K. Carpenter.

## References

[1]H. W. Carpenter, et al., "High Temperature Coal Fired Combustion With Ceramic Heat Exchangers for CCGT Systems," Paper No. 80–GT–155 presented at the Gas Turbine Conference, New Orleans, March 10–13, 1980, DOE Contract DE–Ac21–80–ET15020, Final Report in publication.

[2]H. W. Carpenter, Rocketdyne ITR–80–039, Oct. 10, 1980.

[3]D. L. Mann and K. H. Styhr, "Preliminary Evaluation of Tubular Ceramic Heat Exchanger Materials," Final Report 77–14570, Contract No. EX–76–C–01–2131, Dec. 8, 1977.

# Evaluation of Ceramics for Condensing Heat-Exchanger Applications

I. SEKERCIOGLU, R. RAZGAITIS, AND J. LUX

Battelle Columbus Labs
Columbus, OH 43201

The overall thermal efficiency of domestic space heating units fired by oil or gas can be improved significantly (up to 94 to 96%) by extracting the latent heat of combustion products using condensing heat exchangers. As a result, however, sulfur and chlorine impurities present in the fuels form HCl and $H_2SO_4$ acids, which condense along with water vapor to form corrosive aqueous solutions. In addition, nitrogen oxides can also form in the flame. These oxides then form nitric and nitrous acids in the condensate. Because of the acidic character of the condensate, conventional heat-exchanger materials such as mild steel are subject to rapid corrosion. Since 1979, Battelle has been involved in various aspects of condensing heat-exchanger technology including testing, evaluation, and screening of metals, plastics, polymer coatings, and ceramics using condensate oil and natural-gas-fired heating systems. All the data reported here on ceramics were obtained using oil-fired condensate. As part of the ceramic task, five types of ceramic materials and one acid-resistant enamel were identified and evaluated in the corrosion rig for 900 h. The ceramic materials included a cordierite glass-ceramic, a mullite, a sintered silicon carbide, a reaction-bonded silicon nitride (RBSN), and reaction-bonded silicon carbides (RBSC) from three sources. The RBSC ceramics exhibited the highest corrosion resistance (average weight loss is about 0.01%), whereas the cordierite glass-ceramic showed the least resistance (the average weight loss is about 0.22%). These data translate into average corrosion rates of about 2.0 and 58.0 $\mu$m per year.

The increasing cost of gas and fuel oil has created a need for operating residential and commercial heating systems at the highest possible efficiency. The traditional heat-exchanger and furnace design practice has been to limit the flue gas temperatures to about 230° to 315°C (450°–600°F). This places an upper limit on the steady-state thermal efficiency of about 80% for the furnace and boiler. It is quite possible to increase the efficiency to about 95% by further cooling the flue gas to about 40°C (100°F) using condensing heat exchangers. A small portion of the increase in efficiency comes from the sensible heat of the combustion products. The major portion comes from the latent heat of the condensible water vapor which forms as a result of burning hydrocarbon fuels. Figure 1 contrasts the operational differences between the conventional and condensing heat-exchanger systems.

The increase in the furnace efficiency, however, is not without its penalty. There are two major problems which must be addressed if this technology is to be adopted and utilized more widely. The first problem is that of corrosion; the second is the heat-transfer analysis and design of the hardware. Both of

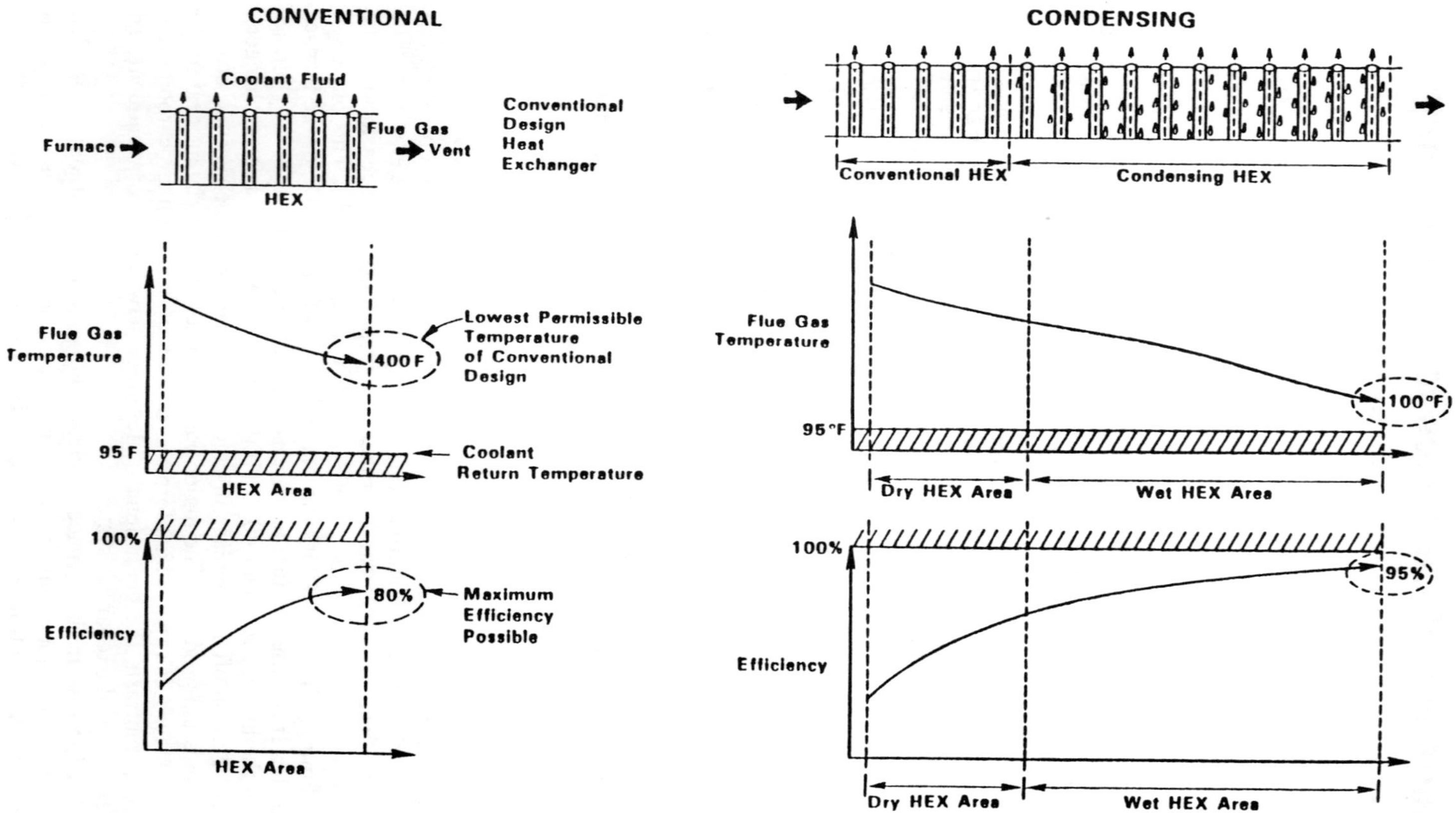


Fig. 1. Contrast between conventional and condensing furnace operation.

these problems must be solved in such a manner that the condensing heat exchangers can be affordable. The corrosion problem arises due to the presence of sulfur and chlorine impurities in the fuels. These impurities form corrosive $H_2SO_4$, $H_2SO_3$, and HCl acids and condense along with the water vapor. In addition, nitric and nitrous acids also form and condense, primarily as a result of the formation of nitrogen oxides in the combustion flame. Other impurities carried into the furnace by the combustion air also contribute to the corrosiveness of the condensate chemistry. As a result, conventional heat-exchanger materials such as mild steel are subject to rapid deterioration because of this corrosive environment. The second problem arises because of the small temperature difference ($\Delta T$) between the two heat-exchanging streams, as shown in Fig. 2. Consequently, it is important to minimize the overall thermal resistance ($R$) of the system so that the required amount of energy ($Q^1$) can be transferred through the smallest possible surface area. Therefore, the two areas of concern in commercializing high-efficiency condensing heat-exchanger furnaces are materials technology data base and thermal design technology. The first concern involves the option and tradeoffs for selecting the candidate materials with regard to their cost and corrosion characteristics in real condensate environments. The second concern involves understanding of the principles of the condensing heat exchangers and minimizing the thermal resistance of the system. In designing the heat exchangers from ceramics, the brittle nature and fabricability of the ceramics must be also considered.

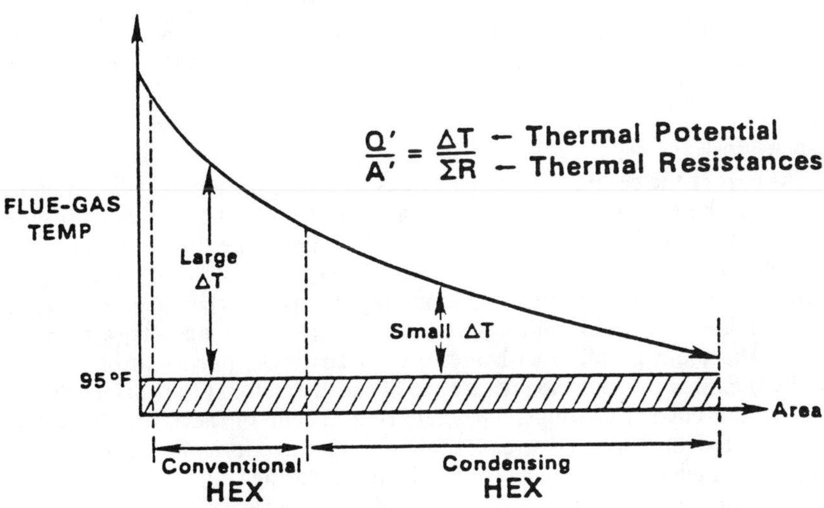


| DESIGN | ΔT | RESULT |
|---|---|---|
| Conventional | Large | Large Q′ possible with given A′ |
| Condensing | Small | Much larger A′ needed for same Q′ |

Fig. 2. Implications of condensing operation.

## Background

### Program Background

In late 1979, a series of research programs was initiated to assess the durability of various materials for condensing applications and other factors relating to the use of heat exchangers in residential and commercial heating equipment. One of the programs conducted at Battelle, sponsored by the U.S. Department of Energy through Brookhaven National Laboratory, concentrated on oil-fired equipment and is the basis of the data reported in this paper. The other program, sponsored by the Gas Research Institute, has focused on metallic candidates with gas-fired equipment.

The objective of the multiphase, multitask Brookhaven/Battelle program is to provide supporting research to aid in the development of oil-fired residential and commercial equipment that will operate in a condensing mode. Since the beginning of the program, various aspects of the condensing heat-exchanger technology have been investigated in different tasks, including review of background information, evaluation of codes and standards, characterization of condensates from oil-fired systems, one- and two-dimensional heat-transfer analyses, development of computer programs and experimental verification of the analyses by testing, designing, and fabricating a number of corrosion test rigs to simulate the condensing heat-exchanger environment, corrosion testing and screening a large number of metallic alloys including brasses and solders, a number of polymeric materials and polymer coatings, and some ceramic materials in Battelle's corrosion test rig. Due to the fact that vast areas are covered in the Brookhaven/Battelle program, the scope of this paper will be limited to the testing and evaluation of the candidate ceramics, the condensate analysis, and the relevant aspects of the corrosion test rig.

### Literature Review

As part of the ceramics task, an extensive literature survey was conducted to identify the available information regarding the testing and evaluation of ceramics in condensing heat-exchanger environments. The survey revealed very little information about testing of ceramics and even less information on quantitative corrosion data. However, in a number of studies, glass-enamel coatings were evaluated[1-5] in acid solutions and in simulated environments, with mixed results. The main problem with the enamel coatings appears to be not necessarily the inherent corrosion resistance of the enamel, but the various defects found in the enamel coatings. These defects appear in the form of bubbles, lightly coated or uncoated areas arising from the fabrication process, and chipped enameled areas caused by accident or during use.

Borosilicate glass is also a good candidate. It has excellent acid-corrosion resistance. Thus, it is used extensively as a laboratory ware as well as for flue liners[6] for power stations. In a recent symposium, Ruston[7] described the successful use of borosilicate glass tubes as condensing heat-exchanger components in the beer-brewing industry[8] where corrosive sulfuric acid forms during malting.

Lithium aluminosilicate (LAS) ceramics have a very low thermal expansion coefficient, which results in excellent thermal shock resistance. Because of this property, they were one of the first ceramics developed for gas-turbine honeycomb rotary (heat) regenerators. Later it became evident in field tests that these materials were susceptible to chemical attack due to the condensation of sulfuric acid on the cold face of the regenerator. Extensive studies conducted by McCollister[9] showed that the hydrogen ions from the acid are exchanged with the lithium ions of the LAS solid solutions. This resulted in the formation of hydrogen aluminosilicate solids solutions (HAS) with a smaller crystal cell structure. The cell contraction then causes the formation of micro- and macrocracks in the regenerator. In addition to sulfuric acid, the LAS ceramics regenerators were found to be susceptible to sodium attack from road salts. In this case, the salt was deposited on the hot face and the Na ions were exchanged with the Li ions of the LAS ceramic. Such an exchange reaction resulted in the formation of high thermal expansion materials at the reaction sites. As a result of the differential expansion, the regenerator deformed and fractured. Consequently, more acid-resistant keatite solid solutions [10] and more recently, cordierite, MAS (magnesium aluminosilicate) ceramics[11,12] were developed. Another good candidate material is mullite (an aluminosilicate ceramic) that has a relatively good thermal shock resistance and excellent acid-corrosion resistance. Since it can be formed from very-low-cost clay materials it offers a potential of fabricating condensing heat exchangers at low cost. Both silicon carbide and silicon nitride ceramics have excellent acid-corrosion resistance, thermal shock resistance, and thermal conductivity. For instance, the thermal conductivity of reaction-bonded silicon carbide is more than six times as great as that of stainless steel (109.7 vs 16.3 W/m·K). This suggests that silicon carbide ceramics would be excellent candidates for condensing heat exchangers.

#### Effect of Thermal Conductivity on Heat Transfer

Of all the relevant properties of ceramics, it would appear that the thermal conductivity would be the most important because the envisioned application is the construction of heat-exchanger surfaces, hence the desirability of minimizing the overall thermal resistance. This is especially important for condensing applications because the temperature difference driving the heat transfer is very small since the last available fractions of energy are being extracted from the flue gas.

Surprisingly, however, the thermal resistance of the heat-exchanger surface (defined as the ratio of the wall thickness divided by the thermal conductivity) plays a relatively small role in determining the overall resistance. The reason for this is that the resistances of the two film coefficients on the air and flue gas sides of the heat-exchanger surface are quite high: Typical heat-transfer coefficients are only on the order of 57 $W/m^2 \cdot K$. Thus, when comparisons are made for the overall resistance using various materials and thicknesses, the effect of widely differing material properties is relatively small.

For instance, despite the large difference in conductivity between, say, copper, aluminum, and 316 stainless steel (389 vs 221 vs 16.3 W/m·K), there is no computed difference in overall resistance to three significant figures assuming the two film coefficients are 64.6 and 53.3 $W/m^2 \cdot K$) and that the wall thickness is 0.49 mm in all cases. When a comparison is made with a typical borosilicate glass using a thermal conductivity of only 1.14 W/m·K and a wall thickness of 2.4 mm, the overall resistance increases by only approximately 5%. Such calculations demonstrate that although high-conductivity, thin-walled materials are naturally desirable for heat-exchanger application, in this instance the overall thermal penalty is small and may be overwhelmingly compensated by other highly desirable properties such as corrosion resistance.

The thermal conductivity of ceramics is important from another standpoint: as a factor in determining susceptibility to causing thermal stresses. In general, it is preferable for ceramics to possess high values of conductivity along with small thermal expansion coefficient, high strength and toughness, and low Young's modulus. Thus, even though it is important to have a high conductivity value for this structural reason as well as for thermal purposes, once again other factors are present which have to be taken into account.

## Experimental Study

### Alternate Immersion Test Rig

The purpose of the alternate immersion test rig is to provide a means of exposing the test coupons to the corrosive condensate in an alternating wet and dry mode of operation to simulate an actual furnace operation.

Figure 3 shows the design of the test rig used in this study. Basically, the test rig consists of a conventional residential oil-fired furnace, which is used as a source of flue gas at about 290°C (550°F). The exhaust stack at the exit end of the rig was fitted with a barometric draft damper to maintain constant furnace draft conditions. Before starting the tests, the burner was adjusted to obtain a Bacharach smoke spot reading of less than 0.5 on the Bacharach scale and a $CO_2$ reading of 10% in the flue gas.

The flue gas from the furnace was passed over a condensing heat exchanger that cools the flue gas and at the same time extracts condensate. The condensing heat exchanger was constructed of 88 borosilicate glass* tubes, 1.27 cm (1/2 in.) in diameter by 11.4 cm (4½ in.) long arranged in a crossflow pattern. The flue gas was on the outside, whereas the cooling water was on the inside. The glass-tube surface temperatures were maintained at a desired

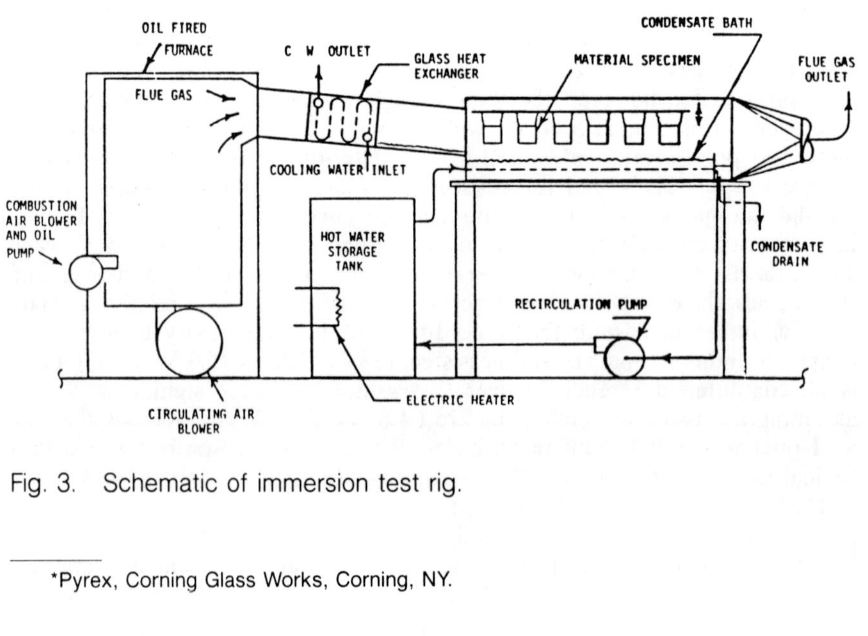


Fig. 3. Schematic of immersion test rig.

*Pyrex, Corning Glass Works, Corning, NY.

level by adjusting the water flow rate through the tubes. Thus, a constant flue gas temperature at about 93°C (200°F), was maintained at the exit of the heat exchanger.

The condensate generated by the heat exchanger was distributed into 13 parallel trays into which the test coupons could be dipped. The trays were isolated from each other by a small fence so that corrosion products from one family of test coupons did not contaminate another family. The temperature of the condensate within the trays was maintained at a constant level of about 41°C (105°F) by circulating heated water through a pair of tubes† immersed in each tray. The condensate from each tray can be drained separately for chemical analysis.

The test rig was essentially fabricated from inert materials. The section of the ductwork between the glass heat exchanger and the condensate was constructed of 304 stainless steel on the top and sides, while the bottom was constructed of CPVC (chlorinated polyvinyl chloride) plastic sheets. This bottom section was cooled with water coils to allow the condensate to run down into the immersion trays without reevaporating. All parts of the immersion bath trays were also fabricated from CPVC plastic.

As shown in Fig. 1, the corrosion test coupons are hung from corrosion-resistant Chromel‡ wires on the hanging racks of the immersion test rig using corrosion-resistant wires. This top section of the rig where coupon racks are attached is built in such a way that it can be raised or lowered pneumatically in a controlled manner. This allows dipping all of the coupons simultaneously into the condensate trays, then raising them back into the passing flue for drying in a cyclic fashion to simulate the operation of a domestic heating system.

#### Fuel and Condensate Properties

As indicated previously, the fuel used in the furnace (in this Brookhaven/Battelle study) was No. 2 fuel oil. The chemical analysis of the fuel, made by X-ray fluorescence, showed that it contained about 400 ppm chlorine; the amount of sulfur was about 1600 ppm (0.16 wt%), which is typical for a No. 2 fuel oil.

Table I summarizes the condensate composition that was used during this study (Phase III). Comparable data are also provided for similar tests per-

Table I. Measured Condensate Properties of Certain Laboratory Studies

| Fuel | Investigator | pH | Cl | $SO_4$ | $NO_3$ | $NO_2$ |
|---|---|---|---|---|---|---|
| Natural gas | A.G.A.L. | 3.9 | — | 1 | 1.1 | 5.4 |
| | VEG-G | 3.6 | 0.5 | 6 | 9 | 1.2 |
| | CGRI | 3.8 | — | — | — | — |
| No. 2 fuel oil | Battelle | 1.5 | 5 | 76 | 7 | |
| | | 3.1 | 18 | 563 | 18 | |

All data in ppm. The table is taken from Ref. 13.
Dash indicates "not measured."
Battelle data given are lowest and highest measured values in all work to date.

†Teflon, E. I. du Pont de Nemours Co., Wilmington, DE.
‡Hoskins Mfg. Co., Detroit, MI.

formed by other investigators using natural gas. It is clear that the fuel oil produces a more aggressive and, in many ways, a more realistic environment: It is known that typical residences contain significant concentrations of airborne contaminants, particularly sources of chlorine. As a result, measurements of condensate produced in homes even by natural gas equipment show that a significant number of sites have chloride concentrations greater than 10 ppm in the condensate.

#### Candidate Materials and Sample Preparation

Based on the literature survey, a number of ceramic materials were identified for testing in Battelle's corrosion test rig. They included a cordierite glass-ceramic, a mullite, a reaction-bonded silicon nitride (RBSN), reaction-bonded silicon carbide (RBSC) from three sources, a sintered alpha silicon carbide (SASC), and an acid-resistant glass-enamel coating on steel. Samples for these materials were obtained from commercially available materials. Mullite was in the form of a 2.54 cm diameter (1 in.) tube from which ring-shaped corrosion test coupons were cut. The glass-enamel samples were nominally 2.54 cm. (1 in.) long, 1 cm (0.40 in.) diameter cylinders. One end of the cylinders was not enameled. The unenameled end contained a threaded hole. Two enamel coupons were screwed on opposite ends of a screw and tightened toward each other against a Teflon (PTFE) washer. The PTFE washer prevents the leakage of corrosive liquids into the unenameled areas. It also serves as a convenient hanger for the samples in the corrosion rig. All other samples were received in the form of large plates or bars, too large and heavy to use in the corrosion rig. Therefore, these samples were cut into small, conveniently sized test coupons; then their surfaces were ground parallel with diamond tooling, and at least one surface was polished. In each coupon, one 1.50 mm (60 mil) hole was drilled near the top edge for hanging the coupons. The holes were drilled using an ultrasonic machining technique with $BC_4$ powder.

All of the test coupons were cleaned and degreased before weighing and inserting into the corrosion rig. The cleaning procedure included washing the coupons ultrasonically in a dilute soap solution[§] for 1 min, rinsing with distilled water, then ultrasonically cleaning and rinsing with alcohol twice. The coupons were dried for 2 h at 110°C, cooled, and then stored in a desiccator until weighed. The coupon weights were determined to a ±0.1-mg accuracy using an analytical balance. The dimensions of the coupons were determined to an accuracy of ±1 mil (25.7 μm) prior to cleaning of the coupons using a caliper equipped with a dial gauge to calculate the external coupon surface area and density from the weight and volume of the coupons.

#### Testing and Evaluation of Samples

The ceramic and enamel test coupons (a minimum three specimens from each candidate material), each identified with a numbered Teflon tag, were hung on the racks using corrosion-resistant Chromel wires. The racks were then inserted into the rig. To simulate the operation of a domestic burner, the furnace was operated in a cyclic manner—11 min ON, 4 min OFF—using a mechanical timer. The cyclic immersion of the coupons into the condensate bath, then drying them in the flue, was synchronized with the furnace ON and OFF cycle. That is, the coupons were immersed into the bath for 4 min

---

[§]Alconox, Alconox, Inc., New York, NY.

during the OFF cycle, then removed from the bath for 11 min for drying, which simulates a duty (or exposure) cycle of 27%. Once a week the coupons were observed through the glass observation ports on the sides of the corrosion rig, to ensure that they were securely attached to their hanging racks. During these observations it was noted that the bottom halves of the coupons did not dry completely during the drying cycles.

After 900 h (37.5 days) of cyclic exposure the coupons were removed from the rig for evaluation. It was observed that the coupons were covered with an uneven layer of soot. There was more soot on the bottom half of the coupons than the top half. This was attributed to the fact that the bottom halves remained wet longer. Thus, these sections collected more soot particles which stuck on the wet surfaces. The soot layer, however, appeared to be porous; thus, the condensate could permeate through the soot layer to interact with the coupons. The coupons were cleaned by subjecting them to three successive 3 min of ultrasonic cleaning in a mild Alconox soap solution, isopropanol, and acetone. Between these treatments, the coupons were rinsed and cleaned with a soft brush. This procedure was adequate for removing all of the soot from the enameled specimens. However, it was not adequate for the ceramic specimens. Small streaks of soot could still be observed in isolated areas. The remaining soot from these samples was removed by oxidizing at a high temperature, specifically by heating the samples in a furnace at 700°C for 6 h. This heating caused the originally white cordierite coupons to assume a white and red marbled appearance. This is attributed to the permeation of iron and other ions into the pores and subsequent oxidation into iron oxides. It was also observed that the polished surfaces of the cordierite samples lost some of their luster and reflectivity. The cream-white mullite coupons did not show any discoloration. This was attributed to the lack of porosity and imperviousness of the samples to the condensate. The dense, black silicon carbide samples did not show any visually observable change, nor did the gray RBSN samples show any disernable discoloration, despite the fact that they contained about 20 vol% porosity.

RBSC contains normally between 10 to 20 wt% free metallic silicon phase. Similarly, RBSN also contains unreacted free silicon. The amount and the grain size of the metallic silicon in RBSN, however, are much smaller than that found in RBSC. To observe whether the silicon phase in the RBSC and RBSN samples was affected by exposure to the condensate, the polished and condensate-exposed surfaces of these samples were examined under a metallographic microscope. The observations at magnifications from 100 to 400 times showed that free silicon was present on the exposed surfaces. This indicated that the free silicon was relatively unaffected by the corrosive chemistry of the condensate.

To obtain quantitative results, corrosion depths were calculated (after postcleaning, drying, and oxidizing the adherent soot layer) from weight loss analysis using the following equation:

$$CD=(\Delta W/d)(1/A) \qquad (1)$$

where $CD$ is calculated corrosion depth, $\Delta W$ weight loss determined from the pre- and post-condensate exposure weights of the samples, $A$ external surface area, calculated from the measured dimensions, and $d$ density of the samples determined from the measured weights and dimensions of the samples. Manufacturer's data were used for the enamel samples. The results are summarized

Table II. Ceramics Corrosion Data (for 900-H Exposure)

| Material | Corrosion Depth Data | | Projected Corrosion | |
|---|---|---|---|---|
| | μm | mils | mpy | μpy |
| Cordierite | 9.4 | 0.37 | 2.3 | 58.0 |
| Mullite | 0.85 | 0.034 | 0.32 | 8.2 |
| RBSC-1 | 0.20 | 0.0080 | 0.08 | 1.9 |
| RBSC-2 | 0.16 | 0.0063 | 0.06 | 1.5 |
| RBSC-3 | 0.18 | 0.0069 | 0.07 | 1.7 |
| Sintered α-SiC | 0.37 | 0.014 | 0.14 | 3.5 |
| RBSN | 0.90 | 0.035 | 0.34 | 8.6 |
| Glass-enamel | 1.4 | 0.054 | 0.52 | 13.3 |

in Table II. The last two columns show the projected corrosion rate in terms of mils per year and micrometers per year.

**Conclusions**

Experimental evaluations have shown that the selected candidate materials exhibit very low corrosion in the simulated condensing heat-exchanger environment. Among the ceramics tested, the RBSC and the sintered silicon carbide ceramics had the lowest corrosion rates. The silicon carbide ceramics also have excellent thermal and mechanical properties; therefore, they are excellent candidates for condensing heat-exchanger applications.

The higher corrosion rate of the RBSN ceramic than that of the silicon carbide is attributed to its high porosity. The pores of the RBSN can be sealed by coating or impregnating with a suitable acid-resistant material to reduce its corrosion rate. This should also minimize the probability of its failure under rapid heating conditions due to the formation of vapors within its pores.

Although the cordierite samples exhibited the highest corrosion rates, the weight losses and the calculated corrosion depths (<58 μm per year of exposure) were acceptable and low compared to most metals. With decreasing porosity, the corrosion rate of cordierite (magnesium aluminosilicate ceramics) should decrease. The cordierite ceramics are already being tested and used as honeycomb heat regenerators for gas turbine applications where both high-temperature salt corrosion and low-temperature acid corrosion conditions exist. Therefore, zero- or very-low-porosity cordierite ceramic is a low-cost potential material for condensing heat-exchanger applications.

The impervious mullite samples tested also exhibited very low corrosion rates. Although the thermal shock resistance of the mullite ceramics is not as good as that of the cordierite ceramics, they should have sufficient and acceptable thermal shock resistance at relatively low thermal gradients expected in domestic condensing heat-exchanger environments. The thermal shock resistance of the mullite ceramics can be improved by minimizing the wall thicknesses.

The acid-resistant glass-enamel coupons also had very low and acceptable corrosion rates (8.2 μm per year of exposure time). The samples evaluated in this program did not show any visually or microscopically observable defects, cracks, or failures after 900 h of exposure. The major causes of failure of enamels evaluated in various other programs are attributable to fabrication defects and mishandling of the materials. Most of the defects and defective components can be eliminated by instituting improved processing and fabrication standards and quality control tests on the finished products, as are being performed for enamelware produced for the chemical industry.

Borosilicate glass has excellent acid-corrosion resistance, and high thermal shock resistance. It is already being used as a condensing heat-exchanger material in the brewery industry. Therefore, it is an excellent candidate for domestic condensing heat-exchanger applications. High surface smoothness of both enamels and glass is a desirable characteristic. It makes it easier to clean the surfaces simply by washing and resists fouling better than rough surfaces.

Although it is desirable to have a high thermal conductivity material for heat-exchanger applications, the low thermal conductivity of oxide ceramics (such as mullite and cordierite), glass, and enamel is not a significant handicap since the overall heat-transfer coefficient between two heat-exchanging streams is determined not by the thermal conductivity of the materials but mainly by the film coefficients at the boundaries.

The oxide ceramics, such as mullite and cordierite, are formulated using widely available inexpensive raw materials. They can be formed into shapes using a variety of ceramic forming methods and sintered in air at relatively low temperatures (compared to the silicon nitride and silicon carbide ceramics). Therefore, they represent low-cost materials for condensing heat-exchanger applications.

Although the silicon carbide and silicon nitride ceramics have the best combination of properties for condensing heat-exchanger applications, they require relatively high sintering temperatures and nonoxidizing, inert sintering conditions. They are expected to be costlier than the oxide ceramics.


## Acknowledgment

The authors are pleased to acknowledge the support of the Department of Energy through Brookhaven National Laboratory/BNL—in particular Lawrence Woodworth of BNL, who was the Project Officer, and Danny Luis and John Ryan of the Building Equipment Division of DOE, who overall have program responsibility for heating equipment research at DOE.


## References


[1]T. Lahtvee, S. W. Khoo, and O. O. Schaus, "Study of Materials to Resist Corrosion in Condensing Gas Fired Furnaces," Quarterly Report Number 1, by Canadian Gas Research Institute for the Gas Research Institute, January 1980.

[2]O. O. Schaus and T. Lahtvee, "Study of Materials to Resist Corrosion in Condensing Gas-Fired Equipment"; in Volume II, Proceedings of Symposium on Condensing Heat Exchangers, Atlanta, Georgia, March 3–4, 1982.

[3]E. De Anda, "Heat Exchangers Fouling and Corrosion Evaluation," Final Technical Report 81-18003, by AiResearch Manufacturing Company for DOE, October 1981.

[4]E. Macanga, Jr., "Study of the Effects of Spray Propellants on Gas Appliances," Research Report No. 1521, Sponsored by American Gas Association, August 1978.

[5]A. A. Mozhaeva and I. F. Golouko, "Study of the Chemical Stability of Enamel Coatings," *Chem. Petrol. Eng.* **9–10**, 690 (1969).

[6]D. M. Berger, R. J. Trewella, and C. J. Wummer, "Evaluation of Linings for $SO_2$ Scrubber Service," *J. Coatings Technol.*, **54** [688] 47–55 (1982).

[7]M. J. Ruston, "Borosilicate Glass—Its Use in Waste Heat Recovery"; Advances in Energy Productivity, Proceedings of the 5th World Energy Engineering Congress, Sept. 14–17, 1982.

[8]Anon., "Corning Glass Gets Million-Dollar Order for Glass-Tubed Energy Recovery System," *Industry News, Am. Glass Rev.*, June 1982.

[9]H. L. McCollister, "Chemical Alteration of $Li_2O \cdot Al_2O_3 \cdot nSiO_2$ Ceramics"; Ph. D. Thesis; University of Michigan, 1978.

[10]H. L. McCollister, "Improved Ceramic Heat Exchanger Material," NASA Report No. CR–135262, 1977.

[11]H. W. Rauch, Sr., "Improved Ceramic Heat Exchanger Material," NASA Report No. CR–135292, 1977.

[12]C. A. Fucinari, "Regenerator Matrix Physical Property Data," NASA Report No. CR–159854, 1980.

[13]R. Razgaitis, D. A. Ball, J. J. Lux, E. L. White, R. A. Markle, E. J. Bradbury, I. Sekercioglu, and D. W. Locklin, *Condensing Heat Exchanger Systems for Residential/Commercial Furnaces and Boilers*, Phase III, Draft Final Report, Prepared by Battelle for DOE/Brookhaven National Laboratories, Contract No. DE–AC02–76CH00016, October 1983.

# Author Index

# Subject Index